Springer Theses

Recognizing Outstanding Ph.D. Research

Aims and Scope

The series "Springer Theses" brings together a selection of the very best Ph.D. theses from around the world and across the physical sciences. Nominated and endorsed by two recognized specialists, each published volume has been selected for its scientific excellence and the high impact of its contents for the pertinent field of research. For greater accessibility to non-specialists, the published versions include an extended introduction, as well as a foreword by the student's supervisor explaining the special relevance of the work for the field. As a whole, the series will provide a valuable resource both for newcomers to the research fields described, and for other scientists seeking detailed background information on special questions. Finally, it provides an accredited documentation of the valuable contributions made by today's younger generation of scientists.

Theses are accepted into the series by invited nomination only and must fulfill all of the following criteria

- They must be written in good English.
- The topic should fall within the confines of Chemistry, Physics, Earth Sciences, Engineering and related interdisciplinary fields such as Materials, Nanoscience, Chemical Engineering, Complex Systems and Biophysics.
- The work reported in the thesis must represent a significant scientific advance.
- If the thesis includes previously published material, permission to reproduce this must be gained from the respective copyright holder.
- They must have been examined and passed during the 12 months prior to nomination.
- Each thesis should include a foreword by the supervisor outlining the significance of its content.
- The theses should have a clearly defined structure including an introduction accessible to scientists not expert in that particular field.

More information about this series at http://www.springer.com/series/8790

Andrew Giltrap

Total Synthesis of Natural Products with Antimicrobial Activity

Doctoral Thesis accepted by
The University of Sydney, Sydney, Australia

Author
Dr. Andrew Giltrap
School of Chemistry
The University of Sydney
Sydney, NSW
Australia

Supervisor
Prof. Richard Payne
The University of Sydney
Sydney, NSW
Australia

ISSN 2190-5053 ISSN 2190-5061 (electronic)
Springer Theses
ISBN 978-981-10-8805-6 ISBN 978-981-10-8806-3 (eBook)
https://doi.org/10.1007/978-981-10-8806-3

Library of Congress Control Number: 2018934889

Printed on acid-free paper

This Springer imprint is published by the registered company Springer Nature Singapore Pte Ltd.
part of Springer Nature
The registered company address is: 152 Beach Road, #21-01/04 Gateway East, Singapore 189721, Singapore

Supervisor's Foreword

Natural products represent a unique source of biologically active molecules that have revolutionised modern medicine. One particularly significant class of natural products is nonribosomal peptides which include a number of clinically essential antibiotics such as penicillin and vancomycin. The ability to synthesise these peptide natural products represents the first step in the development of analogues and is essential in order to improve biological activity and medicinal chemical properties. The development of novel antibiotics with new mechanisms of action is desperately needed as bacteria are rapidly developing resistance to the currently used therapies.

The work described in this thesis represents the first total synthesis of two classes of important bioactive peptide natural products: teixobactin and the skyllamycins. The isolation and structure of teixobactin were first reported in 2015 and possess potent activity against a number of clinically relevant pathogenic bacteria, including methicillin-resistant *Staphylococcus aureus*. Using a solid-phase peptide synthesis approach, the total synthesis of teixobactin was successfully accomplished. Importantly, the synthetic natural product possessed potent activity against a number of gram-positive bacteria. This represented an important breakthrough in how to access teixobactin synthetically, and the technology developed is currently being applied to the generation of analogues.

The second part of this thesis involves the synthetic efforts towards the skyllamycins, a family of modified nonribosomal peptides with bacterial biofilm inhibitory activity. These natural products are highly complex and contain a number of synthetic challenges, including the unusual α-hydroxyglycine moiety. While these natural products were first reported in 2001 no total synthesis had been reported. This work describes the synthesis of four simplified analogues as well as the first total synthesis of a family of skyllamycin natural products. The final step of the total synthesis involved the concomitant cyclisation and formation of the unusual α-hydroxyglycine. This work has laid the foundation for further synthetic and biosynthetic investigations into this unusual class of natural products, as well as the development of analogues with improved biological activity.

January 2018 Prof. Richard Payne

Abstract

Natural products are an essential source of many modern medicines. Examples of important natural products include the antibiotic penicillin, the anticancer drug taxol, the immunosuppressant cyclosporine and the antimalarial quinine. One significant class of bioactive natural products is nonribosomal peptides (NRPs), and two prototypical members of this class are the extremely important antibiotics, penicillin and vancomycin. Currently, bacterial resistance to antibiotics, including penicillin and vancomycin, is one of the most pressing global health issues. The need for new antibiotics with novel mechanisms of action is paramount. This thesis describes the total synthesis of the recently isolated antimicrobial NRPs teixobactin and skyllamycins A–C. Chapter 2 of this thesis describes the first total synthesis of teixobactin, a novel cyclic NRP antibiotic isolated in 2015. This was carried out *via* a solid-phase peptide synthesis (SPPS) strategy with a late-stage cyclisation reaction. The synthetic natural product possessed potent activity against a number of clinically relevant gram-positive bacterial pathogens. Chapters 3 and 4 describe investigations towards the total synthesis of skyllamycins A–C, a family of structurally complex cyclic NRPs. These natural products inhibit the growth of bacterial biofilms, a mechanism by which bacteria evade antibiotics. The most unusual feature of these natural products is the presence of an α-OH-glycine (Gly) moiety, which to date has only been found in one other linear peptide natural product. Chapter 3 details the synthesis of the non-proteinogenic amino acids present in the natural products and their incorporation into the synthesis of four skyllamycin analogues that omit the unusual α-OH-Gly residue. These analogues were analysed for their biofilm growth inhibition activity. Chapter 4 describes the completion of the first total synthesis of skyllamycins A–C. This was achieved through a SPPS strategy followed by a late-stage cyclisation and concomitant formation of the unusual α-OH-Gly residue in one step.

Parts of this thesis have been published in the following journal articles:

Giltrap, A. M.; Dowman, L. J.; Nagalingam, G.; Ochoa, J. L.; Linington, R. G.; Britton, W. J.; Payne, R. J. *Org. Lett.* **2016**, *18*, 2788–2791.

Giltrap, A. M.; Haeckl, F. P J.; Kurita, K. L.; Linington, R. G.; Payne, R. J. *Chem. Eur. J.* **2017**, *23*, 15046–15049.

In addition to the statement above, in cases where I am not the corresponding author of the published item, permission to include the published material has been granted by the corresponding author.

Acknowledgements

To begin I would like to thank my supervisor Prof. Richard Payne for his unwavering support and guidance throughout my Ph.D. I have been a part of the Payne group in some capacity since 2010, and Rich has been a tremendous supervisor throughout this entire time. You have an enormous passion for science and you continually work very hard for the best of your students and for this I am extremely grateful.

Many thanks to all members of the Payne group past and present who have made the laboratory a great environment to work in. I have had a lot of fun during my Ph.D. In particular, thanks to Dr. Katie Terrett, my original mentor—thanks for everything you taught me. Thanks to Luke, it has been a privilege teaching you, and I have really enjoyed working on a number of projects with you. To the fellow Ph.D. students in my cohort, Dave and Nabs, it has been a pleasure doing a Ph.D. with you. We have made it! Thanks as well to the many past Payne group members who have become great friends, in particular James, Gaj, Nick, Bhav and Lukas. I would also like to thank the many friends I made in the other groups on level 5—you have made the Cornforth and Robinson laboratories a great place to work.

I would like to express my gratitude to the many members of the University of Sydney, School of Chemistry, who have made the work presented herein possible, in particular: Carlo Piscicelli, Bruce Dellit, Dr. Shane Wilkinson, Dr. Nick Proschogo, Dr. Cody Szczepina as well as my associate supervisor Associate Professor Chris McErlean. I would also like to thank Dr. Ian Luck for being a great boss in the NMR Facility. I have thoroughly enjoyed working for the facility and have learnt a lot in my time. I am also extremely grateful to the Australian Postgraduate Award and The University of Sydney Vice Chancellor's Scholarship for funding. Furthermore, I would especially like to acknowledge Dorothy Lamberton and John A. Lamberton Research Scholarship, for the generous financial support throughout my studies.

I would like to acknowledge my collaborators Dr. Gaya Nagalingham and Prof. Warwick Britton from the Centenary Institute for the TB expertise. I am also particularly grateful for the fruitful collaboration with Associate Professor Roger

Linington and the many hard-working members of his research group at both the University of California Santa Cruz and Simon Fraser University.

Finally, I would like to thank all of my family and friends for their support throughout my Ph.D. Mum and Dad, thanks for being extremely supportive in all things I have chosen to do and backing me enthusiastically at all opportunities. James and Kate, you have both been very supportive and I really appreciate the way you have kept me grounded at all times. Finally, Veronica you have been such a great support to me over my Ph.D. studies in the last nine and half years. You have been understanding when I have been working late or on weekends, encouraging when you know I have lots to get done, provided delicious baked goods that have made me loved by my laboratory mates and you are always keen for a G&T. Thanks for getting me through it.

Contents

Abbreviations

2,2-DMP	2,2-Dimethoxypropane
2-CTC	2-Chlorotrityl chloride
6-APA	6-Aminopenicillanic acid
A	Adenylation (domains)
Ac	Acetyl
Alloc	Allyloxycarbonyl
AMR	Antimicrobial resistance
APCI	Atmospheric pressure chemical ionisation
Ar	Aromatic
BAIB	*bis*-Acetoxyiodobenzene
Bn	Benzyl
Boc	*tert*-Butyloxycarbonyl
C	Condensation (domain)
Cbz	Benzyloxycarbonyl
CDI	Carbonyldiimidazole
CF	Cystic fibrosis
COMU	(1-Cyano-2-ethoxy-2-oxoethylidenaminooxy) dimethylamino-morpholino-carbenium hexafluorophosphate
COSY	Correlation spectroscopy
DABA	Diaminobutyric acid
DCC	*N,N'*-Dicyclohexylcarbodiimide
DIBAL-H	Diisobutylaluminium hydride
DIC	*N,N'*-Di*iso*propylcarbodiimide
DKP	Diketopiperazine
DMAP	*N,N*-Dimethylaminopyridine
DMF	Dimethylformamide
DMSO	Dimethylsulfoxide
DMTMM	4-(4,6-Dimethoxy-1,3,5-triazin-2-yl)-4-methylmorpholinium
DNA	Deoxyribonucleic acid
E	Epimerase (domain)

EC_{50}	50% effective concentration
ESI	Electrospray ionisation
FDA	Federal Drug Administration
Fmoc	9-Fluorenylmethoxycarbonyl
GC-MS	Gas chromatography–mass spectrometry
GPA	Glycopeptide antibiotics
h	Hour(s)
HATU	2-(7-Aza-1H-benzotriazole-1-yl)-1,1,3,3-tetramethyluronium hexafluorophosphate
HFIP	1,1,1,3,3,3-Hexafluoro*iso*propanol
HMBC	Heteronuclear multiple bond correlation
HMPB	4-(4-Hydroxymethyl-3-methoxyphenoxy)butyric acid
HOAt	1-Hydroxy-7-azabenzotriazole
HOBt	1-Hydroxy-benzotriazole
HPLC	High-performance liquid chromatography
HPLC-MS	High-performance liquid chromatography–mass spectrometry
HRMS	High-resolution mass spectrometry
HTS	High-throughput screen
Hünig's base	*N,N*-Di*iso*propylethylamine
HWE	Horner–Wadsworth–Emmons
IC_{50}	Half maximal inhibitory concentration
LiHMDS	Lithium hexamethyldisilazide
LRMS	Low-resolution mass spectra
M	Methylation (domain)
MeCN	Acetonitrile
MRSA	Methicillin-resistant *Staphylococcus aureus*
MS	Mass spectrometry
MSNT	1-(Mesitylene-2-sulfonyl)-3-nitro-1,2,4-triazole
MSSA	Methicillin-susceptible *Staphylococcus aureus*
Mtb	*Mycobacterium tuberculosis*
NMM	*N*-Methylmorpholine
NMR	Nuclear magnetic resonance
NOESY	Nuclear Overhauser effect spectroscopy
NRP	Nonribosomal peptide
NRPS	Nonribosomal peptide synthetase
ORD	Optical rotatory dispersion
Orn	Ornithine
PAL	Peptidyl-α-hydroxylglycine α-amidating lyase
PAM	Peptidylglycine α-amidating monooxygenase
Pbf	2,2,4,6,7-Pentamethyldihydrobenzofuran-5-sulfonyl
PCP	Peptidyl carrier protein
PDGF	Platelet-derived growth factor
PhF	9-(9-Phenylfluorene)
PHM	Peptidylglycine α-hydroxylating monooxygenase
PMB	*para*-Methoxybenzyl

Ppant	4'-Phosphopantetheine
p-TSA	*para*-Toluenesulfonic acid
PyAOP	(7-Azabenzotriazol-1-yloxy)tripyrrolidinophosphonium hexafluorophosphate
PyBOP	(Benzotriazol-1-yloxy)tripyrrolidinophosphonium hexafluorophosphate
rt	Room temperature
SAM	*S*-Adenoyslmethionine
Sec	Selenocysteine
SFU	Simon Fraser University
SPPS	Solid-phase peptide synthesis
Su	Succinimide
TB	Tuberculosis
TBAF	Tetrabutylammonium fluoride
TBS	*tert*-Butyldimethylsilyl
Te	Thioesterase (domain)
TEMPO	2,2,6,6-Tetramethyl-1-piperidinyloxy
TES	Triethyl silyl
Tf	Triflate
Tf_2O	Trifluoromethanesulfonic anhydride
TFA	Trifluoroacetic acid
TfOH	Triflic acid (trifluoromethanesulfonic acid)
THF	Tetrahydrofuran
TIS	Tri*iso*propylsilane (iPr$_3$SiH)
TLC	Thin-layer chromatography
Trt	Triphenyl methyl (trityl)
UHPLC	Ultra-high performance liquid chromatography
UHPLC-MS	Ultra-high performance liquid chromatography–mass spectrometry
VRE	Vancomycin-resistant enterococci
WHO	World Health Organisation
WTA	Wall teichoic acid

Proteinogenic Amino Acids

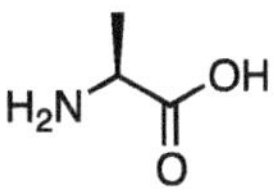

Alanine
Ala, A

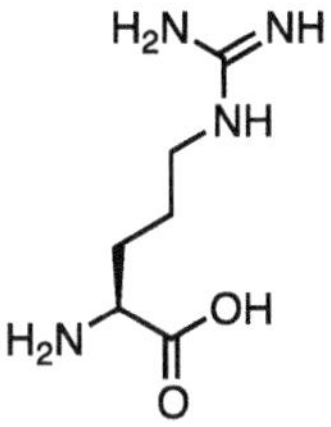

Arginine
Arg, R

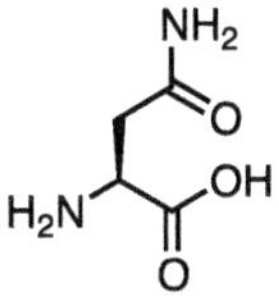

Asparagine
Asn, N

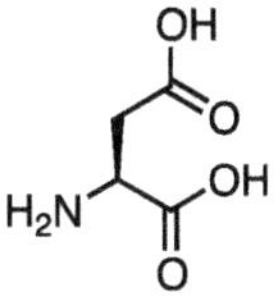

Aspartic Acid
Asp, D

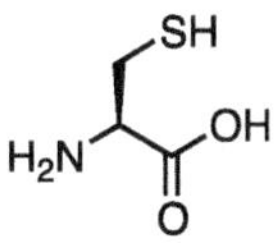

Cysteine
Cys, C

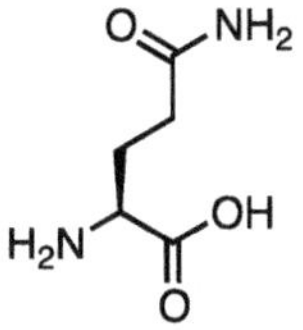

Glutamine
Gln, Q

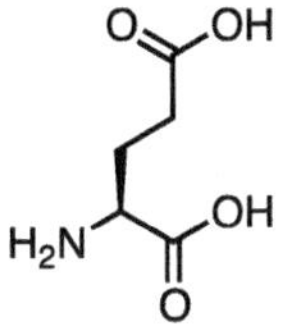

Glutamic Acid
Glu, E

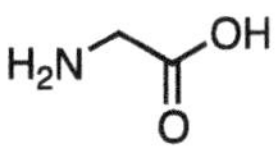

Glycine
Gly, G

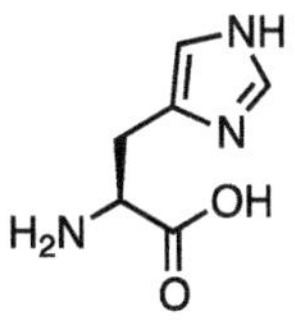

Histidine
His, H

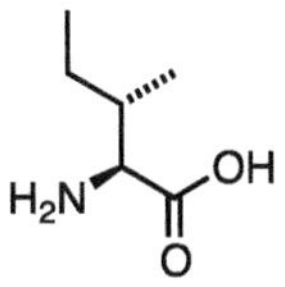

Isoleucine
Ile, I

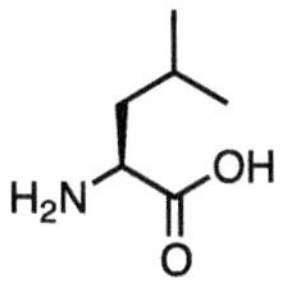

Leucine
Leu, L

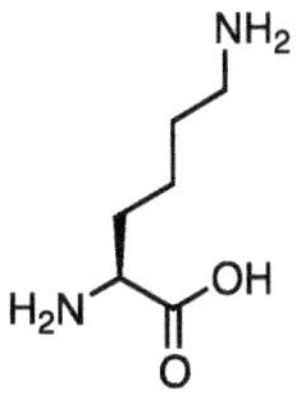

Lysine
Lys, K

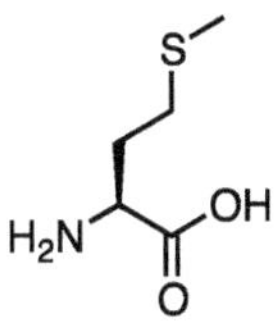

Methionine
Met, M

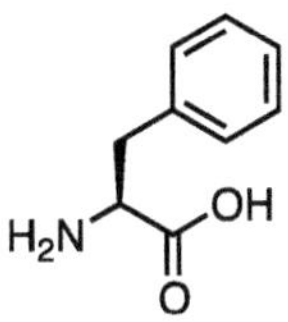

Phenylalanine
Phe, F

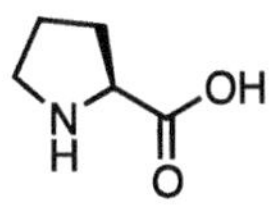

Proline
Pro, P

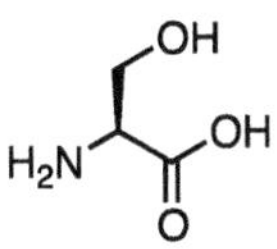

Serine
Ser, S

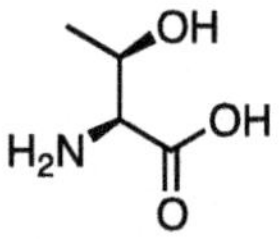

OH
H2N OH
O
Threonine
Thr, T

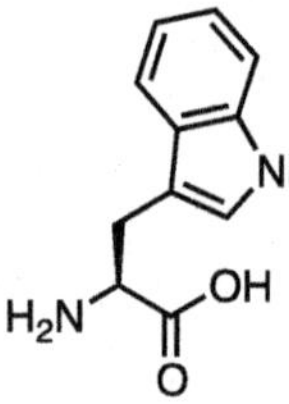

NH
H2N OH
O
Tryptophan
Trp, W

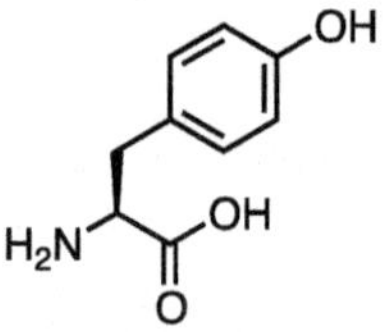

OH
H2N OH
O
Tyrosine
Tyr,T

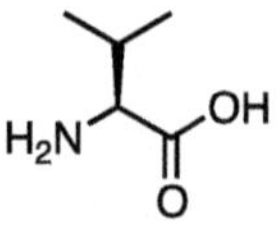

H2N OH
O
Valine
Val, V

Chapter 1
Introduction

1.1 Natural Products as a Source of Drugs

Natural products have been used as therapeutics throughout history, well before the chemistry behind them was understood. In particular, plants formed the basis of early human medicine, however, these were administered without a knowledge of how they exerted a therapeutic effect [1]. For example, since ancient times, the bark of the Willow Tree was known to ease pain and reduce fevers, and was used as a medicine across ancient cultures in Europe, China and North America. It was not until the 1800s that scientists established that these medicinal plants must contain an active molecular constituent which could be isolated and be used to treat disease.

These active constituents (natural products) are secondary metabolites produced by bacteria, algae, corals, sponges, plants and lower animals and are generally not required for growth and reproduction. They do however provide the organism with an evolutionary advantage by interacting with other competing organisms to the producer's benefit [2]. Over millennia organisms have developed specific molecular scaffolds capable of binding to proteins and other biological targets. These natural product structures are extremely diverse and, as such, occupy a large range of biologically relevant chemical space [3]. As scientific understanding advanced, the structure of natural products (including stereochemical elements) could be elucidated and chemists began to synthesise and modify these compounds to understand structure activity relationships. This represented the birth of the field of natural product synthesis, and ultimately the development of new synthetic methods and numerous new medicines.

A. Giltrap, *Total Synthesis of Natural Products with Antimicrobial Activity*,
Springer Theses, https://doi.org/10.1007/978-981-10-8806-3_1

1.1.1 Quinine—The First Anti-malarial

The cinchona tree, native to South America, was known to be medicinally active by the Spanish in the 1600s, and possibly before this by the native South Americans. From the 1600s until the early 1800s an extract of the cinchona bark was used as an effective treatment for malaria, the disease caused by the parasite *Plasmodium sp.* and transmitted via mosquitoes [4]. The active ingredient, the alkaloid quinine (**1**) (Fig. 1.1), was first isolated in 1820 by Pierre Joseph Pelletier and Joseph Caventou and represented the first example of a chemical compound being used to treat an infectious disease. The cinchona extract continued to be the main treatment of malaria until the 1920s, when synthetic alternatives became available. The most notable example is chloroquine (**2**), developed in 1934 by Johann Andersag, which has been used extensively for the treatment of malaria since the 1940s [5].

It was almost 100 years from the initial isolation of quinine until Paul Rabe established the structure of this molecule (1907) [6], with a the partial synthesis reported over a decade later (1918) [7]. R. B. Woodward and Doerring completed the first formal synthesis in 1944 [8], however owing to the chemical complexity of quinine (**1**) it was not until Stork in 2001 [9] that the first asymmetric total synthesis was completed, almost 100 years after its structure was elucidated.

1.1.2 Artemisinin—From Chinese Herbal Medicine to Anti-malarial

The discovery of quinine (**1**), and later the development of synthetic alternatives such as chloroquine (**2**), led to the successful treatment of malaria up until the 1950s. However, at this point, the malaria parasite began to develop resistance to chloroquine, which halted global efforts to eradicate the disease. During the Vietnam War, the fight against malaria became a huge priority for the US military who initiated a drug discovery program, which ultimately led to the discovery of the antimalarial mefloquine [10], however not until 1975.

In 1967, the People's Republic of China established the 523 Office with a mission to find and develop a new anti-malarial. Two years later, in 1969, Youyou Tu was selected to head the Project 523 research group at the Academy of Traditional Chinese Medicine [11]. The group began by reviewing ancient Chinese

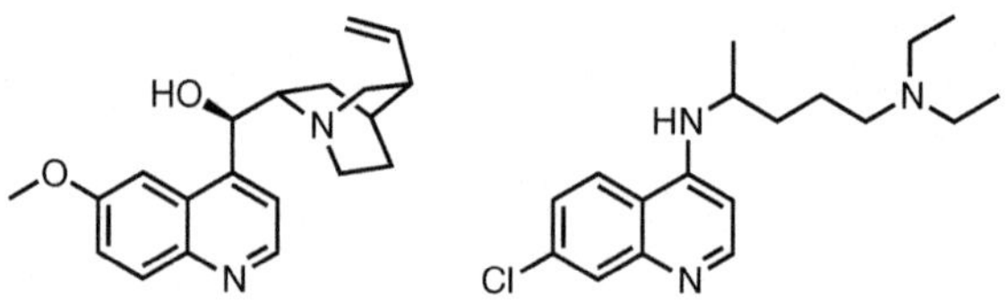

Fig. 1.1 Strucuture of quinine (**1**) and chloroquine (**2**)

HO N O N 1 HN N Cl N 2

Fig. 1.2 Structure of artemisinin (**3**) and dihydroartemisinin (**4**)

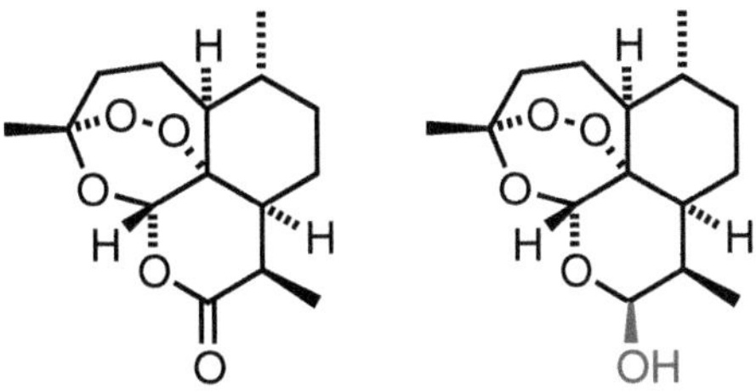

literature and folklore for references to treatments for malaria. The Qinghao family of herbs were traditionally used to treat a number of illnesses, and ancient literature as early as 350 A.D. described the use of the plants to treat symptoms of malaria. Initial extracts were unsuccessful in treating the disease, however in October 1971 an extract was found to be effective in clearing malaria from a rat. The first clinical trial was carried out a year later, with the extract found to be effective in clearing the infection in humans [11].

While the active constituent was purified and named artemisinin (**3**) (Fig. 1.2) in November 1972, it was a further 5 years until the structure of the natural product was published in 1977. Artemisinin (**3**) is a sesquiterpene lactone natural product which contains an unusual endo-peroxide moiety. With the structure determined, guided chemical modifications were next performed. Reduction of the lactone to the hemiacetal (red in Fig. 1.2), to form dihydroartemisinin (**4**), resulted in higher compound stability and improvement in the antimalarial activity of more than an order of magnitude compared to the parent natural product [12]. The discovery of artemisinin and its use (together with structural analogues) for the treatment of malaria was a hugely important scientific breakthrough. As a result, Youyou Tu, who was responsible for the early isolation work on the natural product, was awarded a share of the 2015 Nobel Prize in Medicine and Physiology [11].

1.1.3 Penicillin—From Mould to World Changing Antibiotic

The development of anti-malarials was a significant breakthrough in the treatment of human infectious disease. However, arguably of even greater significance for both medical and human histroy, was the discovery of antibiotics. Prior to the development of antibiotics minor infections, now considered easily treated, were often fatal. One of the key discoveries was that of penicillin (**5**) (Fig. 1.3). In 1928 Alexander Fleming found that the *Penicillium* mould was capable of preventing the growth of bacteria in a Petri dish [13]. He reasoned that this must be due to a substance (i.e. a natural product) produced by the fungus. The true breakthrough came in 1940 when a team led by Howard Florey at Oxford University isolated

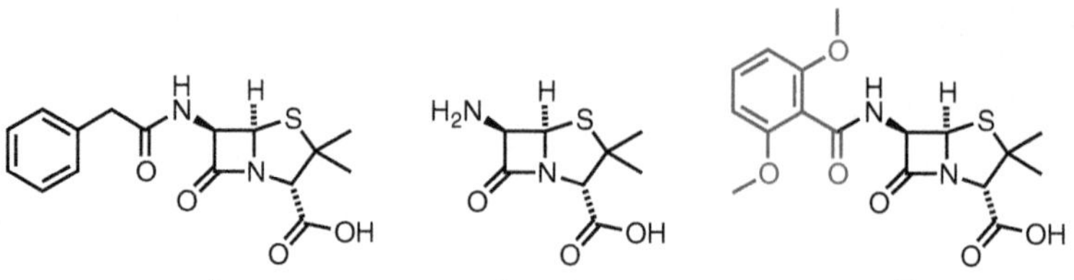

Fig. 1.3 Structure of penicillin G (5), 6-APA (6) and methicillin (7)

penicillin and demonstrated its in vivo antibiotic activity in mice [14]. This was followed, a year later, by a clinical trial in 10 humans suffering from a *Staphylococcus aureus* infection [15]. While a significant achievement, only a very small amount of purified penicillin was produced. This discovery led to significant efforts to produce penicillin (**5**) on a large enough scale to be utilised by the allies in World War II. Numerous different structures were proposed for penicillin, however it was not until 1945 that Dorothy Crowfoot Hodgkin conclusively determined the structure of the natural product by X-ray crystallography [16]. Both Fleming and Florey received the Nobel Prize in Medicine in 1945 along with their colleague Ernst Chain.

With the structure established, extensive chemical studies were carried out, most notably in the laboratory of John Sheehan at MIT. Due to the unusual and reactive β-lactam ring it was commonly thought that the synthesis of penicillin (**5**) would be impossible. Indeed, R. B. Woodward, the Chemistry Nobel Laureate in 1965, described penicillin as being made up of "diabolic concatenations of reactive grouping" [17]. However after nine years of work the Sheehan completed the total synthesis in 1957 [18]. In order to overcome the challenge they developed new chemistry, including the phthalamide and *tert*-butyl protecting groups, and aliphatic carbodiimides as a method for the synthesis of amide bonds, that are still in use today.

Importantly, during both synthetic [19] and biological [20] studies *en route* to the natural product, 6-aminopenicillanic acid (6-APA) (**6**) was developed. This molecule contains the key bioactive β-lactam core of the penicillin family, with an amine handle which can be functionalised to produce analogues possessing different properties. Of particular importance was the development of methicillin (**7**) in 1960 [21]. At the time of the isolation of penicillin it was observed that some bacteria produced an enzyme (β-lactamase) that was capable of breaking down penicillin [22]. Furthermore, during their initial clinical trial Florey et al. observed resistance to penicillin by *Staphylococci* cultures in vitro [15]. As the use of penicillin increased in the clinic, resistance became more common in *S. aureus* infections. However, it was discovered that the incorporation of the dimethoxybenzoic acid moiety (red in Fig. 1.3) onto 6-APA (**6**), in methicillin (**7**) conferred activity against penicillin resistant *S. aureus* [21]. However, resistance to methicillin is now also common, and the strain is commonly dubbed methicillin resistant *S. aureus* (MRSA), or 'Golden Staph'.

These historical examples highlight the importance of natural products for the development of drugs, particularly as antimicrobials. The examples also showcase the power of chemical synthesis in ensuring that the challenge of antibiotic resistance can be met through the generation of synthetic analogues.

1.1.4 The Current State of Affairs—Natural Products and Drug Development

While natural products, or derivatives thereof, were historically the major source of drugs, in the 1990s the pharmaceutical industry turned to combinatorial chemistry as a way to develop extensive small molecule libraries (>1 million) to screen against a particular target [23]. Unfortunately these libraries yielded few hits due to a lack of chemical diversity and drug likeness. Unlike natural products, these molecules often lack complexity and a large number of stereocentres and, as such, there has been a shift towards smaller, focused libraries which contain more 'natural product like' molecules. In their comprehensive review, *Natural Products as Sources of New Drugs 1981–2014*, Newman and Cragg analysed the origin of new drugs approved each year [24]. Their analysis demonstrated that over the 34 year period, an average of 33% of new small molecule drugs approved per year are either natural products or derived from a natural product (Fig. 1.4).

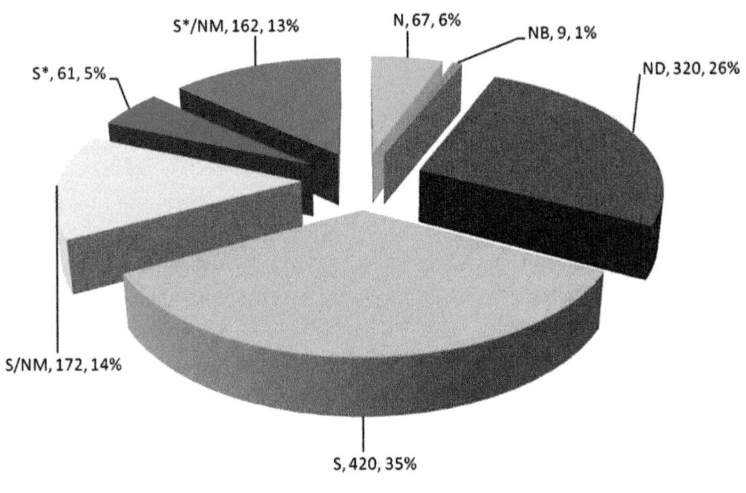

Fig. 1.4 All small molecule drugs approved between 1981–2014 (ND = derived from natural product, NB = natural product botanical drug, N = natural product, NM = natural product mimic, S* = natural product within pharmacophore of synthetic drug, S = synthetic drug) [24]

The oceans provide a large repository of natural products. Indeed, in 2015 alone, 1340 new compounds were isolated from marine sources such as microorganisms, algae molluscs and plants [25]. While this is a significant number of novel molecules, a large portion of 'Nature's treasure trove of small molecules' remains unexplored [24, 26]. With their often unique and privileged chemical structures, which falls in biologically relevant chemical space [3], natural products have the potential to again served as an essential source of new medicines, particularly in the fight against antibiotic resistance [27].

1.2 The Problem of Antimicrobial Resistance

The discovery of penicillin heralded a new method for treating common infections that were previously life threatening. As early as 1951 Sir Frank MacFarlane Burnet, winner of the 1960 Nobel Prize in Medicine, wrote the following:

> If one looks around the medical scene in North America or Australia, the most important change he sees is the rapidly diminishing importance of infectious disease ... With the full knowledge we already possess, the effective control of every important infectious disease, with the one outstanding exception of poliomyelitis, is possible.

The focus of medicine moved from preventing infectious disease to cures for chronic illnesses.

Unfortunately this prediction proved incorrect. In 2015 Dr. Margaret Chan, the Director-General of the World Health Organisation (WHO), wrote that "antimicrobial resistance threatens the core of modern medicine and the sustainability of an effective, global public health response to the enduring threat from infectious disease" [28].

1.2.1 O'Neill Report

In 2014 Jim O'Neill (Baron of Gately), a widely renowned British economist, was commissioned by the then Prime Minister David Cameron to chair a report on global antimicrobial resistance (AMR) which was published in 2016 [29]. AMR is the phenomen in which infectious microbes (bacteria, fungi, parasites, viruses and others) become resistant to the drugs used to treat them. Examples, discussed in the previous sections, include chloroquine-resistant malaria and methicillin-resistant *Staphylococcus aureus* (MRSA). The report found that, in 2014, 700,000 deaths per year are due to resistant strains of common bacteria, with 200,000 of these caused by *Mycobacterium tuberculosis*, the etiological agent of tuberculosis (TB) (Fig. 1.5). It was estimated that currently there are 2 million cases of AMR infections per year in the US alone, which costs the US health system in excess of 20 billion USD.

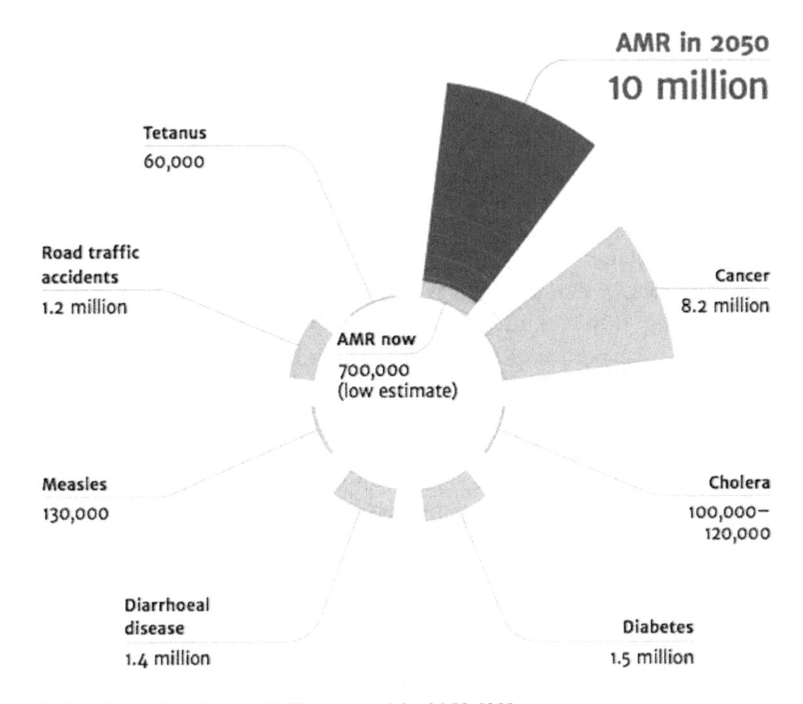

Fig. 1.5 Deaths attributable to AMR now and in 2050 [29]

More alarming were the predications made in the report for the future global impact of AMR. It suggests that, unless action is taken, by 2050 AMR will cause 10 million deaths per year (more than cancer currently causes) and burden the global economy to the scale of 100 trillion USD (Fig. 1.5). A further unquantifiable burden will be placed on healthcare systems. Current treatments for other diseases, such as cancer, require invasive surgery, organ transplant and chemotherapy, all of which leave the patient at a high risk of bacterial infection. With an increase in AMR, the routine nature of these types of procedures will be put at risk, greatly reducing the health outcomes we currently experience.

1.2.2 Mechanisms of Antibiotic Resistance

The widespread utilisation of antibiotic therapy in medicine provides an enormous selective pressure for bacteria to develop resistance. However, antibiotic resistance is a phenomenon that predates the introduction of antibiotics into the clinic. Recent work from Wright et al. [30]. identified genes encoding resistance to a number of currently used antibiotics classes [β-lactam, tetracycline and glycopeptide antibiotics (GPAs)] in a 30,000-year-old permafrost sample. The prototypical GPA is vancomycin, often used as a treatment of last resort for patients with MRSA, and to which there is now associated resistance in enterococci [vancomycin resistant enterococci (VRE)]. In this study, the authors were able to express the protein involved in vancomycin resistance and demonstrate, via X-ray crystal structure analysis, its homology to the modern day resistance protein.

Bacteria are the source of many of our antibiotics. They are produced to confer a selective advantage to the producer by inhibiting the growth of competitive bacteria. As such, to ensure they do not harm themselves, the producing organisms have developed cellular mechanisms to ensure immunity to the antibiotics they create. For example, a number of GPA producing organisms, such as the vancomycin producing *Amycolatopsis orientalis*, contain three of the five genes which are the cause of resistance in VRE [31], suggesting that the origin of VRE may be GPA producing bacteria themselves. Indeed a recent study [32] demonstrated that genes for resistance to all major classes of antibiotics found in non-pathogenic soil bacteria share DNA sequence identity to human pathogenic bacteria.

While there are a number of mechanisms by which bacteria can develop resistance to antibiotics there are three major strategies (Fig. 1.6). Firstly, reducing the concentration of the antibiotic in the cytoplasm. This is achieved by the expression

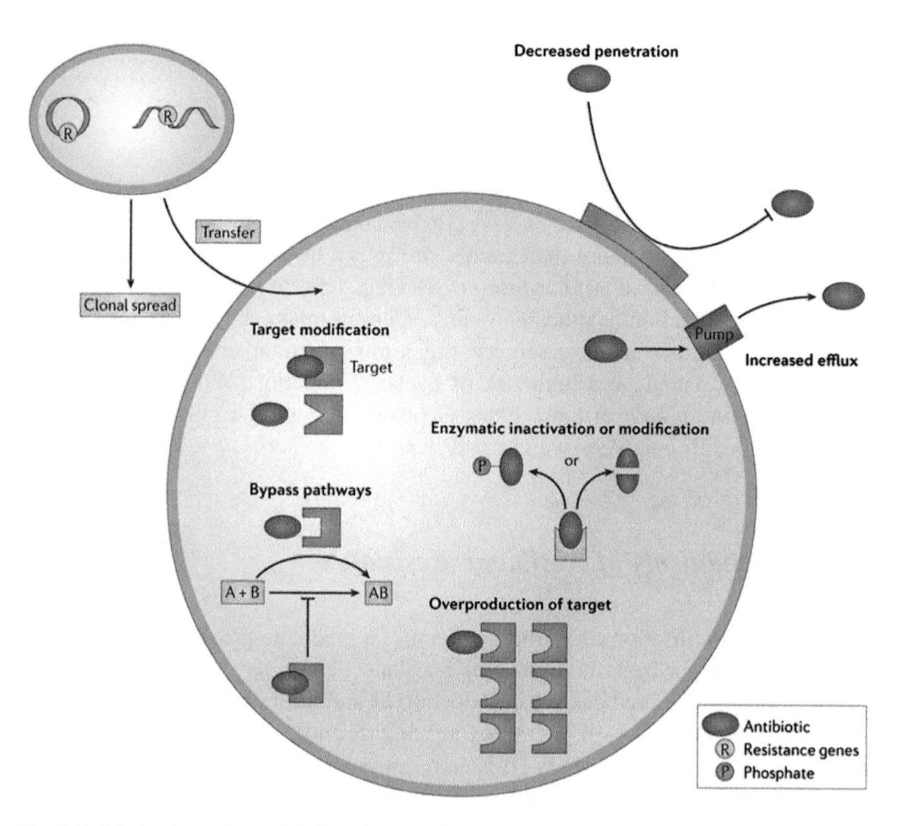

Fig. 1.6 Mechanisms for antibiotic resistance [34]

of efflux pumps, large membrane-spanning protein channels that actively transport a wide variety of chemical entities from the bacterial cell. These pumps are known to confer multi-drug resistance to bacteria, and are associated with clinically resistant pathogens [33]. The second mechanism involves modification of the antibiotic's target. This prevents the antibiotic from effectively carrying out its function and is the mechanism behind the resistance of VRE to vancomycin (see Sect. 1.3.3). The third mechanism is inactivation of the antibiotic. This involves the production of enzymes by the bacteria which are capable of modifying the antibiotic, thus rendering it ineffective. The most widely studied case are the β-lactamases, which hydrolyse β-lactam antibiotics such as penicillin thus preventing them from inhibiting their molecular target [33].

Another mechanism that bacteria use to evade the effect of antibiotics is through the production of biofilms; in which they grow on a surface and encase themselves in an extracellular matrix of polysaccharides and protein. Biofilm infections are commonly associated with medical devices such as catheters and persistent lung infections in cystic fibrosis patients [35]. Biofilm infections are extremely resistant to conventional chemotherapy for three main reasons: the matrix provides a physical barrier preventing antibiotics rapidly diffusing to bacterial cells; the altered microenvironment in the matrix can deactivate antibiotics and also cause the bacterial cells to enter a non-growing state, rendering antibiotics which inhibit growth ineffective; and finally a small portion of the bacterial cells (persisters) enter a protected phenotypic state which allows them to survive an antibiotic attack and re-establish the population when the challenge stops [35, 36] (Fig. 1.7).

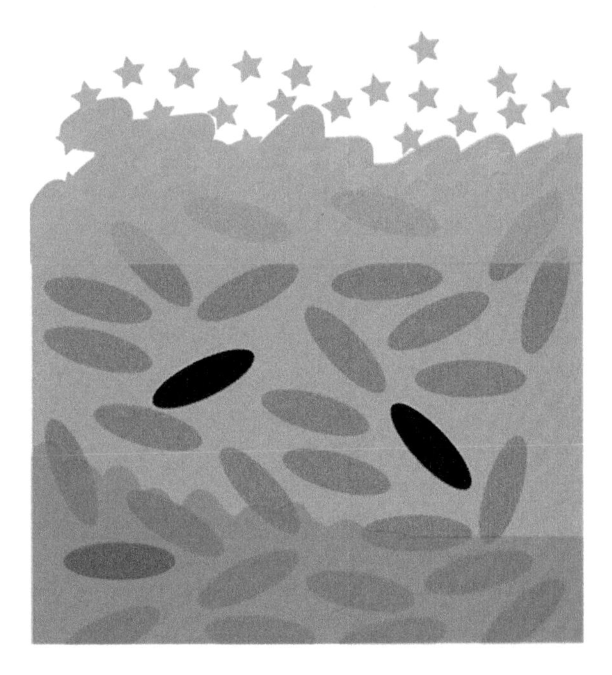

Fig. 1.7 Mechanisms for antibiotic resistance in biofilms: reduced penetration of antibiotics (green), altered microenvironment (blue) and generation of persisters (black). Stars = antibiotics

Interestingly, when bacterial cells break free of the biofilm they are usually highly susceptible to antibiotics. This presents an opportunity for a new type of therapy—molecules which cause the biofilm to break apart, in combination with conventional antibiotics. While bacteria developed resistance to antibiotics thousands of years ago, the emergence of resistant clinically relevant pathogens is rendering therapeutic antibiotics currently employed ineffective. As such, new treatments to treat resistant infections are urgently required.

1.2.3 The Challenge of Antibiotic Drug Discovery

The so called golden era of antibiotic drug discovery (following the discovery of penicillin) involved the screening of microorganism fermentation broths and extracts against bacterial growth in vitro, without an initial understanding of mechanism of action [37]. Selman Waksman of Rutgers University pioneered a hugely successful screening platform for testing extracts produced by soil microbes, particularly streptomycetes [38]. In 1943 this yielded streptomycin, the first known treatment for TB, for which Waksman was awarded the 1952 Nobel Prize in Medicine. This approach was adopted in the pharmaceutical industry and yielded numerous new antibiotics, however unfortunately after a short amount of time the same compounds were continually reisolated and new hits became scarce [38]. Subsequent efforts focused on the modification of known antibiotic scaffolds and yielded several generations of each antibiotic class able to kill resistant bacteria. This led to what is known as the discovery gap or innovation void (Fig. 1.8), in which no new classes of antibiotic were introduced to the clinic between 1962 and 2000 [39].

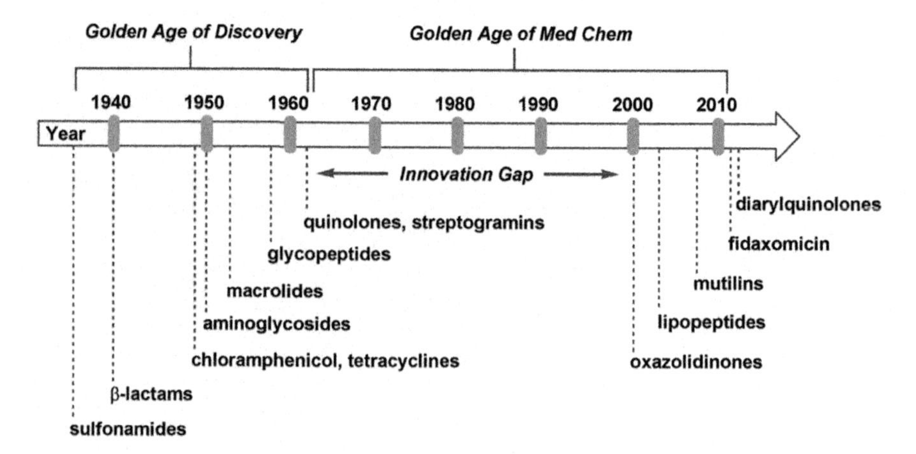

Fig. 1.8 Timeline of introduction of new classes of antibiotics [39]

Pharmaceutical companies began to move away from these programs and focused instead on target-based antibiotic discovery which was being applied to other areas of medicine, i.e. synthesis of potent inhibitors of essential enzymes [37]. The sequencing of the first bacterial genome [40] allowed researchers to determine essential gene products required for bacterial survival. High-throughput-screens (HTS), the development of large compound libraries and activity screening, could then be carried out against newly discovered antibacterial drug targets. GlaxoSmithKline ran an antibacterial discovery program between 1995–2001 completing 67 target based HTS which delivered only 5 leads with none ultimately progressing to trials [41]. Indeed while this was a successful approach for many other types of medicines, this approach is yet to deliver an antibiotic after more than 20 years [27].

One of the main problems with this approach to antibiotic drug discovery is the lack of chemical space interrogated through compounds generated by combinatorial chemistry (Fig. 1.9) [41]. Most HTS libraries fall into Lipinski's 'rule of five' [42]. These rules were developed as a means for assessing whether a molecule is likely to have the solubility and permeability properties to be a successful orally available drug candidate, based on a large data set of drugs that had progressed to clinical trials. These parameters are optimised for drugs to treat human disease, such as drugs for central nervous system conditions. Importantly, in the study Lipinski states that the classes of drugs which violate the 'rule of 5' include antibiotics. This is most likely because antibiotics are required to penetrate complex bacterial cell walls. Kim Lewis, a pioneer of antibiotic natural product discovery, has suggested that 'rules of penetration' based on known antibiotics should be developed to further aid future development of clinical drugs [38].

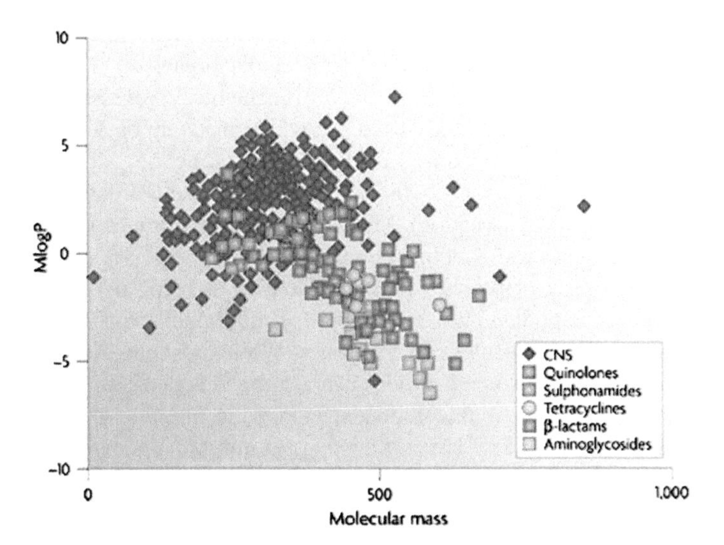

Fig. 1.9 Calculated logP versus molecular mass comparing central nervous system (CNS) drugs which follow Lipinski's 'rule of 5' and various classes of antibiotics [41]

1.2.4 New Antibiotics Are Urgently Needed

In order to fight the rising problem of antibiotic resistance, new drugs are urgently needed. Antibiotic drug development is extremely difficult (vide supra) and new strategies are required. Future efforts must focus on expanding chemical diversity and natural product likeness of chemical libraries—ideas pioneered by Schrieber [43] and Waldmann [23] among others. Indeed of all antibiotics approved, between 1981 and 2014, 73% are natural products or synthetic derivatives of natural products [24]. Other approaches include targeting previously uncultivable bacteria as well as extremophiles living in harsh natural environments. It is estimated that there are between 10^5 and 10^6 microbial species globally, however only a few thousand have been cultured in the lab. Some important developments have been made so that these traditionally uncultivable bacteria can be isolated in the lab [44]. These bacteria are likely a rich source of novel natural products and potential therapeutic antibiotics. Natural products have been the major source of antibiotics throughout history, and with so much biodiversity still left uncovered there remains reason to believe that novel antibiotics are still to be found.

1.3 Therapeutic Peptides

1.3.1 Peptides and Proteins

Proteins are one of the three major classes of biopolymers, along with nucleic acids and carbohydrates. They are made up of amino acids joined together by amide bonds and fall into two main classes: structural and functional. Structural proteins, for example collagen, provide rigidity to multicellular organisms. Functional proteins, for example enzymes, act as catalysts for biochemical processes essential for life. Proteins are synthesised on the ribosome and are made up of 20 proteinogenic amino acids. The sequence of amino acids (primary structure), determines local organisation (secondary structure), which subsequently results in an overall structure of the protein (tertiary structure). Multiple subunits can make up a functional protein and this organisation is defined as the quaternary structure (Fig. 1.10).

Peptides are a subset of proteins normally defined as being no longer than 50 amino acid units. Peptides play a number of important physiological roles, including as signalling molecules and hormones. While most peptides and proteins are synthesised on the ribosome, peptides can also be generated without the aid of the ribosome but rather by multicomponent enzymatic machinery and are called nonribosomal peptides (NRPs). These often contain non-proteinogenic amino acids and possess considerable structural complexity (see below) [45].

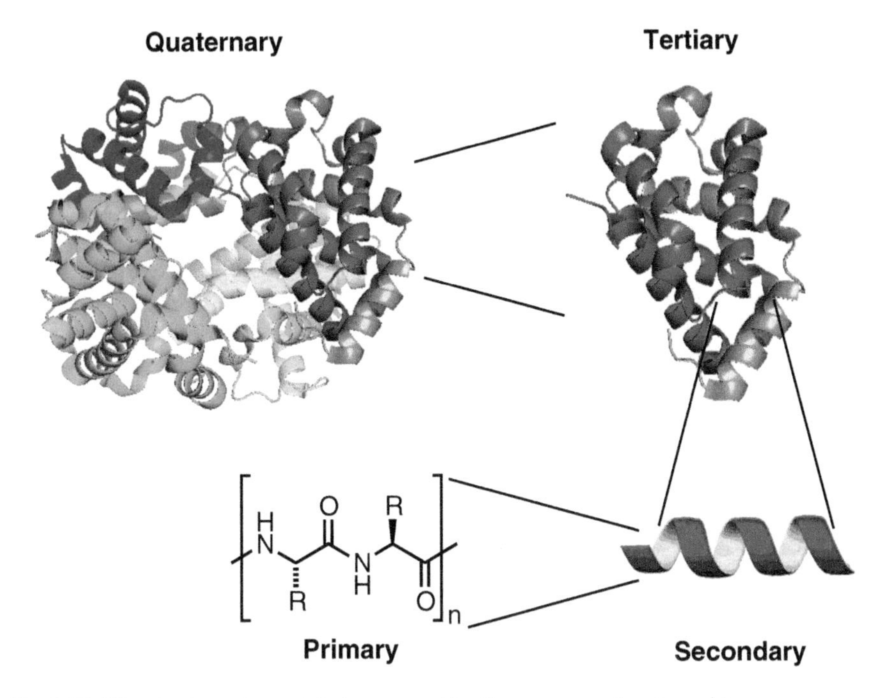

Fig. 1.10 The structure of a protein is governed by the sequence of amino acids

1.3.2 Therapeutic Peptides and NRPs

Over the last decade peptides have became a major source of new drugs. Indeed, there are currently 60 peptides approved by the federal Drug Administration (FDA), and the global value of peptide-based drugs is predicted to increase from 14.1 billion USD (2011) to 25.4 billion USD in 2018 [46]. Therapeutically, peptides are generally used to treat metabolic diseases, including diabetes and obesity, as well as cancer. Nonribosomal peptides are also rich source of therapeutics; of currently marketed drugs approximately 30 contain a NRP core and have a combined value in the billions USD. In contrast to peptide drugs, the majority of NRPs are used as antibacterials, including two of the most clinically significant antibiotics, penicillin and vancomycin [45]. Significant medicinal chemistry studies have been carried out on NRPs, with a number of semisynthetic or synthetic derivatives also in the clinic.

1.3.3 Vancomycin—A Game Changing Antibiotic

The discovery of penicillin (**5**) began the golden age of antibiotic drug discovery in which multiple new classes of natural products were discovered and found to be potent antibiotics. One of the most vital is vancomycin (**8**) (Fig. 1.12) which was isolated from *Amylocalatopsis orientalis* a bacterium present in a soil sample from the jungle in Borneo and developed by the Eli Lilly company in 1956 [47]. Vancomycin was given rapid approval by the FDA and is still used today. However, due to toxic side effects it is reserved as a drug of last resort, generally for treating MRSA.

Vancomycin (**8**) is a structurally complex nonribosomal GPA which eluded structural characterisation for a number of years. Many initial studies were carried out via degradation [48], nuclear magnetic resonance (NMR) [49] spectroscopy and X-ray crystallographic analysis [50] before Harris and Harris assigned the structure in 1982 [51]. This key structural information as well the potent activity of vancomycin and interesting mechanism of action made it a prime candidate for total synthesis in the late 1980s. The synthetic efforts were impeded by some unforeseen challenges in the structure, most notably atropisomerism about the biaryl ring junctions. However, the synthesis of vancomycin (**8**) and its aglycone (**9**) was ultimately achieved by Nicolaou [52], Evans [53] and Boger [54] among others at the end of the 1990s.

While resistance to penicillin and methicillin developed rapidly, it took almost 30 years from the time vancomycin was widely adopted in the clinic until resistant bacteria were reported. This first occurred in entercoccal isolates and became a significant problem within hospitals, with the percentage of vancomycin resistance, in isolates from nosicomal infections in intensive care units rising from 0.4% in 1989 to 13.6% in 1993 [55]. Vancomycin exerts its antibiotic affect by binding, via 5 hydrogen bonds, to the D-Ala-D-Ala terminal portion of Lipid II, a key substrate in bacterial cell wall biosynthesis (Figs. 1.11 and 1.12). This prevents critical cross-links from forming and ultimately leads to a weakened cell wall and cell rupture under osmotic pressure. Resistance was slow to develop because vancomycin binds to a substrate, rather than an enzyme, which is significantly harder for a bacterium to mutate.

Pioneering work by Christopher Walsh at Harvard University uncovered the biochemical mechanisms behind resistance to vancomycin, namely five genes and one missing hydrogen bond [56]. The presence of these genes results in the mutation of D-Ala-D-Ala **10** to D-Ala-D-Lac **11** in Lipid II, which prevents one key hydrogen bonding interaction (red arrow, Fig. 1.12), and reduces the binding affinity of vancomycin to its substrate by 1000 fold. Significantly, the synthetic groundwork laid by Boger et al. in their 1999 total synthesis [54] meant a rationally designed chemical analogue **12** which binds to mutated Lipid II could be readily accessed. In this ground breaking work by Boger et al. the introduction of an amidine in place of the natural amide functional group present in vancomycin (Y = NH, Fig. 1.12) is described [57]. This functional group can make hydrogen bonds to both mutated and unmutated lipid II, resulting in a molecule with restored activity against vancomycin resistant enterococci (VRE), and equipotency against the susceptible strain [58].

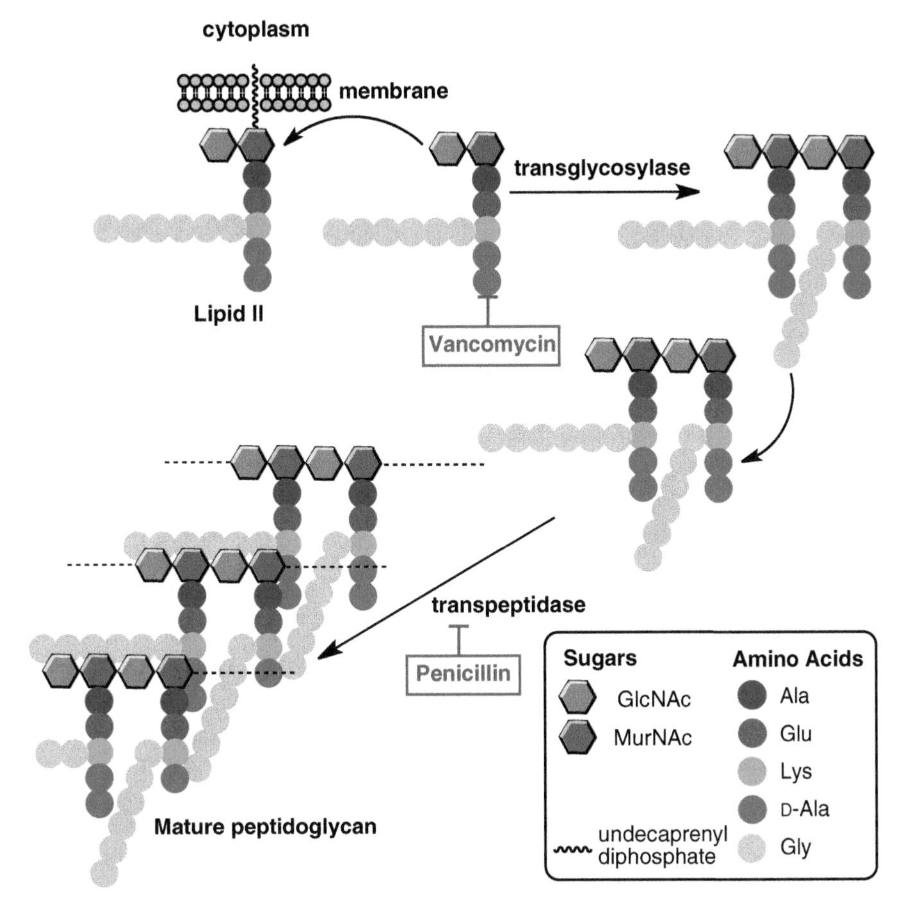

Fig. 1.11 Bacterial cell wall biosynthesis. Vancomycin (**8**) binds to lipid II preventing transglycosylation and transpeptidation, and the β-lactams [such as pencillin (**5**)] inhibit the transpeptidase enzymes preventing the synthesis of mature peptidoglycan

1.3.4 Daptomycin

Another important clinically utilised NRP antibiotic is daptomycin (**13**) (Fig. 1.13) which belongs to the class of cyclic lipopeptide antibiotics. In 1988 scientists at Eli Lilly discovered that a derivative of the antibiotic A21978C, in which the natural fatty acid chain is replaced by a linear C_{10} chain—named daptomycin, had potent antibiotic activity and low mammalian toxicity [59]. Initial clinical trials were carried out by Lilly, however were soon abandoned due to issue with a small therapeutic window between efficacy and safety [60]. It was licenced not long after by Cubist Pharmaceuticals and was eventually approved by the FDA in 2003 as a treatment for MRSA [61]. The introduction of daptomycin represented the first new class of antibiotic to be approved in the clinic for a number of years.

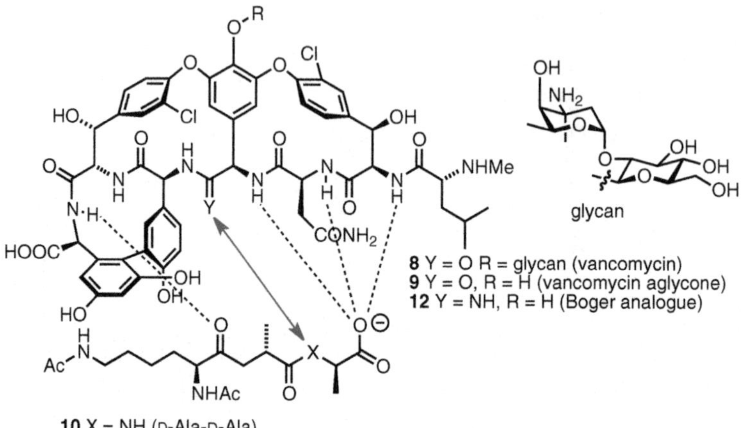

8 Y = O R = glycan (vancomycin)
9 Y = O, R = H (vancomycin aglycone)
12 Y = NH, R = H (Boger analogue)

10 X = NH (D-Ala-D-Ala)
11 X = O (D-Ala-D-Lac)

Fig. 1.12 Structure and binding of vancomycin (**8**) and the Boger analogue **12** to a Lipid II mimic containing an unmutated (D-Ala-D-Ala) **10** or mutated (D-Ala-D-Lac) **11** terminus

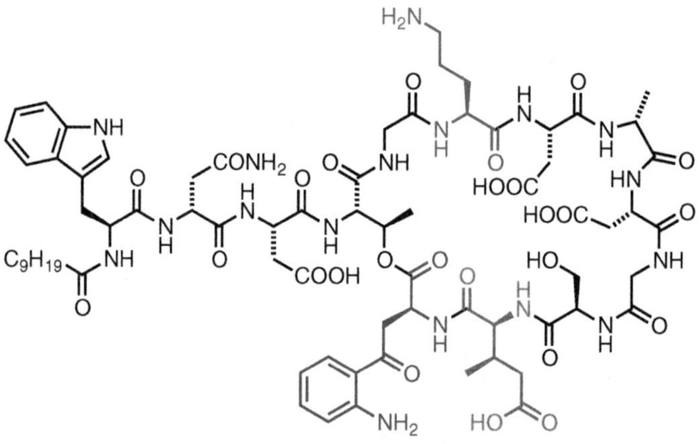

13

Fig. 1.13 Structure of daptomycin (**13**)

Daptomycin (**13**) contains a number of structural features which are common among NRPs. The first of these is the presence of a number of non-proteinogenic amino acids. These include three with D-configuration: D-Ser, D-Ala and D-Asn; the unusual kynurenine (blue), ornithine (red), 3-methyl-glutamic acid (green) and the decanoyl chain (Fig. 1.13). Another common structural feature of NRPs is cyclisation, and daptomycin is cyclised between the Kyn and the side chain of the threonine (Thr) residue.

Compared with other clinical antibiotics, daptomycin has a distinct mechanism of action. In solution, in the presence of calcium ions, daptomycin oligomerises, followed by a dissociation event and insertion into the bacterial cell membrane. Importantly, the ability of daptomycin to insert into the bacterial cell membrane is dependent on the composition of the membrane. Negatively charged phospholipids bind to calcium ions, which facilitate the insertion of daptomycin into the membrane. Once daptomycin inserts in the membrane, leakage of the cytoplasm and depolarisation of the cell occurs resulting in cell death [62]. Due to this novel mechanism of action, there was initially a lot of excitement at the introduction of daptomycin, with reports demonstrating resistance was rare during clinical trials. However, recent reports have found that resistance can be generated by alteration of the phospholipid content in the bacterial cell membrane [33].

Both peptides and NRPs are an important and therapeutically significant source of drugs with NRPs representing a number of classes of clinically employed antibiotics. With the increasing problem of antibiotic resistance and the urgent need for new drugs, novel NRPs represent a class of molecules with great therapeutic potential.

1.4 Biosynthesis of NRPs

Nonribosomal peptides are a structurally diverse and therapeutically relevant class of natural products. They are synthesised almost exclusively by bacteria and fungi, with one of the main producers, the *Streptomyces* genus of actinobacteria, exploited successfully in the Waksman platform. The large multidomain enzymes responsible for their biosynthesis are known as nonribosomal peptide synthetases (NRPSs). The genes that encode for NRPSs are generally found in clusters with other genes required for the synthesis of the mature product such as: the synthesis of amino acid building blocks, peptide modification, self-resistance and export. With improved modern bioinformatic methods, and a better understanding of enzymes involved in the synthesis of NRPs, the so-called 'biosynthetic logic' of NRPSs can now be decoded such that the putative structure of the NRP products can be predicted from the genome [63].

NRPSs produce a vast array of structurally complex molecules. As well as the incorporation of non-proteinogenic amino acids, NRPs are often cyclic and functionalised with a diverse array of chemical modifications. The biosynthetic methods by which they are constructed: synthesis of amino acids, peptide backbone assembly, followed by tailoring and chemical modification, will be discussed below.

1.4.1 Generation of Amino Acid Diversity

Nature has developed a myriad of different structural modifications of amino acid building blocks used in NRPs and this leads to a significant amount of structural diversity. Broadly, they are either carried out prior to loading into the NRPS machinery or while on the NRPS by tailoring enzymes. Two of the most common modifications found in NRPs are methylation and hydroxylation, which are often found at the β-position of the amino acid. These are often generated prior to loading into the NRPS, for example the β-OH of the chlorotyrosine residues present in the GPAs such as vancomycin. Biochemical studies by Wohlleben and co-workers determined the enzymes responsible for the synthesis of the β-OH-tyrosine [64]. Interestingly, one of these encodes a standalone NRPS on which tyrosine is loaded and subsequently β-hydroxylated by the cytochrome P450 monooxygenase enzyme OxyD (Fig. 1.14). Importantly, the authors determined that the chlorination of the β-OH-tyrosine does not occur prior to loading onto the NRPS but must occur while it is loaded on the main NRPS.

Another common modification is methylation. While, dedicated modules on the NRPS (see below) carry out *N*-methylation, β-Me is usually generated in an *S*-adenosylmethinonine (SAM)-dependent manner prior to loading onto the NRPS. As mentioned previously, the potent antibiotic daptomycin contains a β-Me-Glu (**14**). Recently, the biosynthetic pathway to this unusual amino acid was elucidated [65], and interestingly it is not biosynthesised from glutamic acid itself, but rather from α-ketoglutarate (**15**) (Scheme 1.1). The first step is the stereoselective SAM-dependent methylation, followed by transamination of **16** leading to the final β-Me-Glu (**14**), which can then be incorporated into the NRPS en route to daptomycin.

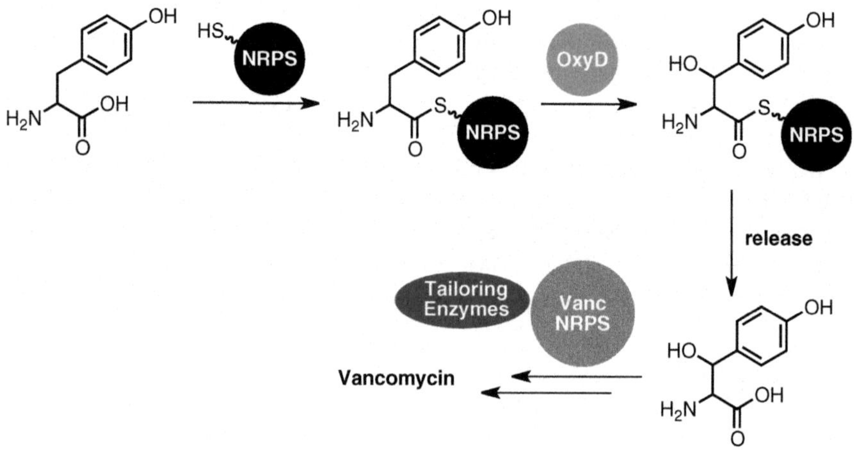

Fig. 1.14 Origin of β-OH-chlorotyrosine present in vancomycin (**8**)

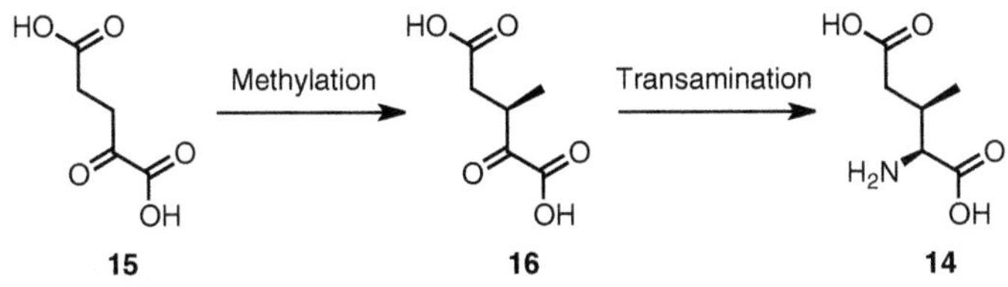

Scheme 1.1 Biosynthesis of β-Me-Glu (**14**) present in daptomycin (**13**)

A detailed discussion of the breadth of Nature's diversity in regard to modified amino acids is beyond the scope of this introduction. However, a number of other important modifications will be discussed in greater detail in Chaps. 2–4.

1.4.2 Peptide Elongation and Release

Once the appropriate amino acids have been synthesised the peptide backbone can be assembled via the NRPS (Fig. 1.15). This enzymatic machinery is extremely complex and years of work by many prominent researchers have gone into elucidating the molecular mechanisms by which they function. Nature employs a modular approach to the synthesis of NRPs. Each amino acid in the mature NRP has an associated module in the NRPS, and each module contains three essential domains: peptidyl carrier protein (PCP), adenylation (A), and condensation (C) domains as well as optional domains such as epimerase (E) and methylation (M), among others. The structure and role of the three essential domains will be discussed in more detail below.

Peptidyl carrier protein

The first step in the activation of a NRPS is the modification of the PCP, also known as the thiolation domain (T), a four-helix bundle. The molecular mechanism of activation is the transfer of a 4′-phosphopantetheine (Ppant) group onto the highly conserved Ser residue of the inactive apo-PCP, resulting in an active holo-PCP (Fig. 1.16). This Ppant moiety bears a terminal thiol which attacks the activated amino acid held in the A domain (see below). The PCP acts as the flexible arm which transfers the amino acid and growing peptide chain between domains for modification (e.g. epimerase), and C domains for construction of the peptide backbone [66]. Another important role carried out by the PCP is the recruitment and binding of enzymes, which modify the growing NRP in *trans* (i.e. they are not part of the NRPS but are recruited separately). These interactions, both with adjacent domains and modifying enzymes, are mainly mediated by two of the four helices forming an interface with the binding partners [66].

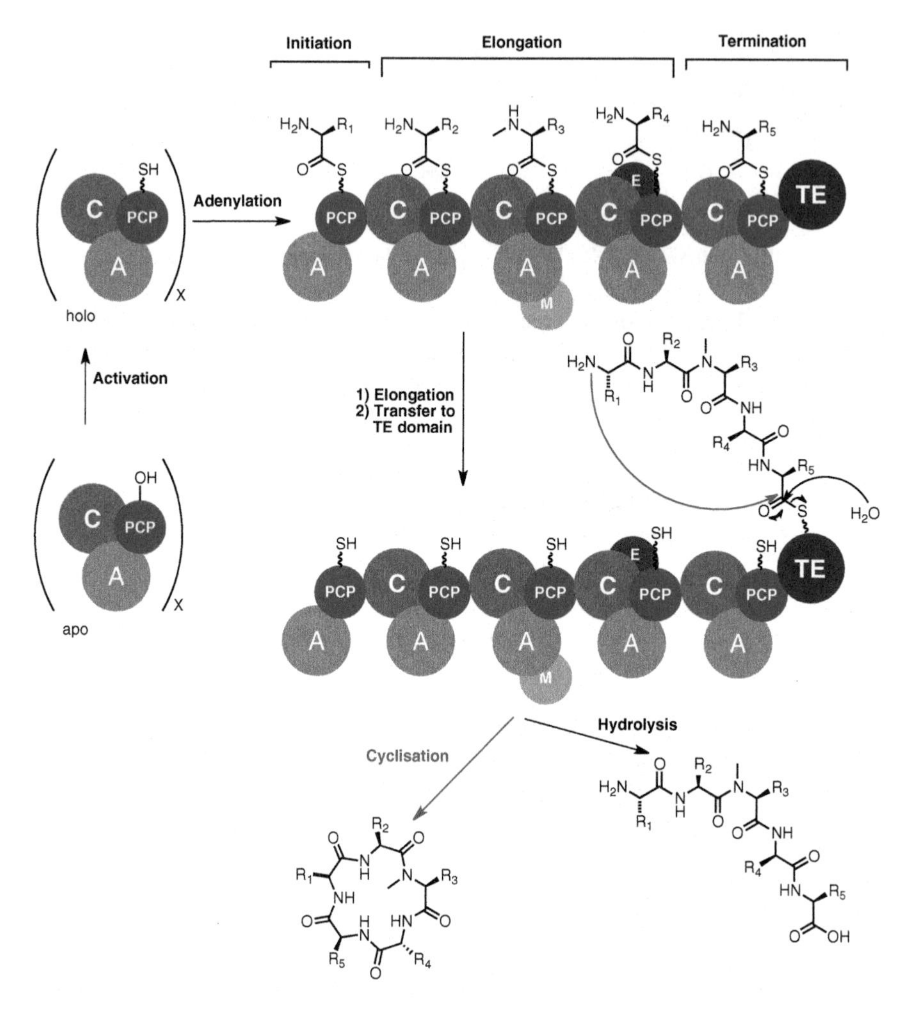

Fig. 1.15 The synthesis of a NRP on a NRPS. The NRPS is first activated, followed by adenylation to form the active amino acid thioesters. Subsequent elongation forms the linear peptide, and release from the TE domain produces the mature NRP

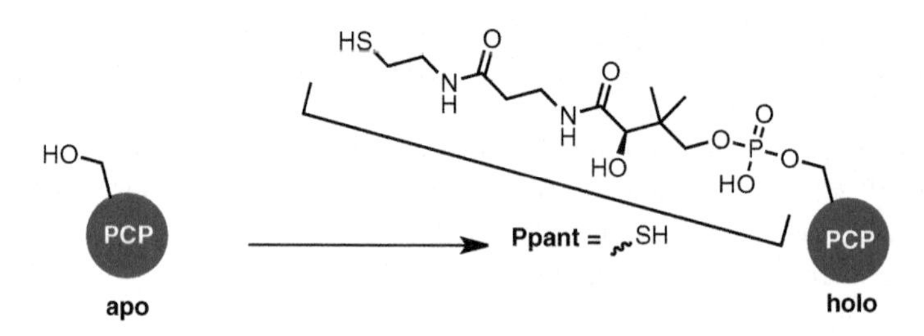

Fig. 1.16 Activation of the PCP domain via PPTase catalysed transfer of Ppant to the conserved Ser residue

Adenylation

The adenylation domain carries out the first step of NRP synthesis, namely selection and activation of the specific amino acid for that module and transfer to the PCP. Adenylation domains are typically 60 kDa and made up of a core (50 kDa) and sub domain (10 kDa) linked by a short hinge region, and function by activating the amino acid as the adenosine monophosphate active ester and shielding the active ester from unproductive hydrolysis by water (Fig. 1.17a). The core domain selects for the specific amino acid (blue) in a binding pocket that contains conserved Asp and Lys residues (red, Fig. 1.17b) which stabilise the amine and carboxylate of the amino acid, respectively, as well as up to eight residues which impart selectivity on the sidechain [67].

Condensation

The C domains carry out the critical role of peptide bond formation (Fig. 1.18). The C domain of a given module catalyses the amide bond formation between the amino acid thioester on the PCP (acceptor) and the amino acid or peptidyl chain from the upstream module (donor). Structurally it is a 50 kDa pseudo dimer

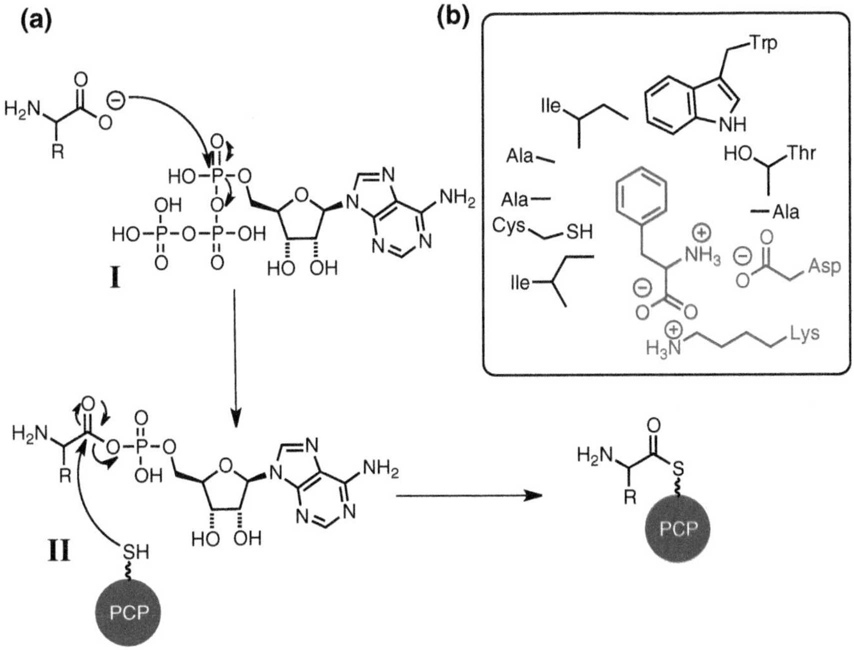

Fig. 1.17 **a** Pathway in which an amino acid is activated in the A domain. I Formation of the adenosine monophosphate active ester. II Loading of the amino acid onto the holo-PCP domain to form the PCP amino acid thioester. **b** A domain binding pocket for selection of the appropriate amino acid (blue), phenylalanine as an example, highlighting conserved Asp and Lys residues (red)

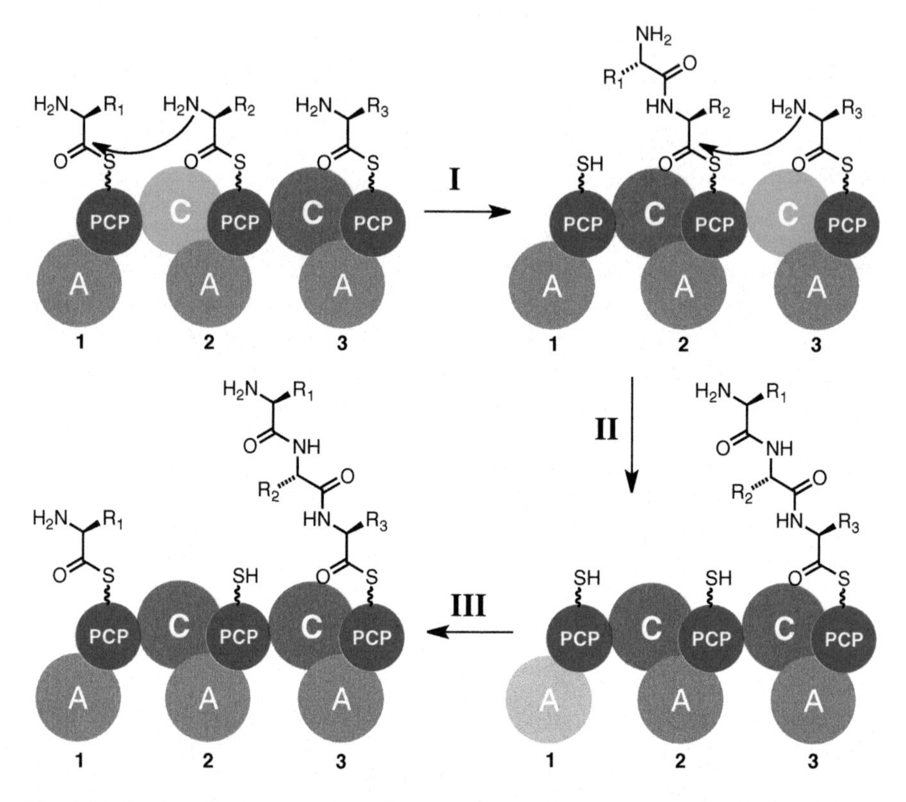

Fig. 1.18 A schematic representation of the iterative condensation events in the NRPS. I The donor amino acid in module 1 is transferred to module 2, catalysed by the module 2 C domain. II The donor dipeptide on module 2 is transferred to acceptor amino acid on module 3, catalysed by the module 3 C domain. III The adenylation domain in module 1 catalyses the formation of the PCP-amino acid thioester

composed of a *C*-terminal and *N*-terminal subdomain which give the protein a V-shaped structure [68]. The Ppant arms of both the donor and acceptor PCP move into the conserved HHxxxDG (H = His, D = Asp, G = Gly) active site, at the central cleft of the C domain. The acceptor site of the C domain also imparts important additional amino acid selectively in case of incorrect amino acid activation at the A domain [69]. The exact catalytic mechanism for peptide bond formation is currently being investigated. It is worth noting that the first module in an NRPS, the initiation module, lacks the condensation domain since the Ppant loaded amino acid is transferred to the downstream C domain for elongation.

Peptide Release

The manner in which the peptide is released from the NRPS is a key generator of chemical diversity. There are numerous methods by which this occurs, generating diverse structures such as linear peptides, macrolactams or macrolactones, and *C*-terminal aldehydes and amides, among others. In the case of linear and cyclic peptides, in most cases, following transfer of the growing peptide chain to the final module PCP, the peptide is transferred to the active site Ser or Cys in the thioesterase (TE) domain [45]. At this point there are a number of alternative mechanisms by which the peptide chain is released. If water acts as a nucleophile, the peptide is cleaved as the linear chain (black arrow, Fig. 1.15), however, as is often the case, the cyclic product is formed by attack of an internal nucleophile (terminal NH_2, sidechain NH_2 or OH) to form the macrolactam (red arrow, Fig. 1.15), or macrolactone. Depending on the NRPS, the TE domain is either accepting of diverse substrates (e.g., tyrocidine [70]), or requires the substrate to contain a specific structural motif (e.g., β-lactam in norcardicin A [71]) before hydrolysis will occur. Interestingly, the TE domain in norcardicin A biosynthesis also acts as an epimerase, further highlighting the diverse catalytic activity of NRPSs.

The complete 3D structure of an intact NRPS, and the phenomenon by which they ensure the correct directionality is not completely understood. However, it seems all three key domains, A, C and PCP work in concert, via conformational changes and binding affinities, to guide the elongating peptide to the correct domains and modules in the right order to produce the desired product [72].

1.4.3 Backbone Modification

NRPs contain numerous modifications that give rise to their chemical diversity, a key factor in the wide bioactivity they exhibit. Two of the most common modifications are epimerisation of the amino acid α-centre and methylation of the backbone amides, carried out by epimerase (E) and methylation (M) domains respectively. Further modifications can be carried out in *trans*, by enzymes interacting with the NRPS bound peptide, while others are carried out once the peptide leaves the NRPS machinery.

Epimerase

One of the most common amino acids found in NRPs are of D-configured amino acid residues. This modification helps to confer resistance to proteolytic breakdown and also greatly increases the chemical space accessible to NRPs. The majority of D-amino acids are incorporated with exquisite selectivity into the peptide backbone owing to the presence of a dedicated epimerase domain present in the module in between the PCP and C domains. As such, the amino acid is epimerised after condensation with the upstream C domain, prior to condensation with the downstream C domain. E domains are structurally very similar to C domains and retain a similar active site structure, including a conserved His residue, indicating a possible common evolutionary origin [73].

Methylation

A further important modification of NRPs is backbone *N*-methylation which has a significant impact on polarity, H–bonding capability, conformation and proteolytic stability. This can lead to an increase in oral bioavailability and cell permeation [74]. The majority of backbone *N*-methylation events are catalysed by an optional methylation (M) domain, integrated into the A domain of the given module, which carries out methylation in the presence of the cosubstrate SAM [75].

The NRPS is an extremely complex molecular machine capable of generating structurally diverse products. Importantly, NRPs have privileged bioactivity, including privileged antibacterial activity. Understanding the biosynthetic mechanism behind NRPSs opens up the possibility of hijacking this machinery to produce natural product analogues, with favourable bioactivity profiles [76]. An important complementary approach is the total chemical synthesis of NRPs. This also opens up the ability to generate analogues, and further understand the biology of these molecules. A beautiful example is Boger's synthesis of the vancomycin analogue described above [57].

1.5 Chemical Synthesis of Peptides

Peptides and proteins are key components of all living organisms. An important area of research within biological and chemical sciences involves understanding how peptides and proteins interact with other proteins and other molecules. One extremely useful method for understanding the structure and function of peptides and proteins is via total chemical synthesis. This provides access to homogenous samples of the biomolecules, often difficult to isolate from a biological source, and allows for chemical modifications to be incorporated.

In 1901 Emil Fischer (winner of the 1902 Nobel Prize in Chemistry) constructed the first synthetic peptide, a glycine-glycine dipeptide [77] which is considered the birth of chemical peptide synthesis research. While a significant advance, the field was hampered by the lack of a suitable protecting group for the α–amine. The benzyloxycarbonyl (Cbz) group **17** was developed by Bergmann and Zervas in 1932 [78], which could be removed easily via catalytic hydrogenation, or by HBr in acetic acid. This was used as the predominant *N*-terminal protecting group, until the introduction of the *tert*-butyloxycarbonyl (Boc) [79, 80] group **18**, which could be conveniently cleaved utilising trifluoroacetic acid (TFA) (Fig. 1.19). An important concept in peptide synthesis is the use of orthogonal protecting groups. All amino acids with reactive side chain functionality, e.g. Lys (NH_2) or Asp (COOH), need to be protected on the side chain to ensure peptide bond formation occurs between the α-amine and the α-carboxylic acid of the two coupling partners, but these protecting groups must remain intact under conditions used to remove protection from the α-amine.

An early elegant example, demonstrating this orthogonality, was the synthesis of adrenocorticotrophic hormone by Schwyzer and Sieber [81] in which they utilised Cbz protection for the α-amine and Boc/*t*Bu (the COOH protecting group equivalent of Boc) for the side chain functionalities. This strategy resulted in a fully elongated protected peptide, which was deprotected using a single TFA treatment.

Fig. 1.19 Common amino acid *N*-terminal protecting groups and their deprotection conditions

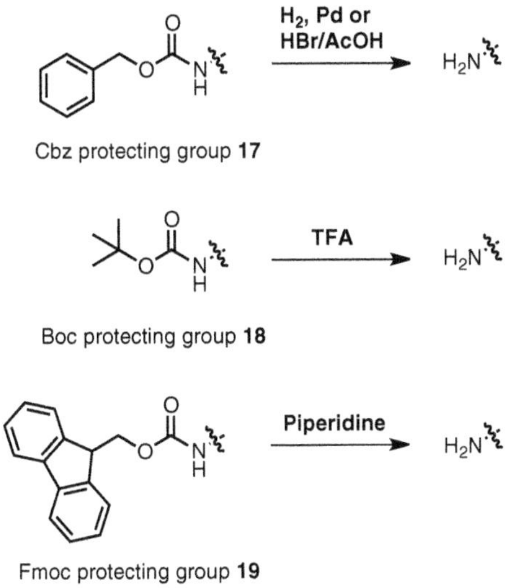

H_2, Pd or HBr/AcOH

Cbz protecting group **17**

TFA

Boc protecting group **18**

Piperidine

Fmoc protecting group **19**

For 61 years peptides were synthesised in solution phase, via the couplings of protected amino acids, highlighted by the total synthesis of the fully functional 124-residue lysozyme enzyme [82]. However this methodology was plagued with inefficiencies including: epimerisation during peptide bond formation, difficulty in purification and analysis, and poor solubility of large protected peptide fragments [83].

1.5.1 Solid-Phase Peptide Synthesis

In 1963 Merrifield [84] introduced the concept of solid-phase peptide synthesis (SPPS) in which an *N*-terminally protected amino acid is loaded onto an insoluble solid support (resin), followed by removal of the protecting group (deprotection) from the α-amine and formation of the peptide bond with an activated amino acid (coupling). The deprotection and coupling steps are repeated iteratively until the desired peptide is generated. The final step involves the cleavage of the complete peptide chain from the solid support and removal of side chain protecting groups (Scheme 1.2). Since the peptide is conjugated to the solid phase, after each reaction the resin can simply be washed to remove any unwanted by-products and reagents. This allows a large excess of amino acids to be used in the coupling step, pushing reactions near to completion resulting in a highly efficient synthesis. This methodology was a huge breakthrough, so much so, that Merrifield was awarded the Nobel Prize in Chemistry in 1984.

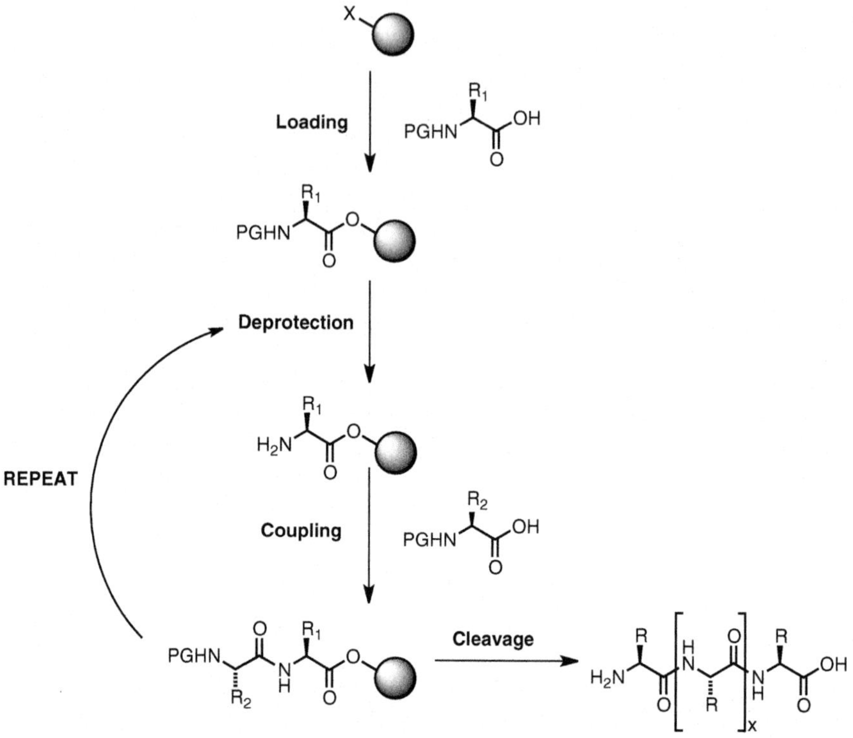

Scheme 1.2 A schematic representation of SPPS. PG = protecting group

1.5.2 Further Developments

Since the development of SPPS, a plethora of synthetic technologies have been developed to allow access to numerous different peptides and proteins. The traditional protecting groups used under the 'Merrifield SPPS protocol' are α-amine Boc and side-chain benzyl protecting groups. This methodology requires the use of TFA for each Boc deprotection step and a final treatment with anhydrous HF, to cleave the peptide from the resin, and remove the side chain protecting groups. A disadvantage of this technique is that the *N*-terminal and side-chain protecting groups are not completely orthogonal, both require acid for removal. As such, over successive TFA treatments, there is a chance side-chain protecting groups will be removed, leading to by-products. A major advancement was the development of the 9-fluorenylmethyloxycarbonyl (Fmoc) *N*-terminal protecting group [85] **19**, and its incorporation into SPPS [86, 87]. This provided a truly orthogonal protecting group strategy; the Fmoc group is deprotected using a mild amine base, such as piperidine, (Fig. 1.19) while side-chains can be protected using Boc/*t*Bu strategy, and removed during a final acidic cleavage from resin using TFA. This methodology removed the requirement for anhydrous HF, which increased the accessibility of peptide synthesis to non-specialist groups, and is now widely utilised alongside the Merrifield protocol.

One of the key aspects of peptide synthesis is the reliable and efficient formation of amide bonds. A key development, in 1955, from Sheehan and Hess was the development of *N,N′*-dicylcohexylcarbodiimide (DCC) [88] (**20**) (Scheme 1.3). This coupling reagent acts by activating a carboxylic acid **21** as an *O*-acylisourea intermediate **23**, allowing an amine to add to the carbonyl carbon, producing an insoluble urea by-product **23**. One of the major problems with this coupling technology is that it can lead to loss of chiral integrity at the α–centre of the amino acids during coupling. Once the *O*-acylisourea intermediate **22** forms, the adjacent carbonyl oxygen can attack the active ester leading to the formation of an oxazolone **24**. The α–H becomes very acidic, and can be deprotonated leading to an sp^2 hybridised carbon atom in **25**, which can then be reprotonated from either face

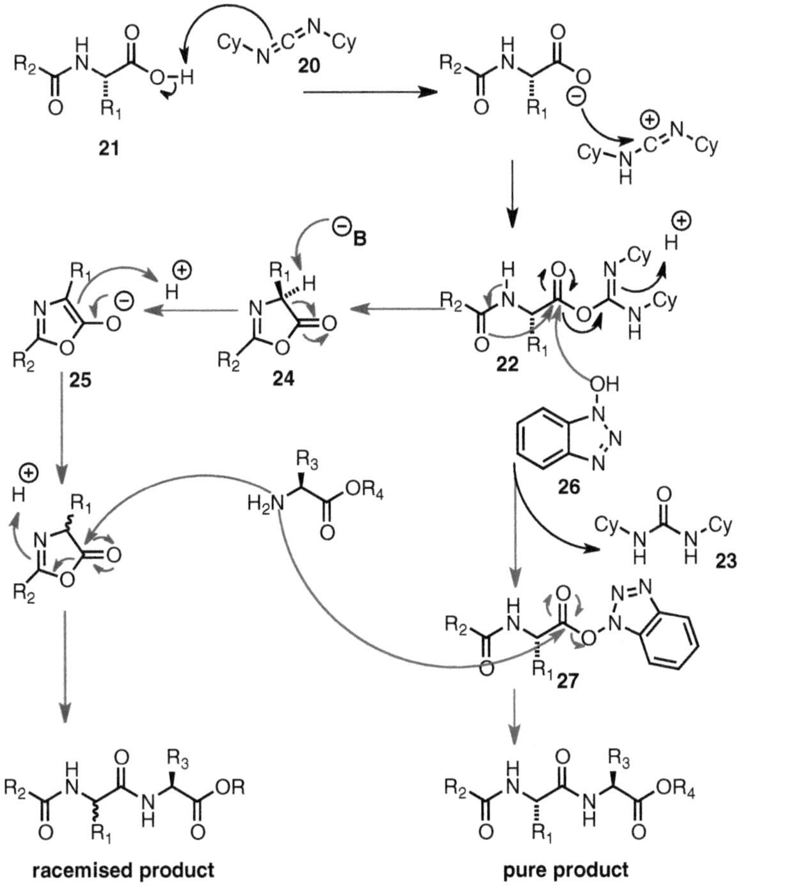

Scheme 1.3 Formation of a peptide bond utilising DCC (**20**) only (blue), or DCC (**20**) and HOBt (**26**) (red)

resulting in racemisation. In 1970 König and Geiger developed an additive, 1-hydroxy-1*H*-benzotriazole (HOBt) [89] (**26**) which reacts with the *O*-acylisourea intermediate, before oxazolone formation occurs. This HOBt active ester **27** is able to react with the amine-coupling partner, but is not reactive enough to induce oxazolone formation, greatly reducing epimerisation. Since the invention of HOBt as a coupling additive, the field of peptide coupling reagents has expanded greatly; there are now numerous coupling reagents, many of which have been optimised for specific purposes [90].

The development of solid-phase peptide synthesis, and further advancements in terms of coupling reagents and resins, have greatly expedited the synthesis of peptides and have allowed for many novel aspects of both chemistry and biology to be explored thoroughly. Furthermore, it also provides an extremely convenient method for the synthesis of NRPs. Due to the complexity of NRPs they were often synthesised via solution phase methods. Recently, however, with improvements in technologies related to SPPS, many NRPs have been synthesised utilising this methodology. These studies often require extensive optimisation to overcome difficulties associated with coupling unusual amino acids, the presence of esters, and presence of *N*-methyl amino acids, or D-amino acids which are not normally encountered during the synthesis of regular peptides. The development of an SPPS-based strategy for the total synthesis of NRPs allows for the rapid synthesis of analogues to further probe important biological functions.

1.6 Aims of Thesis

Antibiotic resistance poses a significant threat to global society, and there is an urgent need for new therapies. Natural products, in particular NRPs, are an essential source of antibiotics. The isolation, structural elucidation and understanding of the biosynthetic pathways that Nature employs is critical. Moreover the total chemical synthesis of natural products plays an essential role in facilitating the generation of analogues with more favourable properties. The aims of the work described in this thesis are to carry out the total chemical synthesis of two NRP natural product classes: teixobactin and the skyllamycins.

- Chapter 2 describes the first total synthesis of teixobactin via a solid-phase peptide synthesis-macrolactamisation approach and subsequent biological evaluation.
- Chapter 3 describes the synthesis of the non proteinogenic amino acids present in skyllamycins A-C and the synthesis of des-hydroxy analogues of the skyllamycins
- Chapter 4 describes the successful total synthesis of skyllamycins A-C utilising and expanding on the chemistry described in Chap. 3.

References

1. D.A. Dias, S. Urban, U. Roessner, Metabolites **2**, 303 (2012)
2. R.A. Maplestone, M.J. Stone, D.H. Williams, Gene **115**, 151–157 (1992)
3. A.L. Harvey, R. Edrada-Ebel, R.J. Quinn, Nat. Rev. Drug Discov. **14**, 111–129 (2015)
4. J. Achan, A.O. Talisuna, A. Erhart, A. Yeka, J.K. Tibenderana, F.N. Baliraine, P.J. Rosenthal, U. D'Alessandro, Malar. J. **10**, 144 (2011)
5. K. Krafts, E. Hempelmann, A. Skórska-Stania, Parasitol. Res. **111**, 1–6 (2012)
6. P. Rabe, Ber. Dtsch. Chem. Ges. **40**, 3655–3658 (1907)
7. P. Rabe, K. Kindler, Ber. Dtsch. Chem. Ges. **51**, 466–467 (1918)
8. R.B. Woodward, W.E. Doering, J. Am. Chem. Soc. **66**, 849 (1944)
9. G. Stork, D. Niu, A. Fujimoto, E.R. Koft, J.M. Balkovec, J.R. Tata, G.R. Dake, J. Am. Chem. Soc. **123**, 3239–3242 (2001)
10. C. Trenholme, R. Williams, R. Desjardins, H. Frischer, P. Carson, K. Rieckmann, C. Canfield, Science **190**, 792–794 (1975)
11. T. Youyou, *Nobel Lecture—Artemisinin—A Gift from traditional Chinese Medicine to the World*, Nobel Media AB 2014 (2015), www.nobelprize.org/nobel_prizes/medicine/laureates/2015/tu-lecture.html
12. Y. Tu, Nat. Med. **17**, 1217–1220 (2011)
13. A. Fleming, Br. J. Exp. Pathol. **10**, 226–236 (1929)
14. E. Chain, H.W. Florey, A.D. Gardner, N.G. Heatley, M.A. Jennings, J. Orr-Ewing, A.G. Sanders, The Lancet **236**, 226–228 (1940)
15. E.P. Abraham, E. Chain, C.M. Fletcher, A.D. Gardner, N.G. Heatley, M.A. Jennings, H.W. Florey, The Lancet **238**, 177–189 (1941)
16. D. Crowfoot, C.W. Bunn, B.W. Rogers-Low, A. Turner-Jones, *Chemistry of Penicillin* (Princeton University Press, 1949), pp. 310-366
17. J.C. Sheehan, K.R. Henery-Logan, J. Am. Chem. Soc. **81**, 3089–3094 (1959)
18. J.C. Sheehan, K.R. Henery-Logan, J. Am. Chem. Soc. **79**, 1262–1263 (1957)
19. J.C. Sheehan, K.R.H. Logan, J. Am. Chem. Soc. **81**, 5838–5839 (1959)
20. F.R. Batchelor, F.P. Doyle, J.H.C. Nayler, G.N. Rolinson, Nature **183**, 257–258 (1959)
21. G.N. Rolinson, F.R. Batchelor, S. Stevens, J.C. Wood, E.B. Chain, The Lancet **276**, 564–567 (1960)
22. E.P. Abraham, E. Chain, Nature **146**, 837 (1940)
23. R. Breinbauer, I.R. Vetter, H. Waldmann, Angew. Chem. Int. Ed. **41**, 2878–2890 (2002)
24. D.J. Newman, G.M. Cragg, J. Nat. Prod. **79**, 629–661 (2016)
25. J.W. Blunt, B.R. Copp, R.A. Keyzers, M.H.G. Munro, M.R. Prinsep, Nat. Prod. Rep. **34**, 235–294 (2017)
26. K.C. Nicolaou, Cell Chem. Biol **21**, 1039–1045 (2014)
27. E.D. Brown, G.D. Wright, Nature **529**, 336–343 (2016)
28. World Health Organisation, *Global Action Plan on Antimicrobial Resistance* (2015), www.who.int/antimicrobial-resistance/global-action-plan/en/
29. J. O'Neill, *Tackling Drug-Resistant Infection Globally: Final Report and Recommendations—The Review on Antimicrobial Resistance* (2016), https://amr-review.org
30. V.M. D'Costa, C.E. King, L. Kalan, M. Morar, W.W.L. Sung, C. Schwarz, D. Froese, G. Zazula, F. Calmels, R. Debruyne, G.B. Golding, H.N. Poinar, G.D. Wright, Nature **477**, 457–461 (2011)
31. C.G. Marshall, I.A.D. Lessard, I.S. Park, G.D. Wright, Antimicrob. Agents Chemother. **42**, 2215–2220 (1998)
32. K.J. Forsberg, A. Reyes, B. Wang, E.M. Selleck, M.O.A. Sommer, G. Dantas, Science **337**, 1107 (2012)
33. J.M.A. Blair, M.A. Webber, A.J. Baylay, D.O. Ogbolu, L.J.V. Piddock, Nat. Rev. Microbiol. **13**, 42–51 (2015)
34. K. Lewis, Nat. Rev. Drug Discov. **12**, 371–387 (2013)

35. P.S. Stewart, J. William Costerton, The Lancet **358**, 135–138 (2001)
36. K. Lewis, Antimicrob. Agents Chemother. **45**, 999–1007 (2001)
37. L.L. Silver, Clin. Microbiol. Rev. **24**, 71–109 (2011)
38. K. Lewis, Nature **485**, 439–440 (2012)
39. C.T. Walsh, T.A. Wencewicz, J. Antibiot. **67**, 7–22 (2014)
40. R. Fleischmann, M. Adams, O. White, R. Clayton, E. Kirkness, A. Kerlavage, C. Bult, J. Tomb, B. Dougherty, J. Merrick et al., Science **269**, 496–512 (1995)
41. D.J. Payne, M.N. Gwynn, D.J. Holmes, D.L. Pompliano, Nat. Rev. Drug Discov. **6**, 29–40 (2007)
42. C.A. Lipinski, F. Lombardo, B.W. Dominy, P.J. Feeney, Adv. Drug Delivery Rev. **23**, 3–25 (1997)
43. S.L. Schreiber, Science **287**, 1964–1969 (2000)
44. T. Kaeberlein, K. Lewis, S.S. Epstein, Science **296**, 1127–1129 (2002)
45. R.D. Süssmuth, A. Mainz, Angew. Chem. Int. Ed. **56**, 3770–3821 (2017)
46. K. Fosgerau, T. Hoffmann, Drug Discov. Today **20**, 122–128 (2015)
47. M.H. McCormick, J. McGuire, G. Pittenger, R. Pittenger, W. Stark, Antibiot. Ann. **3**, 606–611 (1955–1956)
48. F.J. Marshall, J. Med. Chem. **8**, 18–22 (1965)
49. D.H. Williams, J.R. Kalman, J. Am. Chem. Soc. **99**, 2768–2774 (1977)
50. G.M. Sheldrick, P.G. Jones, O. Kennard, D.H. Williams, G.A. Smith, Nature **271**, 223–225 (1978)
51. C.M. Harris, T.M. Harris, J. Am. Chem. Soc. **104**, 4293–4295 (1982)
52. K.C. Nicolaou, H.J. Mitchell, N.F. Jain, N. Winssinger, R. Hughes, T. Bando, Angew. Chem. Int. Ed. **38**, 240–244 (1999)
53. D.A. Evans, M.R. Wood, B.W. Trotter, T.I. Richardson, J.C. Barrow, J.L. Katz, Angew. Chem. Int. Ed. **37**, 2700–2704 (1998)
54. D.L. Boger, S. Miyazaki, S.H. Kim, J.H. Wu, O. Loiseleur, S.L. Castle, J. Am. Chem. Soc. **121**, 3226–3227 (1999)
55. M.N. Swartz, Proc. Nat. Acad. Sci. U.S.A. **91**, 2420–2427 (1994)
56. C.T. Walsh, S.L. Fisher, I.S. Park, M. Prahalad, Z. Wu, Cell Chem. Biol **3**, 21–28 (1996)
57. J. Xie, J.G. Pierce, R.C. James, A. Okano, D.L. Boger, J. Am. Chem. Soc. **133**, 13946–13949 (2011)
58. J. Xie, A. Okano, J.G. Pierce, R.C. James, S. Stamm, C.M. Crane, D.L. Boger, J. Am. Chem. Soc. **134**, 1284–1297 (2012)
59. M. Debono, B.J. Abbott, R.M. Molloy, D.S. Fukuda, A.H. Hunt, V.M. Daupert, F.T. Counter, J.L. Ott, C.B. Carrell, L.C. Howard, L.D. Boeck, R.L. Hamill, J. Antibiot. **41**, 1093–1105 (1988)
60. A. Raja, J. LaBonte, J. Lebbos, P. Kirkpatrick, Nat. Rev. Drug Discov. **2**, 943–944 (2003)
61. B.I. Eisenstein, J.F.B. Oleson, R.H. Baltz, Clin. Infect. Dis. **50**, S10–S15 (2010)
62. S.K. Straus, R.E.W. Hancock, Biochim. Biophys. Acta Biomembr. **1758**, 1215–1223 (2006)
63. D.E. Cane, C.T. Walsh, C. Khosla, Science **282**, 63–68 (1998)
64. O. Puk, D. Bischoff, C. Kittel, S. Pelzer, S. Weist, E. Stegmann, R.D. Süssmuth, W. Wohlleben, J. Bacteriol. **186**, 6093–6100 (2004)
65. C. Mahlert, F. Kopp, J. Thirlway, J. Micklefield, M.A. Marahiel, J. Am. Chem. Soc. **129**, 12011–12018 (2007)
66. T. Kittilä, A. Mollo, L.K. Charkoudian, M.J. Cryle, Angew. Chem. Int. Ed. **55**, 9834–9840 (2016)
67. G.L. Challis, J. Ravel, C.A. Townsend, Cell Chem. Biol. **7**, 211–224 (2000)
68. T.A. Keating, C.G. Marshall, C.T. Walsh, A.E. Keating, Nat. Struct. Mol. Biol. **9**, 522–526 (2002)
69. P.J. Belshaw, C.T. Walsh, T. Stachelhaus, Science **284**, 486–489 (1999)
70. J.W. Trauger, R.M. Kohli, H.D. Mootz, M.A. Marahiel, C.T. Walsh, Nature **407**, 215–218 (2000)
71. N.M. Gaudelli, C.A. Townsend, Nat. Chem. Biol. **10**, 251–258 (2014)

72. M.A. Marahiel, Nat. Prod. Rep. **33**, 136–140 (2016)
73. S.A. Samel, P. Czodrowski, L.-O. Essen, Acta Crystallogr. Sect. D: Biol. Crystallogr. **70**, 1442–1452 (2014)
74. J. Chatterjee, F. Rechenmacher, H. Kessler, Angew. Chem. Int. Ed. **52**, 254–269 (2013)
75. K.J. Labby, S.G. Watsula, S. Garneau-Tsodikova, Nat. Prod. Rep. **32**, 641–653 (2015)
76. A. Kirschning, F. Hahn, Angew. Chem. Int. Ed. **51**, 4012–4022 (2012)
77. E. Fischer, E. Fourneau, Ber. Dtsch. Chem. Ges. **34**, 2868–2877 (1901)
78. M. Bergmann, L. Zervas, Ber. Dtsch. Chem. Ges. **65**, 1192–1201 (1932)
79. L.A. Carpino, J. Am. Chem. Soc. **79**, 4427–4431 (1957)
80. F.C. McKay, N.F. Albertson, J. Am. Chem. Soc. **79**, 4686–4690 (1957)
81. R. Schwyzer, P. Sieber, Nature **199**, 172–174 (1963)
82. G.W. Kenner, Proc. R. Soc. London, Ser. A **353**, 441–457 (1977)
83. S.B.H. Kent, Chem. Soc. Rev. **38**, 338–351 (2009)
84. R.B. Merrifield, J. Am. Chem. Soc. **85**, 2149–2154 (1963)
85. L.A. Carpino, G.Y. Han, J. Am. Chem. Soc. **92**, 5748–5749 (1970)
86. C.D. Chang, J. Meienhofer, Int. J. Pept. Protein Res. **11**, 246–249 (1978)
87. E. Atherton, C.J. Logan, R.C. Sheppard, J. Chem. Soc. Perkin Trans. **1**, 538–546 (1981)
88. J.C. Sheehan, G.P. Hess, J. Am. Chem. Soc. **77**, 1067–1068 (1955)
89. W. König, R. Geiger, Chem. Ber. **103**, 788–798 (1970)
90. A. El-Faham, F. Albericio, Chem. Rev. **111**, 6557–6602 (2011)

Chapter 2
Total Synthesis of Teixobactin

Natural products are an extremely important source of new drug leads. As the number of bacterial strains resistant to current therapeutics increases, the need for molecules with novel modes of action is critical. Almost all the currently used clinical antibiotics are natural products or are derived from natural products, however very few novel classes have been discovered recently. With novel antibiotic scaffolds dwindling, researchers are turning to previously uncultivated bacteria as a source of new natural product scaffolds [1]; a recent example was the discovery of teixobactin (**28**) [2].

2.1 Discovery of Teixobactin

2.1.1 Initial Isolation and Structural Elucidation

Teixobactin (**28**) is a non-ribosomal peptide natural product which was reported by Lewis and co-workers in 2015 [2]. Teixobactin possesses potent antibacterial activity against a number of clinically relevant organisms including *M. tuberculosis*, as well as antibiotic-resistant pathogens such as MRSA, and VRE. The discovery of teixobactin was enabled through the analysis of previously uncultivated soil bacteria for antimicrobial activity. It is estimated that as many as 99% of bacterial species resist cultivation in the lab [3]. A number of technologies have been developed to overcome this problem to enable unstudied bacteria to be cultured in the laboratory. As biological novelty is revealed, it is likely that chemical novelty, i.e. new antibiotic natural product scaffolds, will follow [4].

In the report by Lewis and co-workers [2], a soil sample from Maine, USA was collected and diluted into the iChip, a device previously developed by the authors for the simultaneous isolation and growth of uncultivable bacteria [5]. The iChip is a multi-channel device that partitions a suspension of bacteria into individual channels, resulting in the growth of pure cultures. The device is then returned to the soil so that

A. Giltrap, *Total Synthesis of Natural Products with Antimicrobial Activity*,
Springer Theses, https://doi.org/10.1007/978-981-10-8806-3_2

the bacteria can grow in their natural environment (Fig. 2.1). With a number of bacterial colonies established in the iChip, they were cultured under traditional laboratory conditions then the extracts were screened for activity against *S. aureus*. From the screen the authors identified that a previously uncultivable Gram-negative bacterial species, which they named *Eleftheria terrae*, was a potent inhibitor of *S. aureus* growth. Subsequent isolation, and structure determination of an active natural product led to the discovery of teixobactin (**28**). Teixobactin possesses a number of unusual structural features, namely a cyclic macrolactonised *C*-terminus (green), four D-amino acids, an *N*-terminal *N*-methyl-D-phenylalanine residue (blue) and most interestingly an L-*allo*-enduracididine residue (red) (Fig. 2.2).

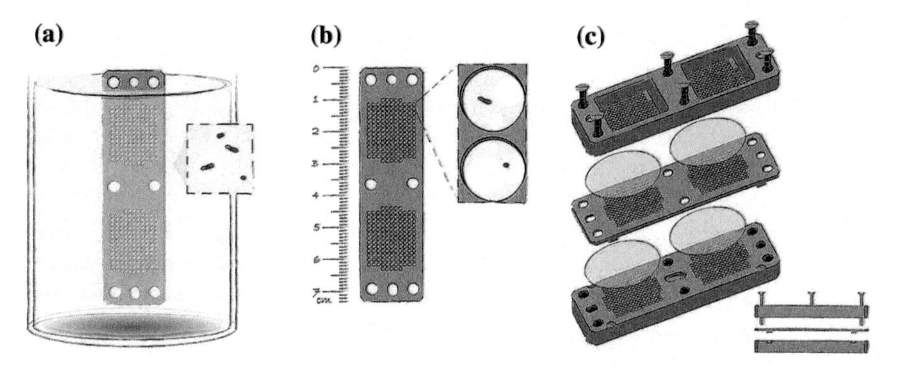

Fig. 2.1 Cultivation of bacteria using the iChip [5]. **a** The iChip central plate is immersed in a mixture of bacteria. **b** On average one bacterial cell is distributed into each well. **c** The iChip is reassembled and returned to the native environment

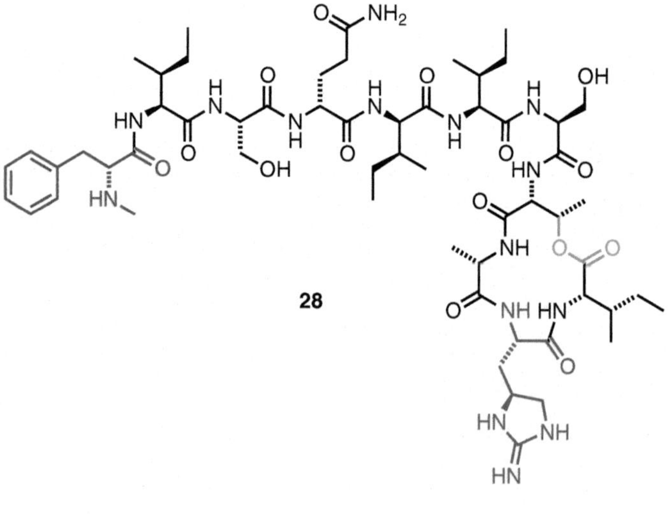

Fig. 2.2 Structure of teixobactin (**28**) highlighting the macrolactone linkage (green), *N*-methyl-D-phenylalanine (blue) and L-*allo*-enduracididine residues (red)

2.1.2 Biosynthesis of Teixobactin

Teixobactin is a NRP natural product. After the structure of the natural product was determined, Lewis and co-workers were able to identify the gene cluster encoding for the biosynthesis of teixobactin [2]. Two key genes, *txo1* and *txo2*, were identified as encoding for the NRPS (Fig. 2.3a). Analysis of these genes revealed 11 modules, and the adenylation domains matched with the amino acids present in mature teixobactin (**28**). The methyl transferase domain in module 1 is responsible for installing the *N*-Me moiety of the *N*-terminal phenylalanine residue (Fig. 2.3). The final thioesterase domain catalyses the formation of the *C*-terminal macrolactone and concomitant release of teixobactin (**28**) from the NRPS.

The unusual amino acid enduracididine **29** occurs in a number of other important NRP antibiotic families, including the mannopeptimycins [6] and the enduracidins [7]. While the biosynthetic pathway for enduracididine produced in *Eleftheria terrae* has not be disclosed, it is presumably very similar to that in the aforementioned natural products. Interestingly, the mannopetimycins contain β-OH-enduracidine **30** and the biosynthesis of this interesting amino acid is shown in Scheme 2.1. To begin, arginine (**31**) undergoes hydroxylation at the γ-position and conversion to the keto-acid **32** catalysed by the enzyme mppP [8]. Subsequent dehydration and cyclisation catalysed by mppR [9] yields **33**, which then undergoes transamination, facilitated by mppQ, to yield enduracididine **29**. The final step in the reaction is the β-hydroxylation, catalysed by mppO which yields the final amino acid **30** which can then be incorporated into the biosynthesis of mannopeptimycin [10].

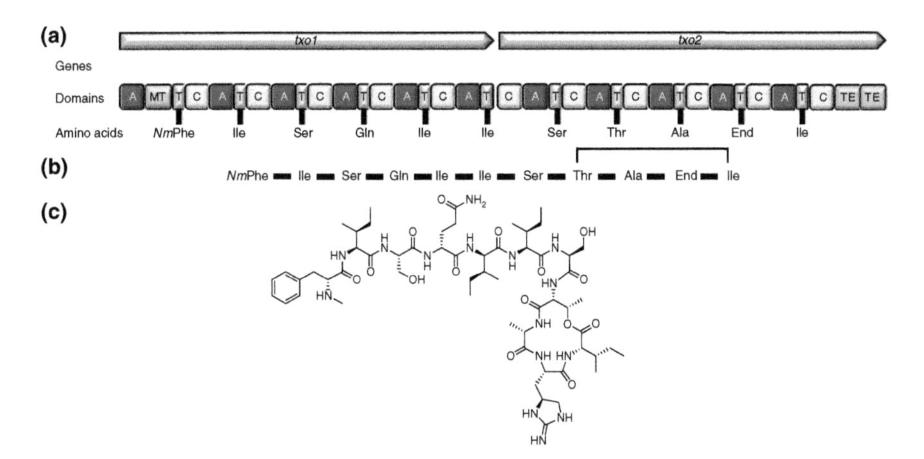

Fig. 2.3 Organisation of the NRPS responsible for the synthesis of teixobactin (**28**). **a** Genes and NRPS organisation. **b** Schematic representation of teixobactin (**28**). **c** Structure of teixobactin (**28**)

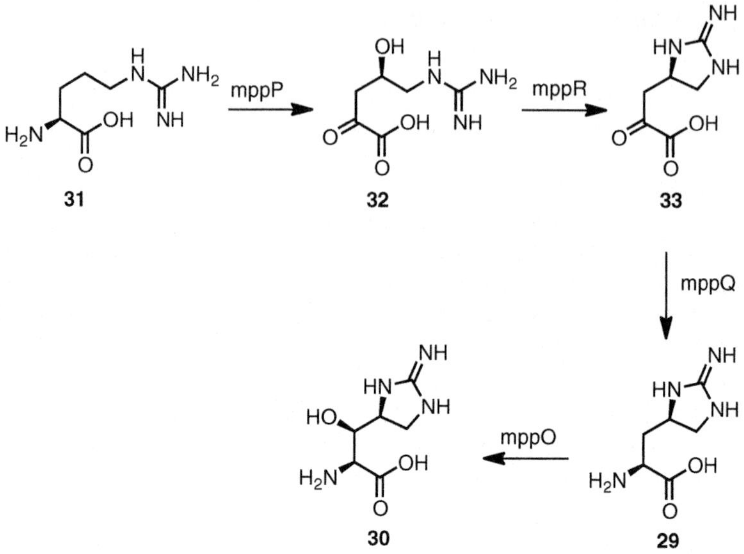

Scheme 2.1 Biosynthesis of enduracididine **29** and β-OH-enduracididine **30**

2.1.3 Antimicrobial Activity and Mechanism of Action

Once isolated and structurally characterised, Lewis and co-workers carried out comprehensive studies to fully characterise its antibiotic activity and the mechanism of action [2]. To begin, teixobactin (**28**) was screened against a number of clinically relevant bacterial strains. Here the natural product was shown to possess potent activity against 'Golden Staph', VRE and *M. tuberculosis* H37Rv, the causative agent of tuberculosis (Table 2.1). Teixobactin was shown to be inactive against the Gram-negative bacteria tested including *Escherichia coli* and *Pseudomonas aeruginosa*. The authors also tried to generate *S. aureus* and *Mtb* strains resistant to teixobactin, by multiple treatments at a sub-lethal dose, however this was unsuccessful. This suggested that the molecular target of teixobactin (**28**) was not an enzyme, which can be mutated easily to afford resistance, but rather an enzyme substrate. Further experiments determined that teixobactin is a bacterial

Table 2.1 Activity of teixobactin (**28**) against a number of clinically relevant pathogens

Organism	Teixobactin (**28**) MIC (μg/mL)
Staphylococcus aureus (MRSA)	0.25
Enterococcus faecalis (VRE)	0.5
Mycobacterium tuberculosis H37Rv	0.125
Escherichia coli	25
Pseudomonas aeruginosa	>32

peptidoglycan synthesis inhibitor. Significantly, teixobactin (**28**) showed no toxicity against mammalian cell lines at 100 μg/mL.

In order to assess the potential of teixobactin (**28**) as a clinical candidate, the authors carried out stability tests in serum and against microsomes. Teixobactin showed low toxicity and good serum and microsomal stability. Specifically, it was shown to be stable in serum for 4 h after an intravenous injection. In vivo efficacy studies were carried out in three different models; with mice infected with either MRSA or *Streptococcus pneumonia*. Importantly teixobactin (**28**) was highly efficacious against both organisms in all three models.

With the potent biological activity and basic mechanism of action determined, the authors carried out further studies to fully elucidate the molecular target of teixobactin (**28**). In vitro experiments revealed that teixobactin (**28**) inhibits multiple reactions in the peptidoglycan biosynthesis pathway, which utilise a number of lipid substrates (lipid I, lipid II and undecaprenyl pyrophosphate). Importantly dosing *S. aureus* with lipid II prevented teixobactin from exerting its antibiotic effect. These results demonstrated that the molecular target of teixobactin is the peptidoglycan lipid precursors rather than the enzymes themselves (Fig. 2.4). As discussed in Sect. 1.3.3 vancomycin (**8**) also exerts its action by binding to lipid II, specifically the terminal D-Ala-D-Ala motif. However, teixobactin (**28**) was able to bind to the mutated lipid II (D-Ala-D-Lac). Furthermore, teixobactin (**28**) also binds to lipid III, a precursor to wall teichoic acid (WTA), a further component of the

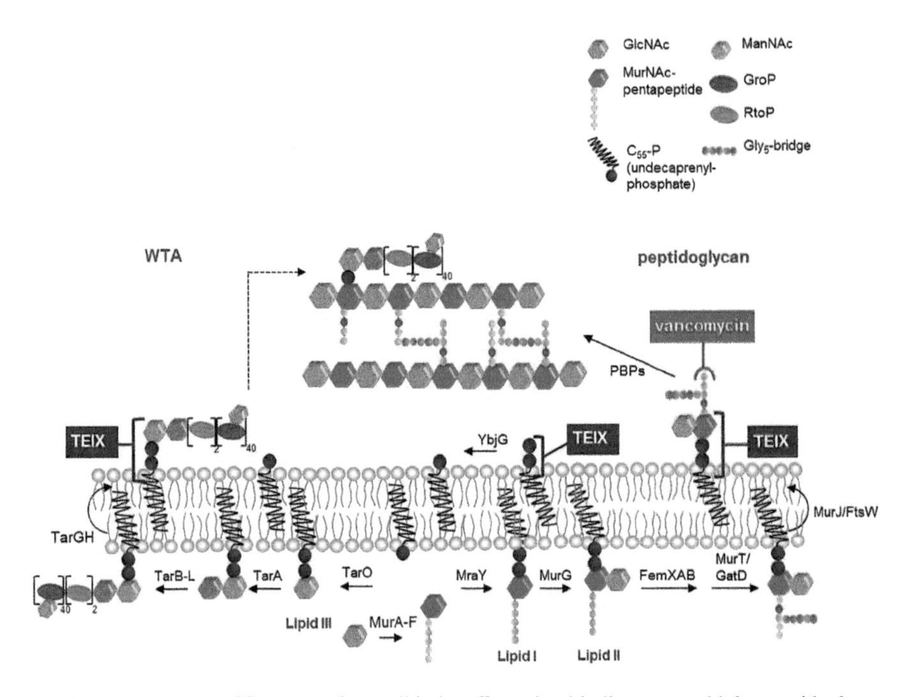

Fig. 2.4 Teixobactin (**28**) exerts its antibiotic effects by binding to multiple peptidoglycan precursors, with a novel binding mode compared to vancomycin (**8**) [2]

peptidoglycan cell wall in Gram-positive bacteria. While WTA is not critical for the viability of bacteria [11] inhibition of its biosynthesis is thought to liberate autolysins, which cause cell-wall lysis and bacterial killing [12]. Further studies carried out by Lewis and co-workers revealed that the potent activity of teixobactin (**28**) could be attributed to inhibition of both peptidoglycan and WTA leading to synergistic bacterial killing [13].

Combining the results together, Lewis and co-workers proposed that teixobactin (**28**) binds primarily to the pyrophosphate and first sugar attached to the lipid in the peptidoglycan precursors. The nature of this sugar is likely to be of low significance, which means teixobactin (**28**) can exert its antibiotic affect against a number of bacteria which utilise different sugars in their cell wall, including the arabinogalactan present in *M. tuberculosis*.

The novel mechanism of action means that resistance to teixobactin (**28**) is likely to be slow to develop. As discussed in the previous chapter, due to the fact that vancomycin (**8**) binds to a substrate, and not an enzyme, resistance to this antibiotic was slow to develop [14]. Furthermore, it is likely that the origin of resistance to vancomycin (**8**) was from bacteria which produce this antibiotic themselves [15]. Teixobactin is produced by *E. terrae* a Gram-negative bacterium. After teixobactin (**28**) is transported out of the bacterial cell, it is likely that the outer cell wall of *E. terrae* prevents teixobactin from coming into contact with the peptidoglycan precursors and exerting its antibiotic action. While it is possible that resistance to teixobactin could develop in the future, it is feasible to suggest that this might take even longer than the 30 years it took for resistance to vancomycin (**8**) to emerge.

2.1.4 Media Attention

Due to its potent and broad-spectrum antibiotic activity, its in vivo efficacy, as well as the low potential for resistance to develop, teixobactin (**28**) received a significant amount of attention in the mainstream media. This included reports from the BBC News—'Antibiotics: US discovery labelled 'game-changer' for medicine'; The Wall Street Journal—'Scientists Discover Potent Antibiotic, A Potential Weapon Against a Range of Diseases'; and Forbes Magazine—'Teixobactin and iChip Promise Hope Against Antibiotic Resistance'. While some of these claims are bold, particularly considering the difficulties associated with getting a drug to market, it highlights the importance of developing new drugs for antibiotic resistant bacteria.

2.1.5 Synthesis of Arginine Analogue 34

Shortly after the isolation and structure of teixobactin (**28**) was reported two independent reports detailing the synthesis of teixobactin analogue **34** were

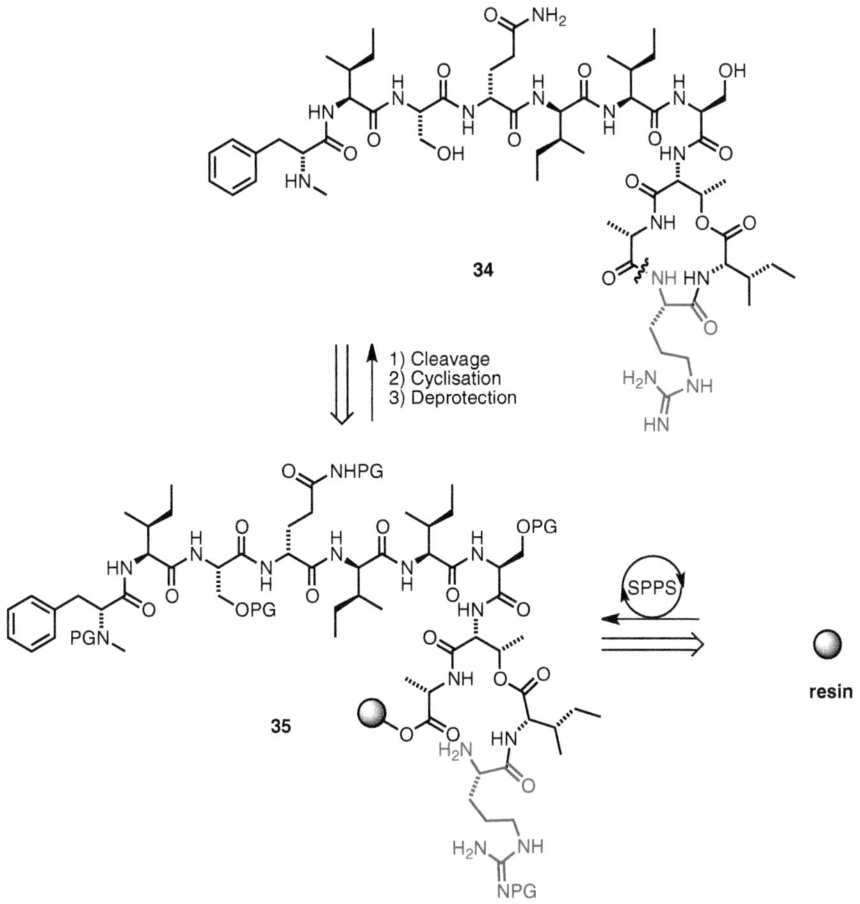

Scheme 2.2 General synthetic scheme for the two independent syntheses of teixobactin analogue **34** [16, 17]. PG = protecting group

published [16, 17] (Scheme 2.2). This analogue **34** is the same as teixobactin in all regards, except for the substitution of non-proteinogenic enduracididine for arginine. The same cyclisation, namely between the Ala and Arg residues was used in both syntheses. Furthermore, both syntheses were completed in one SPPS linear sequence yielding resin-bound peptide **35**, with a final cleavage from resin and cyclisation followed by global deprotection of the side chain protecting groups. In both studies the biological activity of the synthesised analogue was tested and shown to be about an order of magnitude less effective against *S. aureus* than the native natural product. This clearly revealed the importance of the enduracididine residue for the bioactivity of teixobactin (**28**) and further reiterates the importance of a total synthesis of this molecule.

2.2 Retrosynthesis of Teixobactin

Due to the interesting chemical structure, as well as the potent antibacterial activity against a number of clinically relevant pathogens, teixobactin (**28**) was chosen as a target for total synthesis in the Payne research group.

2.2.1 Initial Retrosynthetic Approach

The initial retrosynthesis of teixobactin (**28**) construction through the ligation of two fragments, the cyclic peptide portion and the linear peptide portion through a selenium-mediated peptide ligation between selenoester **36** and diselenide **37** [18] (Scheme 2.3). This could be followed by oxidative deselenisation of the installed selenocysteine residue (blue) to yield the native serine residue at the ligation junction [19]. Both of these technologies were recently developed in the Payne group, and it was envisaged they would expedite the synthesis of teixobactin and the rapid generation of analogues. While it was assumed that the synthesis of linear selenoester **36** could be rapidly generated via SPPS beginning from the resin-bound amino acid **38**, the synthesis of the cyclic peptide **37** would pose more challenges.

2.2.2 Considerations Regarding Cyclic Peptide Synthesis

The formation of peptide macrocycles is not trivial, and there has been a significant body of scientific study devoted to determining the best strategies for their construction. Indeed there are a number of considerations that need to be taken into account when designing a synthetic route towards a cyclic peptide to ensure the greatest chance of a successful synthesis [20]. There are a number of classes of cyclic peptides; one of the most common are cyclic depsipeptides which contain an ester bond in the cycle, normally present at the side chain alcohol of a Ser or Thr residue. The cyclisation is generally carried out via a final macrolactonisation, or macrolactamisation reaction. However, it is preferential to carry out a final macrolactamisation, as this has the advantage of a more nucleophilic amine as a coupling partner, as opposed to an alcohol in a macrolactonisation [21]. In order to prevent oligomerisation (i.e. formation of dimers and trimers etc.) during cyclisation, the reaction needs to be conducted at high-dilution [22].

A further important factor, which dictates the ability of a linear peptide to cyclise, is the ability for the linear peptide to adopt a conformation that brings the reactive ends together. While it is known that large peptides (more than seven amino acids) are relatively easy to cyclise, it can be extremely difficult to affect the cyclisation of smaller peptide chains, particularly those containing all L or D amino acids. In a seminal study, Schmidt and Langner [23] discussed the difficulty in

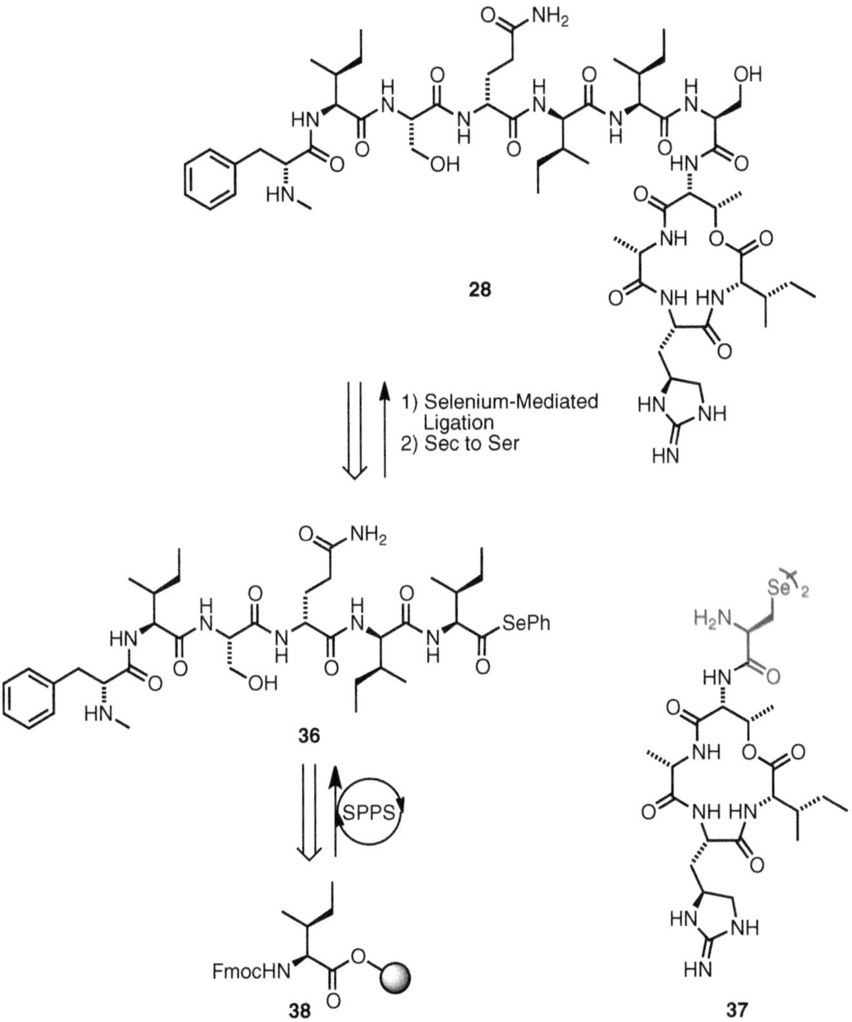

Scheme 2.3 Retrosynthetic analysis of teixobactin (**28**)

preparing an all L tetrapeptide, indeed the only cyclic tetrapeptide formed under their conditions had undergone epimerisation at the *C*-terminus. The authors found they could synthesise the all L pentapeptide Pro-Ala-Ala-Phe-Leu, but only if they cyclised between the Ala-carboxylate and Phe-amine. Indeed, further to this, work carried out by Veber and co-workers [24] revealed that if possible, a cyclisation junction between an L and D amino acid at each termini is most favourable. Furthermore, peptide cyclisations of all L amino acids are known to be facilitated by the presence of a D amino. They are thought to exert a turn-inducing affect on the linear peptide chain. Similarly proline residues are able to form *cis* amide bonds, which also induces a turn in the chain, increasing the propensity for peptide cyclisation [20].

There are a number of other factors which need to be considered when assessing cyclisation junctions [25]. If possible, the cyclisation junction should possess low steric hindrance, namely no β-branched amino acids. Similarly, the absence of *N*-alkylated residues at the *N*-terminus of the linear peptide also decreases steric hindrance at the cyclisation junction. Finally, a *C*-terminal residue that cannot undergo epimerisation such as a Gly residue, or a β-amino acid, is also desirable.

2.2.3 *Retrosynthesis of Cyclic Peptide 37*

The cyclised *C*-terminus of teixobactin (**28**) is a tetrapeptide made up of a D-Thr, Ile, L-*allo*-End and Ala residues with the ester bond formed between the side chain of the D-Thr and the Ile residues. When designing the synthetic strategy to access **37**, it was decided that cyclisation should take place between the D-Thr and Ala residues (Scheme 2.4). This is a favourable D-L junction [24], and the Ala residue is sterically the smallest residue in the cycle. The cyclisation reaction would also be a lactamisation, which is more facile than a lactonisation. It was envisioned the resin-bound linear tetrapeptide **39** could be synthesised via Fmoc-SPPS with a key on-resin esterification step utilised.

2.3 Synthesis of Enduracididine

Before the synthesis of the teixobactin (**28**) could commence an efficient synthesis of the non-proteinogenic amino acid L-*allo*-enduracididine needed to be developed.

Scheme 2.4 Retrosynthesis of cyclic peptide **37**

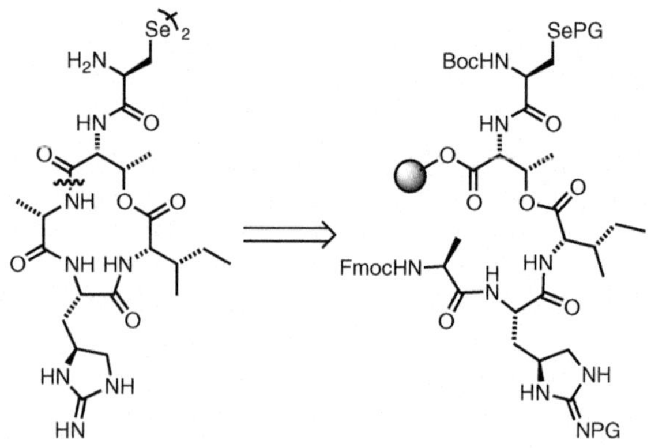

2.3.1 *Previous Syntheses of Enduracididine*

There have been a number of syntheses of enduracididine previously reported in the literature, the first being by Shiba et al. in 1975 [26]. This synthesis began from L-histidine, and while concise, yielded enduracididine as a mixture of diastereomers at the γ-position. No further syntheses were reported until Dodd et al. published a synthesis of both L-enduracididine and L-*allo*-enduracididine [27]. This method relied on a key aziridination reaction, which proceeded in low yield and moderate diastereoselectivity. The discovery of teixobactin reignited the interest in this unusual amino acid. In 2015, Yuan and co-workers reported an elegant synthesis of protected L-*allo*-enduracididine **40** beginning from 4-hydroxyproline (**41**) [28] (Scheme 2.5). The desired stereochemistry at the γ-position was set by mesylation of the alcohol and displacement with azide yielding **42**, followed by oxidation to yield lactam **43**. Reductive ring opening afforded **44**, followed by installation of the guanidine moiety which yielded **45**. The key cyclisation reaction was then performed, followed by final deprotection of the Boc and *t*Bu protecting groups followed by reinstallation of the Boc protecting group to afford protected L-*allo*-enduracididine **40**.

2.3.2 *Retrosynthesis of Protected* L-allo-*Enduracididine 46*

In order to synthesise teixobactin a suitably protected L-*allo*-enduracididine **46** was required (Scheme 2.6). In order to incorporate this building block into Fmoc-SPPS, the *N*-terminus would need to be protected with an Fmoc group. The side chain

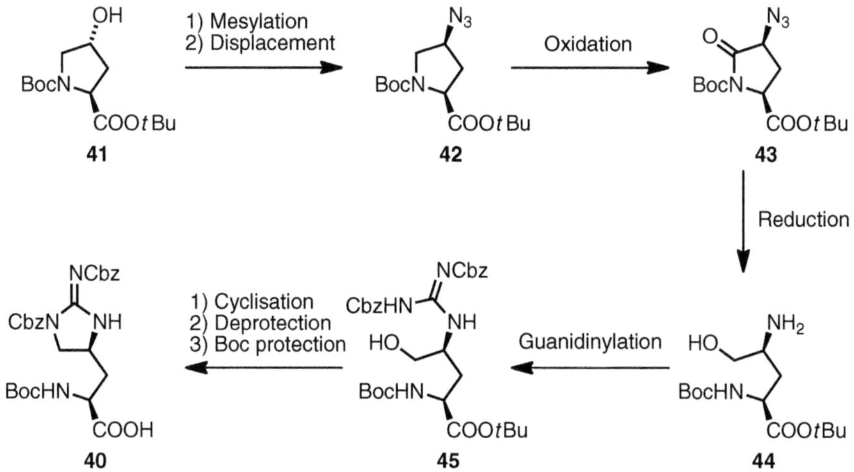

Scheme 2.5 Synthesis of protected L-*allo*-enduracididine **40** by Yuan and co-workers [28]

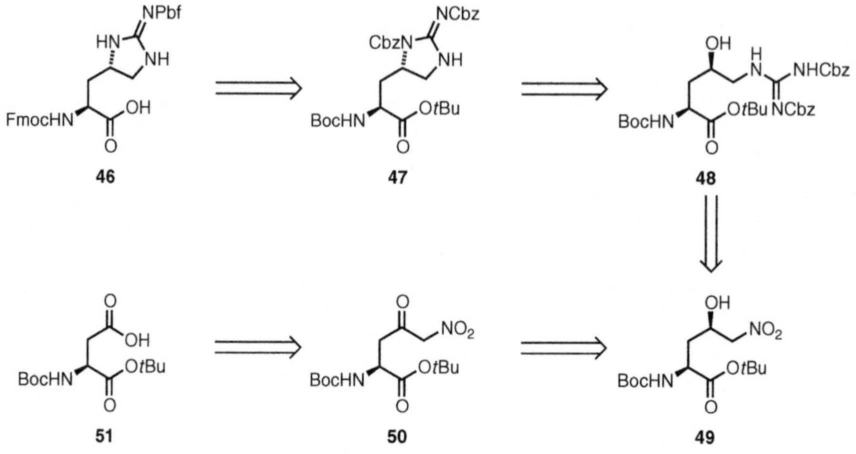

Scheme 2.6 Retrosynthesis of suitably protected L-*allo*-enduracididine **46**

would need to bear an orthogonal acid-labile protecting group such as the 2,2,4,6,7-pentamethyldihydrobenzofuran-5-sulfonyl (Pbf) group, which is used to protect the side chain of arginine. Inspired by the synthesis of enduracididine published in the patent concerning teixobactin [29], we envisioned **46** could be accessed from **47** via protecting group manipulations, with **47** in turn synthesised from the suitably protected linear guanidine **48**. This in turn could be accessed via reduction of the nitro group and guanidinylation from nitro alcohol **49**. It was envisioned this could be synthesised from nitroketone **50**, in turn accessed from protected aspartic acid **51**.

2.3.3 Synthesis of Suitably Protected Enduracididine 58

To begin the synthesis, the formation of the key nitro-alcohol **49** was investigated (Scheme 2.7). To this end, inspired by a report from Rudolph et al. [30], Boc-Asp-O*t*Bu (**51**) was preactivated with carbonyl diimidazole (CDI) in nitromethane before the addition of potassium *tert*-butoxide. A number of reaction conditions were trialled including those utilised by the authors, but the optimal and

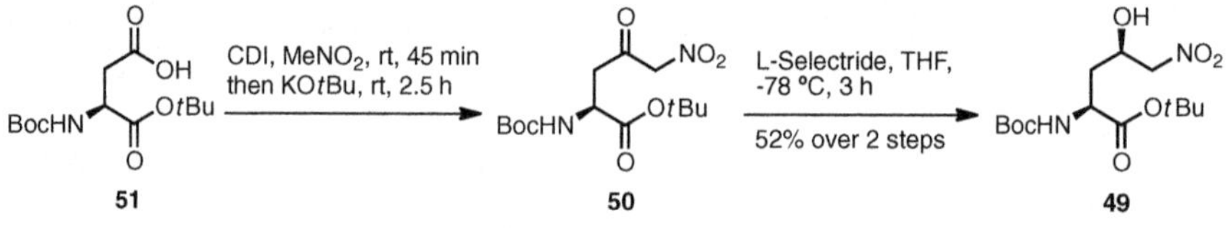

CDI, $MeNO_2$, rt, 45 min then KO*t*Bu, rt, 2.5 h

L-Selectride, THF, -78 °C, 3 h; 52% over 2 steps

Scheme 2.7 Synthesis of alcohol **49**

most reproducible conditions proved to be excess nitromethane and 2 equivalents of KO*t*Bu. Ketone **50** was utilised without further purification, and after work-up was dried thoroughly and reacted with L-Selectride [31] at −78 °C for two hours before quenching. This yielded alcohol **49** as a mixture of diastereomers in a ratio of ~5:1 in favour of the desired (2*S*,4*R*) configuration. Gratifyingly, the diastereomers could be readily separated by silica gel chromatography and, as such, the desired (2*S*,4*R*) alcohol **49** was isolated in 52% yield over the two steps, comparable to that reported by Rudolph et al. [30].

The stereochemistry of this reaction can be rationalised by invoking the Cram-Reetz model for 1,3 asymmetric induction (Scheme 2.8). This model was initially proposed by Reetz and co-workers to explain 1,3-stereoinduction in non-chelating aldol reaction systems [32]. It builds on the polar model proposed by Cram and co-workers [33] in which a polar atom at the chiral β-position, in this case the NHBoc moiety, is oriented anti-periplanar to the carbonyl moiety to minimise the dipole in the transition state **52**. The nucleophile then adds from the least hindered face. Furthermore, Evans and co-workers [34, 35] have extended this model to include torsional effects, which dictate that the transition state **53** adopts a staggered conformation between the forming bond and the substituents α to the carbonyl. To this end, the polar NHBoc moiety is placed pseudo-antiperiplanar to the carbonyl to minimise the dipole in the transition state **53** and then the nucleophile is delivered to the carbonyl at the Burgi-Dunitz angle [36]. Both models provide an effective explanation for the observed stereoselectivity of the reduction of ketone **50** to form nitro alcohol **49**.

With alcohol **49** in hand, the next step was the reduction of the nitro group and installation of the guanidine moiety (Scheme 2.9). Before this reaction could proceed, a suitable method for the installation of the guanidine moiety was required and the Goodman guanidinylating reagent **54** was selected for this purpose [37]. To this end, guanidine hydrochloride (**55**) was reacted with benzylchloroformate to yield the diCbz guanidine **56** in good yield. This was then reacted with trifluoromethanesulfonic anhydride (Tf_2O) under strictly anhydrous conditions to yield the desired guanidine triflate **54** in excellent yield. With Goodman's reagent **54** in hand, nitro-alcohol **49** was reduced using catalytic Pd/C, hydrogen gas and

Scheme 2.8 Cram-Reetz and Evans models for 1,3-asymmetric induction in the reduction of ketone **50** to alcohol **49**

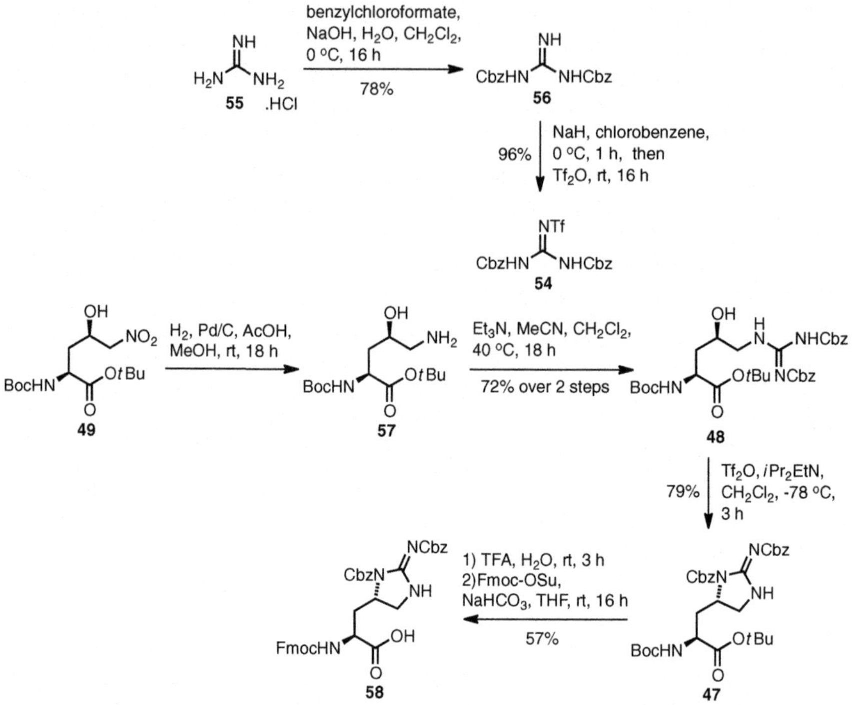

Scheme 2.9 Synthesis of protected L-*allo*-enduracididine **58**

acetic acid to yield the desired amine **57** which was used without purification. Amine **57** was reacted with Goodman's reagent **54** and triethylamine to yield the desired linear guanidine **48**, which proceeded in excellent yield over the two steps. With the guanidine moiety installed, the key transformation, cyclisation of the guanidine unit, was then carried out. As such, linear guanidine **48** was reacted with Tf_2O to activate the secondary alcohol as a triflate, which subsequently underwent rapid intramolecular cyclisation via an S_N2 pathway to yield cyclic guanidine **47** with the desired (2*S*,4*S*) stereochemistry. Pleasingly, this reaction proceeded in good yield, affording the core L-*allo*-enduracididine unit.

The final challenge in the synthesis of the suitably protected enduracididine residue was the manipulation of the protecting groups. To this end, the Boc and *t*Bu protecting groups of **47** were removed using a mixture of trifluroroacetic acid (TFA) and water. The solvent was removed and the crude zwitterion was reacted with Fmoc-succinimide (Fmoc-OSu) in a mixed solvent system to yield the desired Fmoc-protected amino acid **58**. Unfortunately, the yield over these two steps was moderate, possibly due to the reaction of the guanidine moiety with the Fmoc-OSu.

At this point a number of reaction conditions were trialled to substitute the Cbz protecting groups on the guanidine moiety of **58** for an acid labile alternative (Scheme 2.10). To this end, the Cbz protecting groups were removed via a transfer

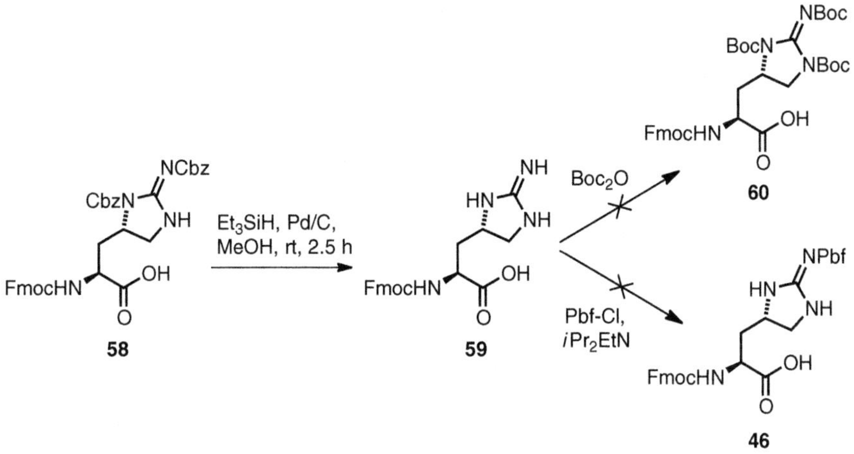

Scheme 2.10 Attempted synthesis of L-*allo*-enduracididine bearing acid labile side-chain protecting groups

hydrogenation employing catalytic Pd/C and triethylsilane as a source of hydrogen [38]. This method is compatible with Fmoc-groups, which in some cases can be labile under standard hydrogenation conditions using Pd/C and H_2 gas [39]. Pleasingly this yielded the desired unprotected guanidine **59** by HPLC-MS analysis. Crude guanidine **59** was reacted with Pbf-Cl and Hünig's base, in order to install the Pbf functionality. Unfortunately this reaction only led to recovered starting material and, as such, an alternative approach was trialled. The crude guanidine **59** was then reacted with four equivalents of Boc-anhydride in an effort to synthesise **60**. Analysis of the reaction by HPLC-MS revealed the formation of multiple products corresponding to one, two or three Boc protecting groups. With both strategies unsuccessful, the protecting group strategy for the guanidine moiety was reconsidered. It was reasoned that the Cbz groups could be removed in the last step of the synthesis employing conditions normally utilised for the HF-free cleavage of protecting groups in Boc-SPPS [40]. To this end, the desired suitably protected enduracididine **58** retained the Cbz protecting groups.

2.4 Initial Efforts Towards Teixobactin

2.4.1 Initial Cyclisation Approaches by Mr. Luke Dowman

While the synthesis of the suitably protected L-*allo*-enduracididine residue **58** was being carried out, an honours student in the Payne laboratory, Mr. Luke Dowman, carried out investigations towards the synthesis of the cyclic peptide segment **37** of teixobactin (**28**) discussed above. To begin, the synthesis a cyclic peptide analogue, containing Arg instead of L-*allo*-enduracididine was attempted. To this end, side chain unprotected Fmoc-D-Thr-OH was loaded to 2-chlorotrityl chloride functionalised polystyrene resin (2-CTC) yielding resin-bound amino acid **61**. (Scheme 2.11). This linker was developed by Barlos and co-workers [41–43] and has a number of favourable characteristics, namely, it is highly acid labile and, as such, the protected peptide can be removed from the resin with the extremely weak acid hexafluoroisopropanol (HFIP). Furthermore, the amino acid is loaded without activation of the *C*-terminus preventing epimerisation. Finally, the steric bulk of the linker prevents the formation of diketopiperazines (DKP) from forming at the dipeptide stage of the peptide synthesis [44].

The next step in the synthesis was Fmoc-deprotection, which was carried out using 10 vol.% piperidine in DMF to reveal the amine of the Thr residue **62**. Next, Boc-Sec(PMB)-OH was coupled utilising 1.2 equivalents of the coupling

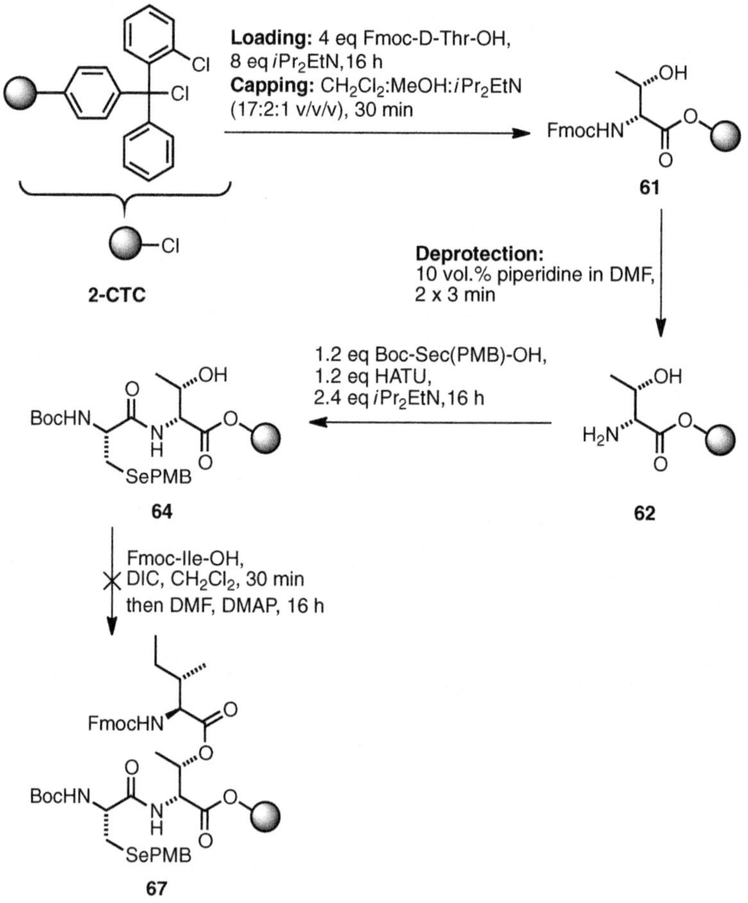

Scheme 2.11 Attempted on-resin esterification reaction

reagent 1-[bis(dimethylamino)methylene]-1H-1,2,3-triazolo[4,5-b] pyridinium hexafluorophosphate 3-oxide (HATU) (**63**) with 2.4 equivalents of Hünig's base, yielding resin-bound dipeptide **64**. As discussed in Sect. 1.5.2, there is a plethora of coupling reagents available for the formation of amide bonds during SPPS. HATU is an extremely active uronium based coupling reagent [45]. It contains 1-hydroxy-7-azabenzotriazole (HOAt) linked directly to the uronium. This acts via the same mechanism to the HOBt additive, to supress epimerisation in coupling reactions. However, the HOAt active ester **65** formed is more reactive than the corresponding HOBt ester [46] (Scheme 2.12). This is due to a two-fold effect; the nitrogen atom at position 7 of the benzotriazole is electron withdrawing, improving the leaving group ability. Furthermore, this nitrogen is in a suitable position to facilitate the coupling, increasing the nucleophilicity of the amine through a neighbouring group effect in the coupling transition state **66** [46].

The next step in the reaction was the key on-resin esterification reaction which was carried out with Fmoc-Ile-OH (Scheme 2.11). This reaction was performed using eight equivalents of Fmoc-Ile-OH and four equivalents of *N*,*N*′-diisopropylcarbodiimide (DIC) to form the symmetric anhydride. This was then added to the resin along with catalytic *N*,*N*-dimethylaminopyridine (DMAP) to form the branched resin-bound peptide **67**. In 1978 Steglich showed that addition of DMAP significantly speeds up the carbodiimide-mediated esterification of an alcohol and carboxylic acid [47]. Unfortunately, after the reaction was analysed by high performance liquid chromatography mass-spectrometry (HPLC-MS), <5% product was observed. The amount of product did not increase after the resin was treated multiple times. It was therefore hypothesised that the bulkiness of the 2-CTC resin prevented the on-resin esterification reaction from occurring. Indeed, Brimble and co-workers encountered a similar issue during their total synthesis of YM-280193 in which a peptide with a *C*-terminal β-OH-Leu loaded to resin did not undergo any esterification [48].

Scheme 2.12 Amide bond formation mediated by HATU (**63**)

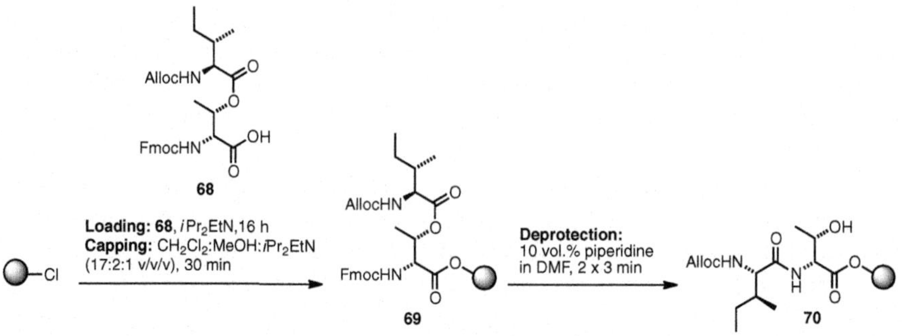

Scheme 2.13 Unwanted *O* to *N*-acyl shift after loading dipeptide **68** to 2-CTC resin

To circumvent this issue, further efforts involved the in-solution synthesis of dipeptide **68** and subsequent loading to 2-CTC to yield resin-bound dipeptide **69** (Scheme 2.13). With this in hand, the Fmoc-group was removed in preparation for coupling of Boc-Sec(PMB)-OH. However, HPLC-MS analysis of the coupling reaction revealed no coupling product at all. It is presumed this was caused by an *O* to *N*-acyl shift at the dipeptide stage yielding **70**. Indeed, removal of the allyloxycarbonyl (Alloc) protecting group and coupling resulted in a tripeptide, providing further evidence for the above. Due to this unexpected reaction the route was abandoned.

2.4.2 Optimisation of the Key On-resin Esterification Reaction

Due to time constraints in his honours project, Mr. Luke Dowman was unable to carry out any further work towards the total synthesis of teixobactin (**28**). Furthermore, due to the difficulties described above, a new method to synthesise the key ester bond was required. As discussed above, it was presumed that bulkiness of the 2-CTC linker prevented the on-resin esterification reaction from proceeding. As such, it was envisioned that loading Fmoc-D-Thr-OH onto a less bulky resin could circumvent this issue. After searching the literature, 4-(4-Hydroxymethyl-3-methoxyphenoxy) butyric acid (HMPB) functionalised NovaPEG resin was selected as a potential alternative (Fig. 2.5). The HMPB linker [49] is used for the

Fig. 2.5 The structure of the HMPB linker

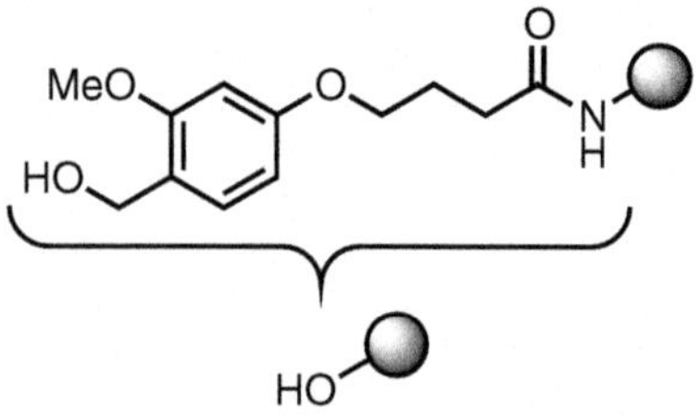

synthesis of protected peptide fragments. Accordingly, the peptide can be cleaved from the resin with 1 vol.% TFA/CH_2Cl_2 which leaves the side chain protecting groups intact. Moreover, this resin is significantly less bulky than the 2-CTC linker used previously, possibly allowing for the esterification reaction to take place. Finally, the NovaPEG resin is a polyethylene glycol based resin which has superior swelling properties to the standard polystyrene resin traditionally used for SPPS.

The HMPB linker attaches to the first amino acid via a benzylic alcohol therefore the first amino acid needs to be loaded utilising an esterification reaction. With this in mind, the side chain alcohol of Fmoc-D-Thr-OH required protection before loading could be carried out to prevent unwanted oligomerisation of the amino acid. To begin the allyl protecting group, which can be cleaved orthogonally to the base labile Fmoc, and acid labile Boc groups, was chosen for the side chain of threonine. To this end, H_2N-D-Thr-OH **71** was Boc-protected before being reacted with NaH and allylbromide to yield the Boc-protected allyl ether **72** as the exclusive product (Scheme 2.14). This was subsequently subjected to protecting group manipulations, namely removal of the Boc group by treatment with HCl in dioxane, followed by removal of the solvent and subsequent protection with Fmoc-OSu to yield the desired amino acid **73**.

Protected Thr **73** was loaded to HMPB functionalised resin using 1-(Mesitylene-2-sulfonyl)-3-nitro-1,2,4-triazole (MSNT) and *N*-Me-imidazole as the base, a standard method for loading hydroxyl functionalised resins [50, 51] (Scheme 2.15). This yielded the desired resin-bound, orthogonally protected Thr residue **74**. At this point, deprotection of the allyl group was attempted. After optimisation, it was found that repeated treatment of the resin with palladium(0) tetrakistriphenylphosphine and phenylsilane in a mixture of CH_2Cl_2 and MeOH was partially effective [52]. To this end, four successive treatments removed most of the allyl protecting yielding HMPB-bound Fmoc-Thr **75**, however approximately 20% starting material was present. This was then treated with Fmoc-deprotection conditions followed by the coupling of Boc-Sec(PMB)-OH to yield resin-bound dipeptide **76**. At this point the key on-resin esterification reaction was carried out. Gratifyingly, the reaction proceeded smoothly after one treatment to yield the desired branched resin-bound peptide **77**. This supports the hypothesis that the 2-CTC linker was too bulky for on-resin esterification to occur, whereas it can proceed in the presence of the less bulky HMPB resin.

1) Boc_2O, $NaHCO_3$, THF, rt, 16 h
2) NaH, AllylBr, DMF, 0 °C to rt, 3 h
69%

1) HCl in dioxane, CH_2Cl_2, rt, 2 h
2) Fmoc-OSu, $NaHCO_3$, THF, rt, 16 h
82%

71 **72** **73**

Scheme 2.14 Synthesis of Fmoc-D-Thr(Allyl)-OH **73**

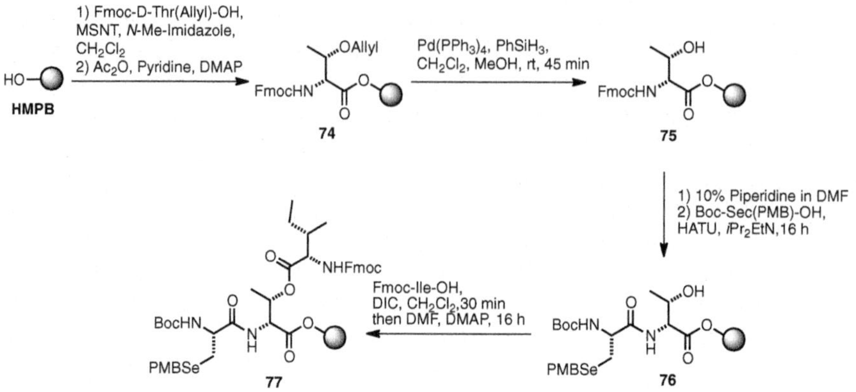

Scheme 2.15 Successful on-resin esterification using HMPB resin

While excited that the key on-resin esterification reaction was successful, the difficulty in removing the allyl protecting group meant a new protecting group strategy was considered.

2.4.3 *Towards the Synthesis of Cyclic Peptide 37*

As discussed above, the protecting group on the side chain alcohol of threonine needed to be stable to basic Fmoc-deprotection, but could be removed without concomitant deprotection of a Boc group, or cleavage from the resin. To this end, the silyl protecting group *tert*butyldimethylsilyl (TBS) was trialled. As such, Fmoc-D-Thr-OH (**78**) was protected with TBS-Cl, under conditions developed by Liu and co-workers [53], to yield the desired Fmoc-D-Thr(TBS)-OH **79** in good yield (Scheme 2.16).

With **79** in hand, the synthesis of the target cyclic peptide was again trialled. After loading **79** to HMPB resin under the same conditions described above affording **80**, the TBS group was removed with tetrabutylammonium fluoride (TBAF) in THF, which proceeded smoothly to yield the desired resin-bound free alcohol **75** (Scheme 2.17). Fmoc deprotection and coupling of Boc-Sec(PMB)-OH yielded the desired resin-bound dipeptide **76**. The next step was the on-resin esterification, which again proceeded smoothly under the conditions described

Scheme 2.16 Synthesis of Fmoc-D-Thr(TBS)-OH **79**

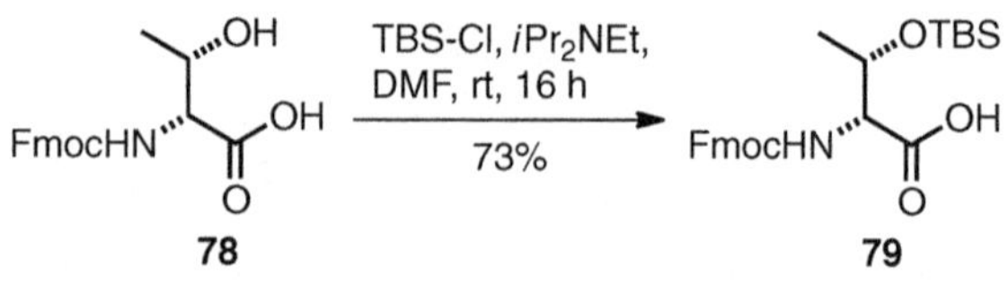

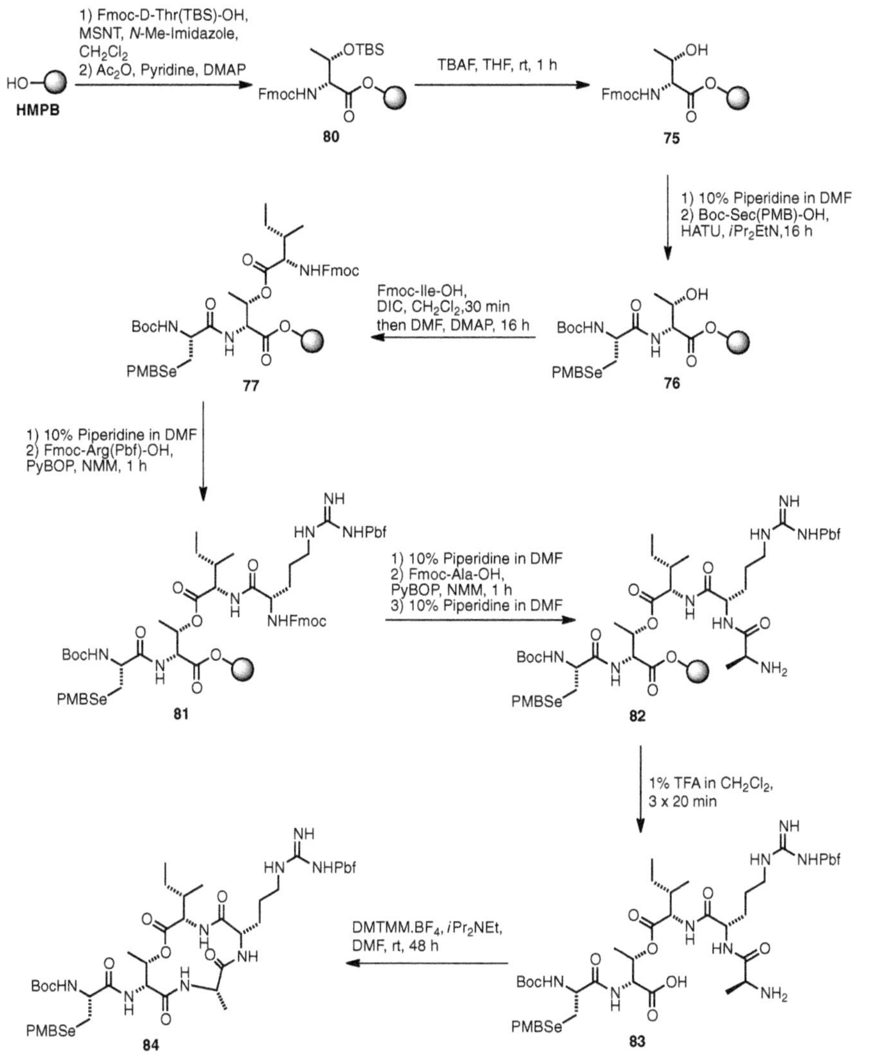

Scheme 2.17 Synthesis of protected cyclic peptide **84**

previously to yield the branched tripeptide **77**. The synthesis was continued, with Fmoc-deprotection followed by coupling of Fmoc-Arg(Pbf)-OH, using the coupling reagent (Benzotriazol-1-yloxy) tripyrrolidinophosphonium hexafluorophosphate (PyBOP) with *N*-methylmorpholine as the base affording **81**. These coupling conditions were the standard coupling conditions for proteinogenic amino acids employed in the Payne research group, and are used as the standard conditions throughout this thesis. Fmoc-Arg(Pbf)-OH was coupled, rather the precious protected enduracididine **58** in order to validate the synthetic viability of this route to teixobactin (**28**). Deprotection and coupling of Fmoc-Ala-OH, followed by final Fmoc-deprotection yielded the desired resin-bound linear peptide **82**. The resin was then treated with 1 vol.% TFA in CH_2Cl_2 to yield the protected linear peptide **83**.

At this point the key cyclisation reaction was attempted. To this end, **83** was reacted with the coupling reagent 4-(4,6-Dimethoxy-1,3,5-triazin-2-yl)-4-methylmorpholinium tetrafluoroborate (DMTMM.BF_4) and Hüning's base in DMF at high dilution for 48 h. HPLC-MS analysis of the reaction pleasingly revealed the presence of cyclised peptide **84**. This coupling reagent was initially chosen based on its ability to affect cyclisation more effectively than standard coupling reagents in a report by Butler and Jolliffe [54]. However, when this crude cyclic peptide was subjected to side-chain deprotection conditions, namely treatment with TFA:*i*Pr_3SiH:H_2O (9:0.5:0.5, v/v/v) an unidentified by-product was generated. Intriguingly, treatment of the crude cyclic peptide **84**, or linear peptide **83** with higher acid concentration than 1 vol.% TFA in CH_2Cl_2 resulted in the formation of the same by-product.

At this point, due to this unexpected issue, the route to teixobactin (**28**) was abandoned. However, some important discoveries were made, namely the optimisation of the on-resin esterification conditions by utilising the HMPB resin linker and the successful cyclisation of linear peptide **83**. With the viability of these two key transformations established, a new route to the important antibiotic natural product teixobactin (**28**) was next investigated.

2.5 Total Synthesis of Teixobactin

2.5.1 Revised Retrosynthesis

Due to the difficulties encountered in the synthesis of cyclic peptide **84** bearing a pendant selenocysteine residue, a new approach to the total synthesis of teixobactin (**28**) was developed (Scheme 2.18). Buoyed by the successful cyclisation reaction observed above, the cyclisation disconnection was not changed. However, it was envisioned that the cyclisation followed by a global deprotection step could be carried out on the complete protected linear peptide **85**. In turn this protected linear peptide could be generated from resin-bound linear peptide **86**, in turn synthesised on resin via SPPS again beginning with HPMB-bound Fmoc-D-Thr(TBS)-OH **80**.

2.5.2 Synthesis of Alloc-Ile-OH

For the strategy described above to be effective, an orthogonally protected Ile building block was required. This would allow for the generation of the branched peptide, followed by the elongation of the linear portion of teixobactin by

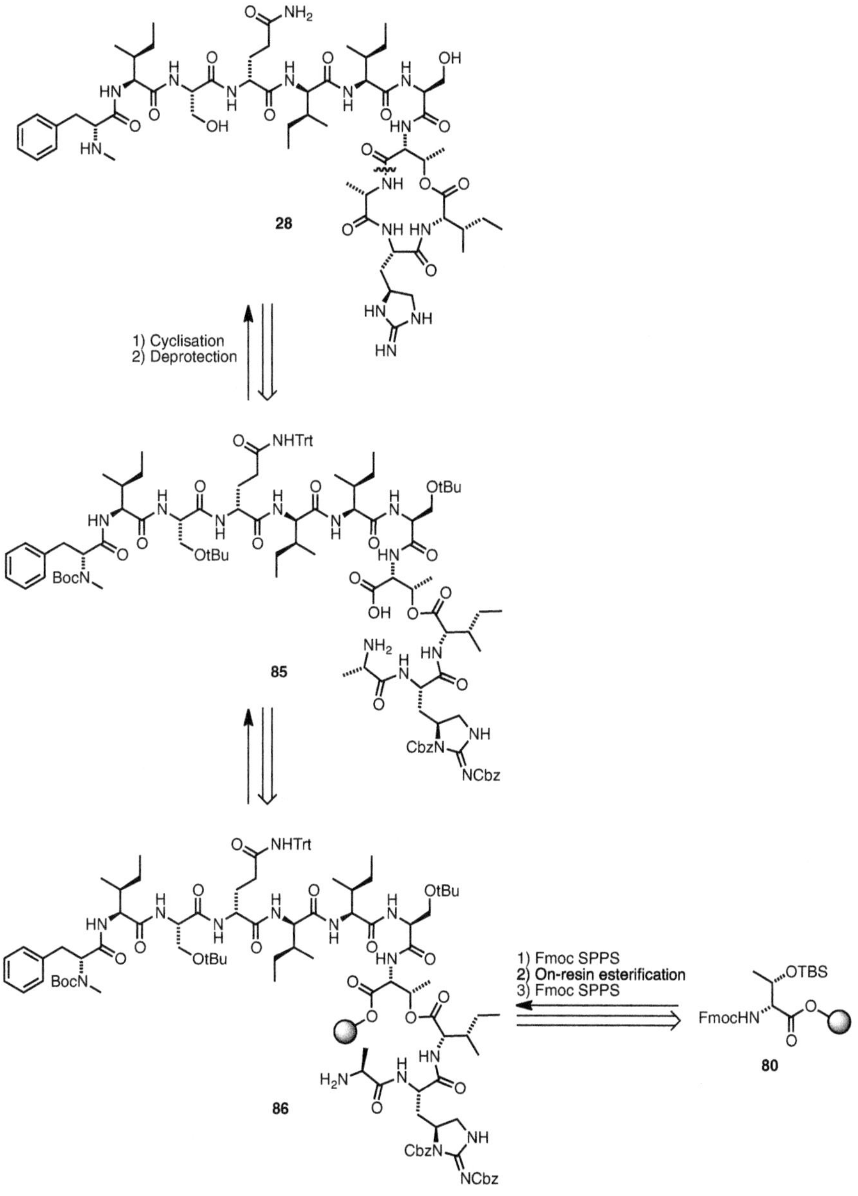

Scheme 2.18 Revised retrosynthesis of teixobactin (28)

Fmoc-SPPS. To this end the Alloc protecting group was chosen. This protecting group is stable to Fmoc deprotection conditions and can be easily cleaved with catalytic $Pd(PPh_3)_4$ and a suitable scavenger via the intermediate π-allyl-Pd species [52, 55]. To this end, L-isoleucine (**87**) was reacted with allylchloroformate in a

Scheme 2.19 Synthesis of Alloc-Ile-OH (**88**)

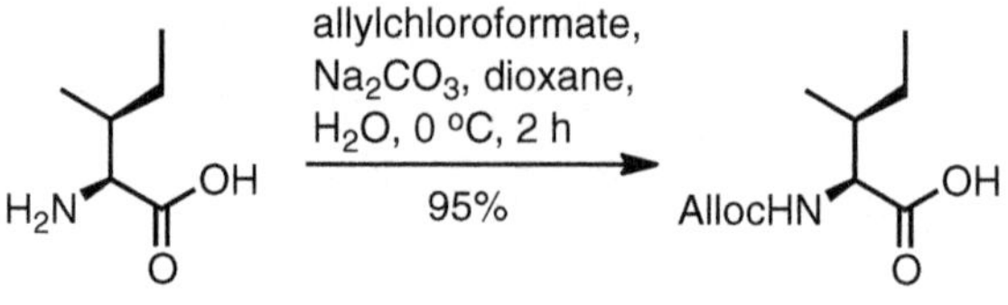

H_2N OH O **87** → allylchloroformate, Na_2CO_3, dioxane, H_2O, 0 °C, 2 h; 95% → AllocHN OH O **88**

mixed solvent system to afford the desired Alloc-Ile-OH (**88**) in excellent yield (Scheme 2.19).

2.5.3 *Synthesis of Fmoc* D*-Thr(TES)-OH*

During the attempted synthesis of teixobactin (**28**) utilising the new strategy described above, the removal of the TBS group using TBAF proved to be problematic, and not reproducible. Indeed, in one instance, a batch of resin required nine treatments with 20 equivalents of TBAF. A decrease in the loading of peptide on the resin, was observed after multiple treatments with unbuffered TBAF solution, presumably via base-catalysed cleavage of the ester bond linking the Thr residue to the HMPB linker. Due to these problems a further revision was made to the protecting group strategy used for the side chain alcohol of the D-Thr residue. Specifically, the more fluoride labile triethyl silyl (TES) protecting group was trialled. The synthesis was carried out in a manner analogous to the synthesis of Fmoc-D-Thr(TBS)-OH **79** described above, namely Fmoc-D-Thr-OH (**78**) was reacted with TES-Cl in the presence of Hünig's base to yield the desired TES protected amino acid **89** (Scheme 2.20). Unfortunately, this reaction proceeded in poor yield, presumably due to instability of the TES ether to the mildly acidic silica gel chromatography. However, enough **89** was obtained for the synthesis of teixobactin.

2.5.4 *Synthesis of Teixobactin*

With TES-protected Thr **89** in hand, the synthesis of the natural product could commence with the loading of this amino acid to resin (Scheme 2.21). This was

Scheme 2.20 Synthesis of Fmoc-D-Thr(TES)-OH (**89**)

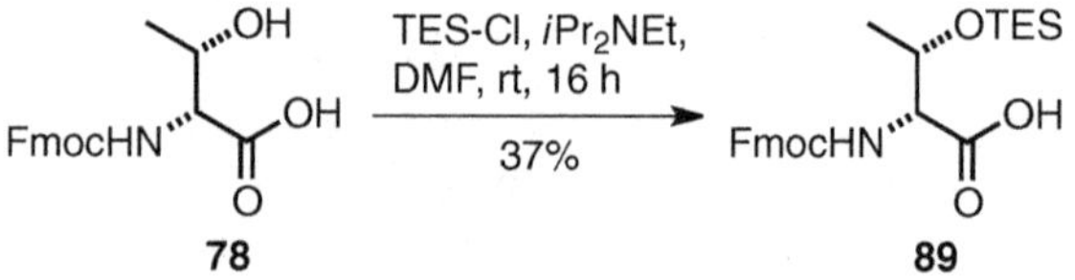

FmocHN OH OH O **78** → TES-Cl, $i$$Pr_2NEt$, DMF, rt, 16 h; 37% → FmocHN OTES OH O **89**

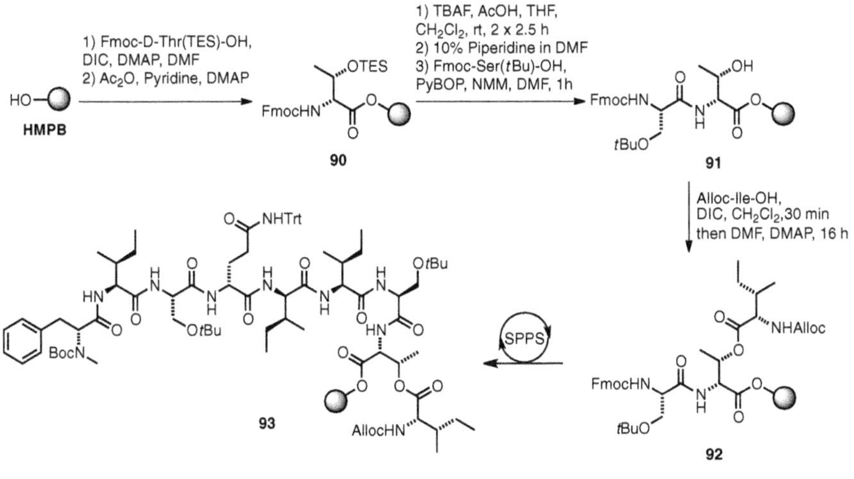

Scheme 2.21 Synthesis of resin-bound peptide **93**

carried out via the formation of the symmetric anhydride, akin to the on-resin esterification reaction. Specifically, **89** was allowed to react with DIC for 30 min to pre-form the symmetric anhydride. After the solvent was removed this was re-dissolved in mixture of DMF and CH_2Cl_2 and added to the HMPB resin with catalytic DMAP and allowed to react overnight yielding **90**. Capping of the resin followed by Fmoc-deprotection and two treatments of the resin with equimolar TBAF and AcOH in a mixture of THF and CH_2Cl_2 yielded the desired deprotected Thr attached to resin. Next, Fmoc-Ser(*t*Bu)-OH was coupled using standard PyBOP coupling conditions affording **91**, followed by on-resin esterification with Alloc-Ile-OH under the previously described conditions to yield resin-bound tripeptide **92**. With the branched resin-bound tripeptide **92** in hand the linear portion of the peptide was next extended. To this end, iterative Fmoc-SPPS, using PyBOP couplings, was carried out with the appropriate L- and D-amino acids to the *N*-terminal *N*-methyl-Boc-D-Phe residue, which yielded resin-bound peptide **93**.

The next step was the unmasking of the branched Alloc-Ile residue, which was carried out using $Pd(PPh_3)_4$ and $PhSiH_3$ as a scavenger, which proceeded smoothly affording **94** (Scheme 2.22). Fmoc-End(Cbz)$_2$-OH (**58**) was then coupled using the coupling reagent HATU, as well as the additive HOAt and Hünig's base for 20 h to yield the desired resin-bound peptide **95**. At this point the Fmoc group was removed under the standard conditions, 10 vol.% piperidine in DMF (2 $\times$ 3 min), to yield the desired free amine. HPLC-MS analysis of a small amount of cleaved resin revealed the presence of unesterified linear peptide. This suggested the formation of a diketopiperizine by which the α-amine of the enduracidine had cyclised onto the α-carboxylate of the preceding Ile residue [56]. To reduce the incidence of this unwanted by-product, the deprotection conditions were modified to 10 vol.% piperidine in DMF (1 $\times$ 40 s) and after washing, the resin was immediately exposed to a pre-activated solution of the final amino acid, Fmoc-Ala-OH, PyBOP

Scheme 2.22 Synthesis of resin-bound peptide **86**

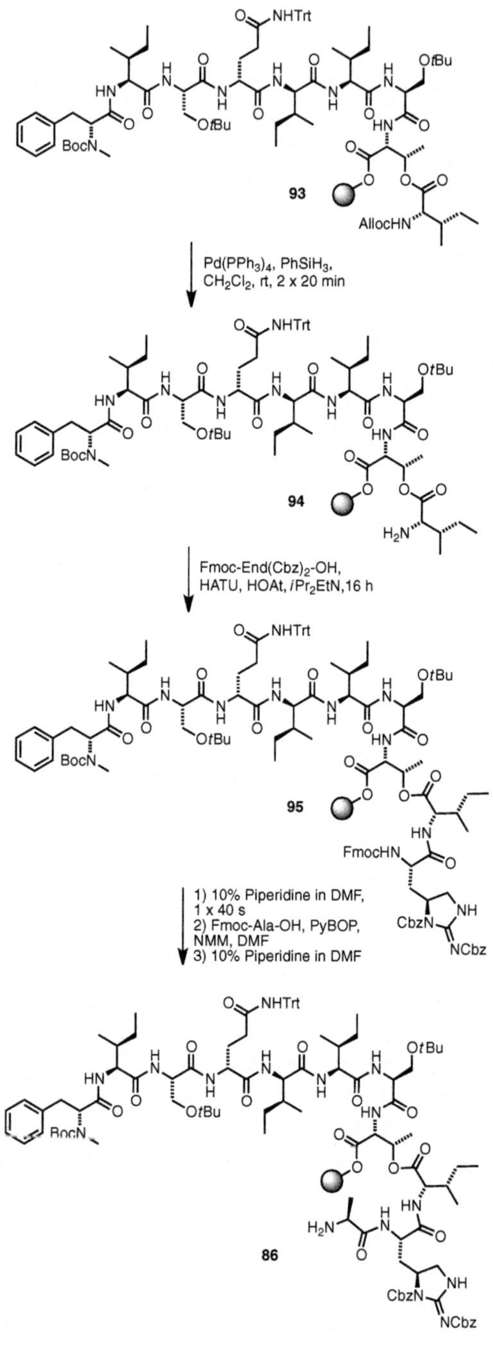
O NHTrt
HN
OtBu
BocN
93
AllocHN
Pd(PPh3)4, PhSiH3,
CH2Cl2, rt, 2 x 20 min
94
H2N
Fmoc-End(Cbz)2-OH,
HATU, HOAt, iPr2EtN,16 h
95
FmocHN
CbzN
NH
NCbz
1) 10% Piperidine in DMF,
1 x 40 s
2) Fmoc-Ala-OH, PyBOP,
NMM, DMF
3) 10% Piperidine in DMF
86

and NMM. The final Fmoc group was removed to yield the desired resin-bound linear peptide **86**.

Resin-bound linear peptide **86** was then cleaved from the resin with multiple treatments of 1 vol.% TFA in CH_2Cl_2 to yield protected linear peptide **85** (Scheme 2.23). It was important to dry the linear peptide thoroughly after resin cleavage as the presence of residual TFA lead to the formation of a TFA amide by-product during the final cyclisation reaction. Without purification, protected peptide **85** was subjected to cyclisation conditions described previously, namely DMTMM.BF_4, Hünig's base at low concentration (10 mM) in DMF. Pleasingly, UHPLC-MS analysis of the crude reaction mixture showed complete consumption of the starting material and formation of the cyclised material (Fig. 2.6b).

The final step remaining in the synthesis was the deprotection of side-chain protecting groups under acidic conditions. After the removal of the solvent the crude cyclic peptide **96** was treated with an acidic cocktail of TFA:TfOH:thioanisole:*m*-cresol (70:12:10:8, v/v/v/v). As discussed above, this cocktail was developed as an alternative to HF for the global deprotection of peptides synthesised using Boc-SPPS chemistry [40]. An ether precipitation to remove the organic soluble protecting groups and scavengers was subsequently carried out to aid in the solubilisation and analysis of the final product. The crude deprotected cyclic peptide was analysed by UHPLC-MS, and gratifyingly the deprotection was shown to have proceeded efficiently (Fig. 2.6c). In the final step, the crude material was subject to HPLC purification to yield teixobactin (**28**) as a TFA salt (Fig. 2.7). In the initial isolation reported by Ling et al. [2], teixobactin was characterised as the *bis*-HCl salt. To this end, the TFA salt was lyophilised multiple times in the presence of 5 mM HCl to afford teixobactin (**28**) in 3.3% yield based off the initial resin loading. Whilst this overall yield appears low, this corresponds to an average of 87% per step over the 24 steps carried out during the synthesis.

2.5.5 Characterisation of Teixobactin

Teixobactin (**28**) was next thoroughly characterised by mass spectrometry and NMR spectroscopy. To begin, the high-resolution mass-spectrometry confirmed that the synthetic teixobactin had the same molecular formula as the isolated natural product. Next, the NMR spectra obtained for synthetic teixobactin were analysed and compared to the data reported for isolated teixobactin [2]. Pleasingly, synthetic teixobactin was consistent with the isolated material, with notable regions including the resonances corresponding to aliphatic side-chains (2.2–0.6 ppm), which showed strong similarity to the isolated material (Fig. 2.8a). Furthermore, the resonances corresponding to the key α-protons (5.0–3.5 ppm) and α-carbons (60–50 ppm) of the synthetic material, as observed in the ^{1}H-^{13}C HSQC spectrum, were consistent with those reported for the isolated natural product (Fig. 2.8b). Overall the ^{1}H chemical shifts (Fig. 2.9a) for these key resonances differed by less than 0.1 ppm and less than 1 ppm on average for the corresponding ^{13}C chemical shifts

Scheme 2.23 Completion of the synthesis of teixobactin (**28**)

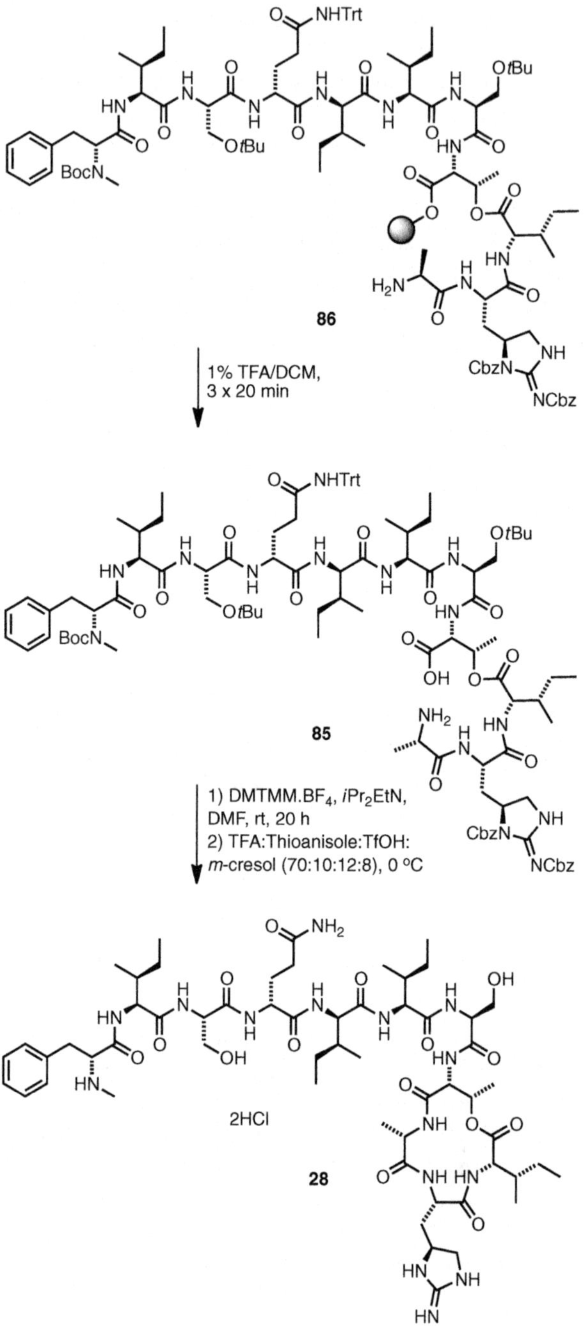
O NHTrt
OtBu
HN
BocN
OtBu
HN
86
1% TFA/DCM,
3 x 20 min
CbzN
NH
NCbz
OH
NH2
85
1) DMTMM.BF4, iPr2EtN,
DMF, rt, 20 h
2) TFA:Thioanisole:TfOH:
m-cresol (70:10:12:8), 0 °C
O NH2
OH
2HCl
28
HN
NH

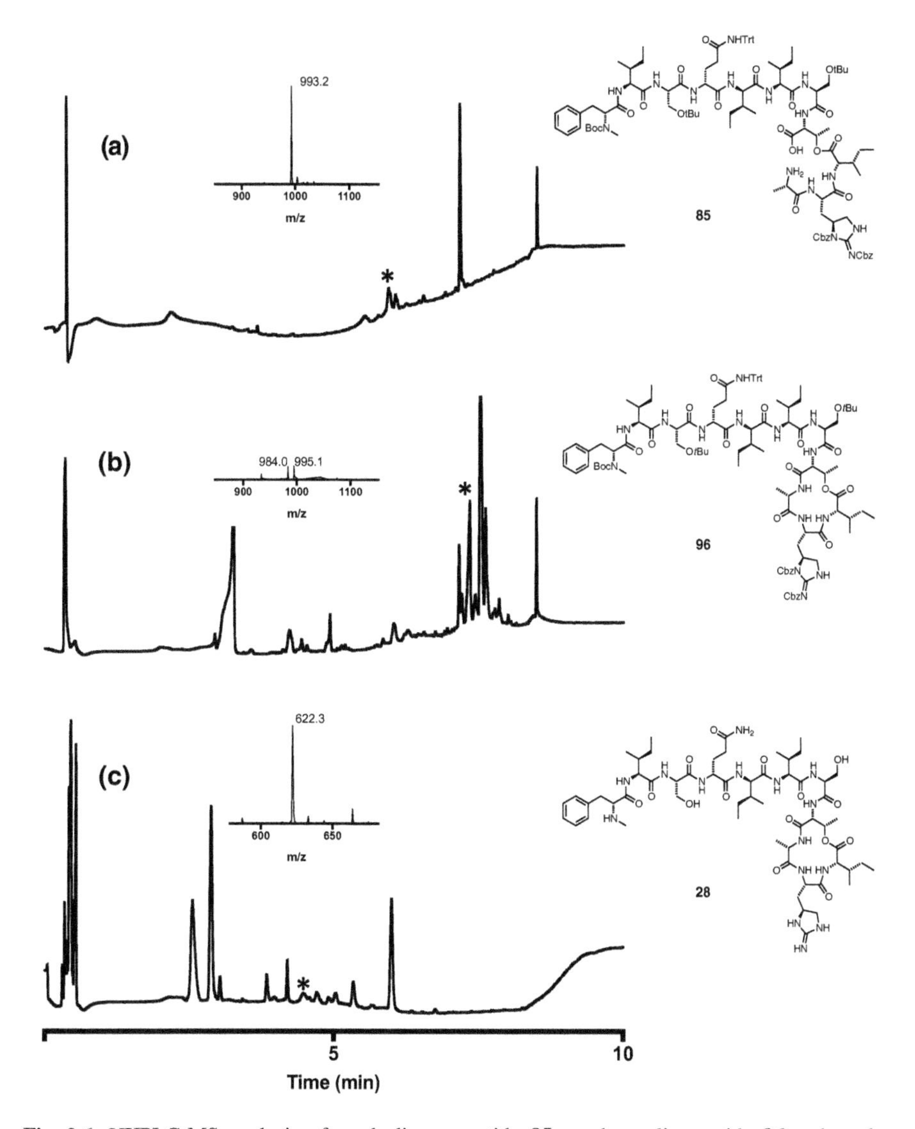

Fig. 2.6 UHPLC-MS analysis of crude linear peptide **85**, crude cyclic peptide **96** and crude teixobactin **28**. All chromatograms: 0–100% MeCN (0.1% formic acid) in H_2O (0.1% formic acid) over 8 min λ = 280 nm. **a** Crude linear peptide **85**; *m/z* 992.6 $(M+2H)^{2+}$; **b** crude cyclic peptide **96**; *m/z* 983.5 $(M+2H)^{2+}$, 994.5 $(M+H+Na)^{2+}$; **c** crude teixobactin (**28**); *m/z* 621.9 $(M+2H)^{2+}$. * Corresponds to the desired product in each trace

(Fig. 2.9b). There are some small differences in the chemical shifts in the ^{1}H and ^{13}C-NMR spectra between synthetic and isolated teixobactin (**28**), however these can be attributed to differences in concentration and pH in the acquisition of NMR data, as well as some difficulties in assignments due to peak overlap (see

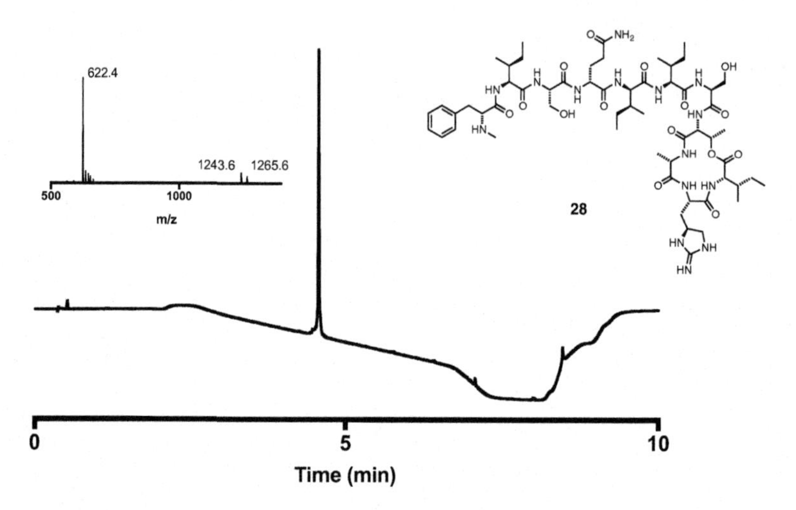

Fig. 2.7 UHPLC-MS analysis of purified teixobactin (**28**). Chromatogram: 0–70% MeCN (0.1% TFA) in H_2O (0.1% TFA) over 5 min λ = 210 nm. Teixobactin (**28**); *m/z* 621.9 $(M+2H)^{2+}$, 1242.7 $(M+H)^+$, 1264.7 $(M+Na)^+$

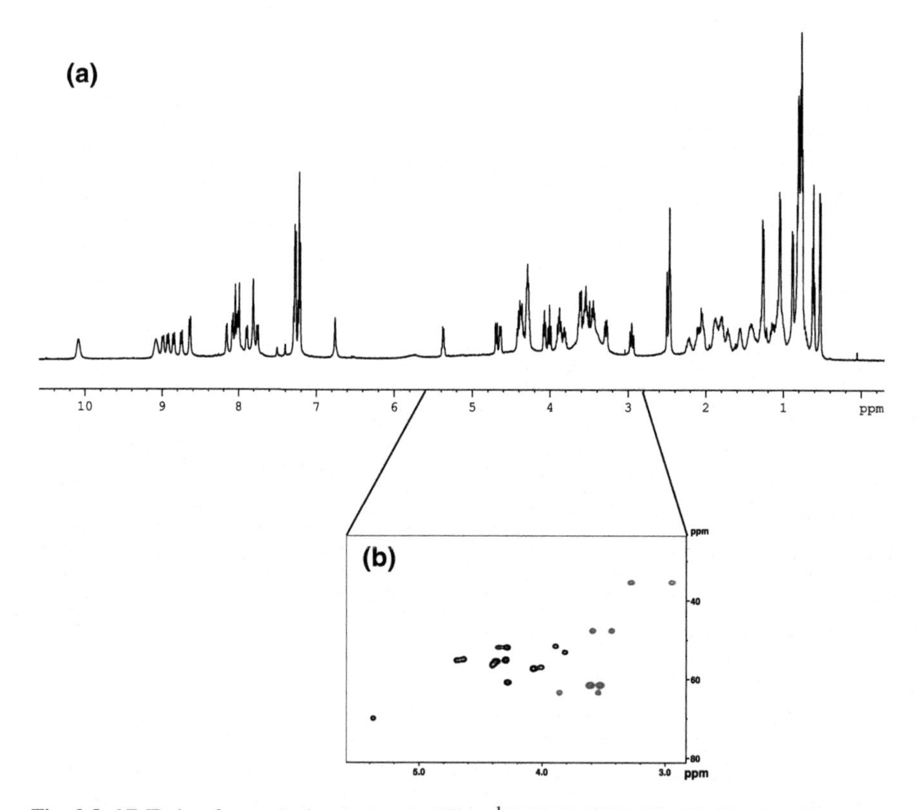

Fig. 2.8 NMR data for synthetic teixobactin (**28**). **a** ^{1}H-NMR (500 MHz, DMSO-d_6) of synthetic teixobactin. **b** ^{1}H-^{13}C HSQC spectrum of region containing key signals including α-carbons (black—CH, blue—CH_2)

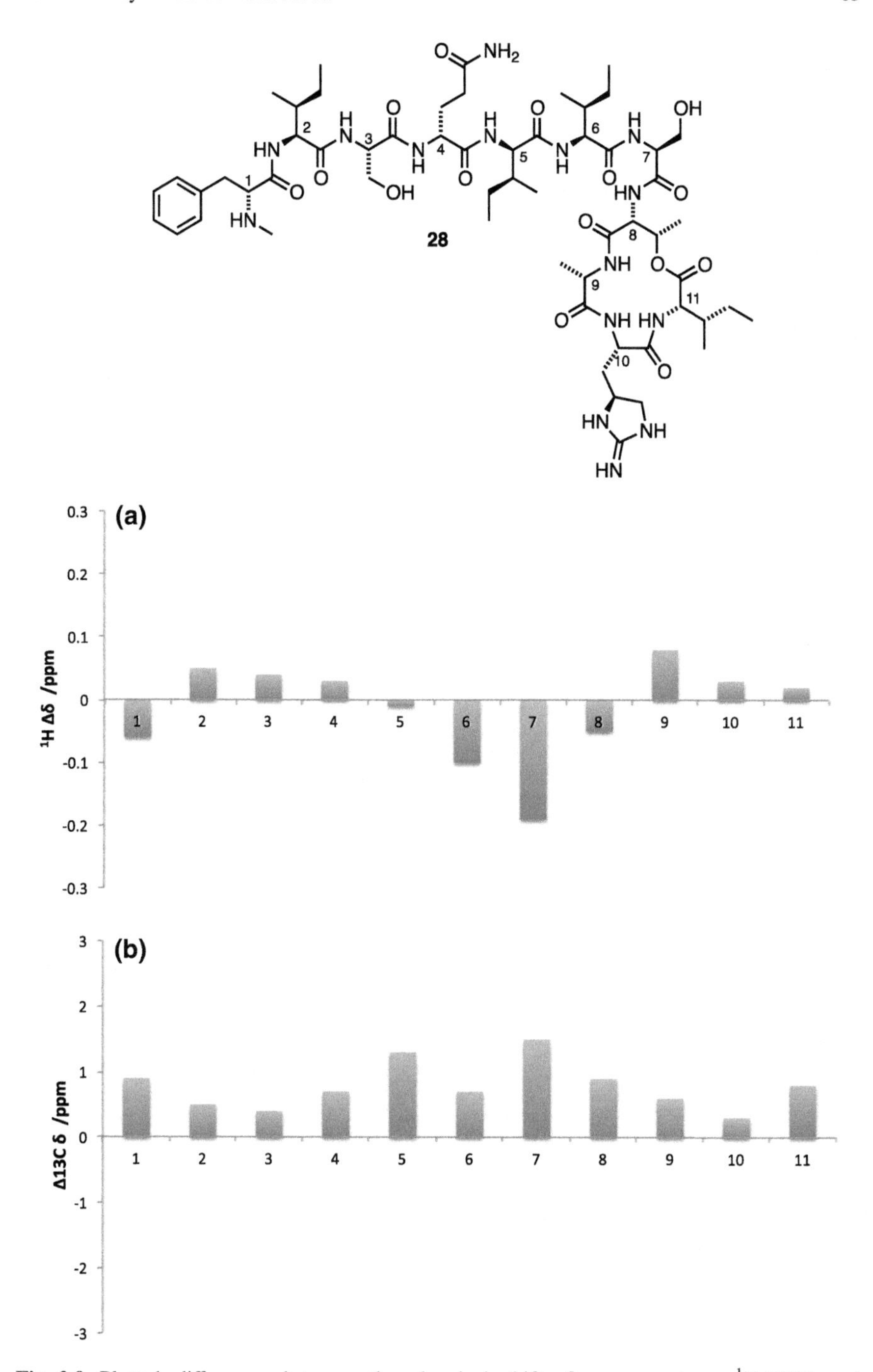

Fig. 2.9 Plotted differences between the chemical shifts for **a** α-protons, ^{1}H-NMR and **b** α-carbons, ^{13}C-NMR of synthetic and isolated teixobactin (**28**) reported by Ling et al. [2]

experimental section for more detail). Therefore, this constituted the first reported total synthesis of this highly significant molecule [57].

2.5.6 Total Synthesis of Teixobactin by Li and Co-workers

Shortly after our publication on the total synthesis of teixobactin (**28**), Li and co-workers also published a total synthesis of teixobactin (**28**) [58]. The strategy employed was similar to the initial strategy investigated in Sect. 2.4, albeit using methodology developed by Li, namely the serine-salicyaldehyde ester ligation [59]. Cyclic peptide **97** (17% yield) and linear salicyaldehyde ester **98** (23% yield) were synthesised by Fmoc- and Boc-SPPS respectively in moderate yield (Scheme 2.24). The final ligation reaction proceeded in 37% yield to give teixobactin (**28**). Together this corresponds to a 6% yield in the longest linear sequence.

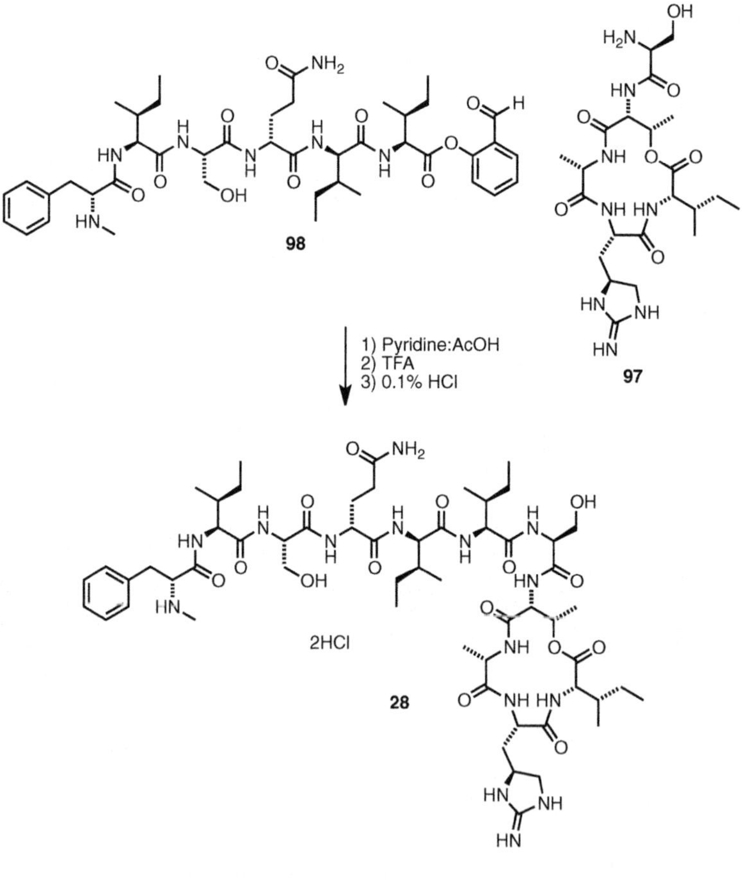

1) Pyridine:AcOH
2) TFA
3) 0.1% HCl

Scheme 2.24 Total synthesis of teixobactin (**28**) by Li and co-workers [58]

Table 2.2 Activity of synthetic teixobactin (**28**) against a panel of pathogenic bacteria. *MIC against *Mtb* determined in a separate assay

Organism	Synthetic teixobactin (**28**) MIC (μM)
Staphylococcus aureus (MSSA)	1.1
Staphylococcus aureus (MRSA)	1.1
Escherichia coli	>27
Bacillus subtilis 168	0.21
Providencia alcalifaciens	>27
Ochrobactrum anthropi	>27
Enterobacter aerogenes	>27
Actinetobacter baumannii	>27
Vibrio cholerae	>27
Salmonella typhimurium	>27
Pseudomonas aeruginosa	>27
Yersinia pseudotuberculosis	>27
Mycobacterium tuberculosis H37Rv*	1.5

2.6 Biological Activity

Synthetic teixobactin (**28**) was next screened against a panel of pathogenic bacteria by our collaborator Associate Professor Roger Linington at the University of California Santa Cruz using a high-throughput screening platform [60]. Pleasingly synthetic teixobactin showed comparable biological activity to the natural product, with strong growth inhibition of Gram-positive bacteria, including MSSA and MRSA (MIC = 1.1 μM), as well as *Bacillus subtilis* (MIC = 0.21 μM) (Table 2.2). Synthetic teixobactin (**28**) was not active against the Gram-negative pathogens tested, which is unsurprising considering the mechanism of action of teixobactin (**28**). Synthetic teixobactin (**28**) also showed strong growth inhibition of the virulent H37Rv strain of *Mtb* (MIC = 1.5 μM), which was determined using a resazurin-based assay [61] by Professor Warwick Britton and Dr. Gaya Nagalingam at the Centenary Institute, The University of Sydney.

2.7 Conclusions and Future Directions

This chapter described the first reported total synthesis of teixobactin (**28**). In order to complete the total synthesis of this important natural product the unusual amino acid L-*allo*-enduracididine first required synthesis. This was successfully achieved in 17% overall yield in 7 steps. Key steps in the synthesis included a stereoselective reduction of nitroketone **50** and a guanidinylation step, followed by protecting group manipulations to yield the suitably protected Fmoc-End($Cbz)_2$-OH **58**. With this key building block in hand the solid-phase assembly of teixobactin was commenced. Initial efforts involved a two-component ligation strategy, however difficulties in the synthesis of cyclic peptide **37** meant this was abandoned. Importantly, during these efforts, the critical on-resin esterification step was optimised by the use of HMPB linked resin. A new strategy, namely the synthesis of the full-length linear peptide followed by a final cyclisation and deprotection was devised. The side-chain protecting group for Thr was optimised to allow for the most efficient removal. Ultimately, the total synthesis of teixobactin (**28**) was achieved via an iterative SPPS protocol, followed by a final cyclisation and global deprotection to afford the natural product in 3.3% overall yield. When assessed for its activity against a panel of pathogenic bacteria, synthetic teixobactin (**28**) showed comparable activity to the isolated natural product, with potent inhibition of MRSA and *Mtb*.

Future work in this project will involve the synthesis of novel teixobactin analogues to further probe the SAR of the natural product framework. The generation of analogues containing Arg instead of End revealed that the End residue is critical for biological activity [16, 17]. Recently there have been a number of reports on the synthesis of teixobactin analogues [62–66]. Interestingly, all the reports have substituted the End residue, presumably due to the difficult synthesis of the building and incorporation into the peptide. To date, an analogue of teixobactin more potent than the natural product **28** is yet to be found. This further reinforces the importance of the End residue to the bioactivity of teixobactin (**28**). As such, rather than replacing this residue, analogues containing altered linear peptide portions could be synthesised. In another drug discovery program in the Payne lab, the replacement of naturally occurring amino acids with synthetic amino acids has proved extremely successful and this approach could also be applied to teixobactin analogues [67]. Furthermore, the replacement of the metabolically labile ester bond in the peptide cycle, resulting in analogue **99** could be investigated. It is envisioned that the incorporation of a 2,3-diaminobutyric acid (DABA) residue **100** in place of the Thr

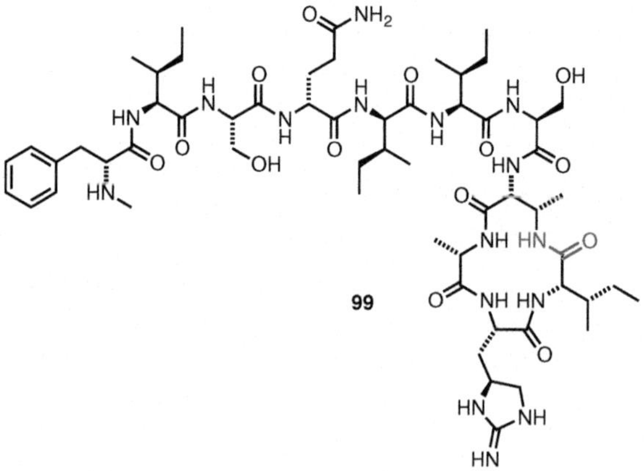

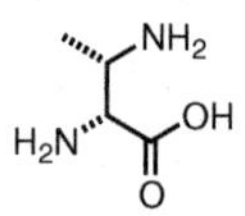

Fig. 2.10 Structure of proposed teixobactin analogue **99** and DABA amino acid **100**

residue would result in a more stable amide bond (red, Fig. 2.10). These structural variations could result in the development of teixobactin analogues with potent bioactivity.

References

1. M.C. Wilson, T. Mori, C. Ruckert, A.R. Uria, M.J. Helf, K. Takada, C. Gernert, U.A.E. Steffens, N. Heycke, S. Schmitt, C. Rinke, E.J.N. Helfrich, A.O. Brachmann, C. Gurgui, T. Wakimoto, M. Kracht, M. Crusemann, U. Hentschel, I. Abe, S. Matsunaga, J. Kalinowski, H. Takeyama, J. Piel, Nature **506**, 58–62 (2014)
2. L.L. Ling, T. Schneider, A.J. Peoples, A.L. Spoering, I. Engels, B.P. Conlon, A. Mueller, T.F. Schaberle, D.E. Hughes, S. Epstein, M. Jones, L. Lazarides, V.A. Steadman, D.R. Cohen, C. R. Felix, K.A. Fetterman, W.P. Millett, A.G. Nitti, A.M. Zullo, C. Chen, K. Lewis, Nature **517**, 455–459 (2015)
3. T. Kaeberlein, K. Lewis, S.S. Epstein, Science **296**, 1127–1129 (2002)
4. K. Lewis, S. Epstein, A. D'Onofrio, L.L. Ling, J. Antibiot. **63**, 468–476 (2010)
5. D. Nichols, N. Cahoon, E.M. Trakhtenberg, L. Pham, A. Mehta, A. Belanger, T. Kanigan, K. Lewis, S.S. Epstein, Appl. Environ. Microbiol. **76**, 2445–2450 (2010)
6. H. He, R.T. Williamson, B. Shen, E.I. Graziani, H.Y. Yang, S.M. Sakya, P.J. Petersen, G.T. Carter, J. Am. Chem. Soc. **124**, 9729–9736 (2002)
7. S. Horii, Y. Kameda, J. Antibiot. **21**, 665–667 (1968)
8. L. Han, A.W. Schwabacher, G.R. Moran, N.R. Silvaggi, Biochemistry **54**, 7029–7040 (2015)
9. A.M. Burroughs, R.W. Hoppe, N.C. Goebel, B.H. Sayyed, T.J. Voegtline, A.W. Schwabacher, T.M. Zabriskie, N.R. Silvaggi, Biochemistry **52**, 4492–4506 (2013)
10. N.A. Magarvey, B. Haltli, M. He, M. Greenstein, J.A. Hucul, Antimicrob. Agents Chemother. **50**, 2167–2177 (2006)
11. M.A. D'Elia, K.E. Millar, T.J. Beveridge, E.D. Brown, J. Bacteriol. **188**, 8313–8316 (2006)
12. G. Bierbaum, H.-G. Sahl, Arch. Microbiol. **141**, 249–254 (1985)
13. T. Homma, A. Nuxoll, A.B. Gandt, P. Ebner, I. Engels, T. Schneider, F. Götz, K. Lewis, B. P. Conlon, Antimicrob. Agents Chemother. **60**, 6510–6517 (2016)
14. C.T. Walsh, S.L. Fisher, I.S. Park, M. Prahalad, Z. Wu, Cell Chem. Biol **3**, 21–28 (1996)
15. K. Lewis, Antimicrob. Agents Chemother. **45**, 999–1007 (2001)
16. Y.E. Jad, G.A. Acosta, T. Naicker, M. Ramtahal, A. El-Faham, T. Govender, H.G. Kruger, B. G. de la Torre, F. Albericio, Org. Lett. **17**, 6182–6185 (2015)
17. A. Parmar, A. Iyer, C.S. Vincent, D. Van Lysebetten, S.H. Prior, A. Madder, E.J. Taylor, I. Singh, Chem. Commun. **52**, 6060–6063 (2016)
18. N.J. Mitchell, L.R. Malins, X. Liu, R.E. Thompson, B. Chan, L. Radom, R.J. Payne, J. Am. Chem. Soc. **137**, 14011–14014 (2015)
19. L.R. Malins, N.J. Mitchell, S. McGowan, R.J. Payne, Angew. Chem. Int. Ed. **54**, 12716–12721 (2015)
20. C.J. White, A.K. Yudin, Nat. Chem. **3**, 509–524 (2011)
21. J.S. Davies, J. Pept. Sci. **9**, 471–501 (2003)
22. G. Illuminati, L. Mandolini, Acc. Chem. Res. **14**, 95–102 (1981)
23. U. Schmidt, J. Langner, J. Pept. Res. **49**, 67–73 (1997)
24. S.F. Brady, S.L. Varga, R.M. Freidinger, D.A. Schwenk, M. Mendlowski, F.W. Holly, D.F. Veber, J. Org. Chem. **44**, 3101–3105 (1979)
25. J.M. Humphrey, A.R. Chamberlin, Chem. Rev. **97**, 2243–2266 (1997)
26. T. Shinichi, K. Shoichi, S. Tetsuo, Chem. Lett. **4**, 1281–1284 (1975)
27. L. Sanière, L. Leman, J.-J. Bourguignon, P. Dauban, R.H. Dodd, Tetrahedron **60**, 5889–5897 (2004)

28. W. Craig, J. Chen, D. Richardson, R. Thorpe, Y. Yuan, Org. Lett. **17**, 4620–4623 (2015)
29. A.J. Peoples, D. Hughes, L.L. Ling, W. Millett, A. Nitti, A. Spoering, V.A. Steadman, J.Y.C. Chiva, L. Lazarides, M.K. Jones, K.L. Poullenec, K. Lewis, Patent Number: WO 2014/089053 AI (2014)
30. J. Rudolph, F. Hannig, H. Theis, R. Wischnat, Org. Lett. **3**, 3153–3155 (2001)
31. H.C. Brown, S. Krishnamurthy, J. Am. Chem. Soc. **94**, 7159–7161 (1972)
32. M.T. Reetz, K. Kesseler, A. Jung, Tetrahedron Lett. **25**, 729–732 (1984)
33. T.J. Leitereg, D.J. Cram, J. Am. Chem. Soc. **90**, 4019–4026 (1968)
34. D.A. Evans, M.J. Dart, J.L. Duffy, M.G. Yang, J. Am. Chem. Soc. **118**, 4322–4343 (1996)
35. D.A. Evans, J.L. Duffy, M.J. Dart, Tetrahedron Lett. **35**, 8537–8540 (1994)
36. H.B. Bürgi, J.D. Dunitz, J.M. Lehn, G. Wipff, Tetrahedron **30**, 1563–1572 (1974)
37. K. Feichtinger, H.L. Sings, T.J. Baker, K. Matthews, M. Goodman, J. Org. Chem. **63**, 8432–8439 (1998)
38. P.K. Mandal, J.S. McMurray, J. Org. Chem. **72**, 6599–6601 (2007)
39. J. Martinez, J.C. Tolle, M. Bodanszky, J. Org. Chem. **44**, 3596–3598 (1979)
40. Z.P. Gates, B. Dhayalan, S.B.H. Kent, Chem. Commun. **52**, 13979–13982 (2016)
41. P. Athanassopoulos, K. Barlos, D. Gatos, O. Hatzi, C. Tzavara, Tetrahedron Lett. **36**, 5645–5648 (1995)
42. K. Barlos, D. Gatos, J. Kallitsis, G. Papaphotiu, P. Sotiriu, Y. Wenqing, W. Schäfer, Tetrahedron Lett. **30**, 3943–3946 (1989)
43. K. Barlos, D. Gatos, S. Kapolos, G. Papaphotiu, W. Schäfer, Y. Wenqing, Tetrahedron Lett. **30**, 3947–3950 (1989)
44. N. Bayó-Puxan, A. Fernández, J. Tulla-Puche, E. Riego, C. Cuevas, M. Álvarez, F. Albericio, Chem. Eur. J. **12**, 9001–9009 (2006)
45. A. El-Faham, F. Albericio, Chem. Rev. **111**, 6557–6602 (2011)
46. L.A. Carpino, J. Am. Chem. Soc. **115**, 4397–4398 (1993)
47. B. Neises, W. Steglich, Angew. Chem. Int. Ed. **17**, 522–524 (1978)
48. H. Kaur, P.W.R. Harris, P.J. Little, M.A. Brimble, Org. Lett. **17**, 492–495 (2015)
49. B. Riniker, A. Flörsheimer, H. Fretz, P. Sieber, B. Kamber, Tetrahedron **49**, 9307–9320 (1993)
50. B. Blankemeyer-Menge, M. Nimtz, R. Frank, Tetrahedron Lett. **31**, 1701–1704 (1990)
51. R. Frank, R. Döring, Tetrahedron **44**, 6031–6040 (1988)
52. M. Dessolin, M.-G. Guillerez, N. Thieriet, F. Guibé, A. Loffet, Tetrahedron Lett. **36**, 5741–5744 (1995)
53. Y.C. Huang, Y.M. Li, Y. Chen, M. Pan, Y.T. Li, L. Yu, Q.X. Guo, L. Liu, Angew. Chem. Int. Ed. **52**, 4858–4862 (2013)
54. S.J. Butler, K.A. Jolliffe, Org. Biomol. Chem. **9**, 3471–3483 (2011)
55. P.D. Jeffrey, S.W. McCombie, J. Org. Chem. **47**, 587–590 (1982)
56. B.F. Gisin, R.B. Merrifield, J. Am. Chem. Soc. **94**, 3102–3106 (1972)
57. A.M. Giltrap, L.J. Dowman, G. Nagalingam, J.L. Ochoa, R.G. Linington, W.J. Britton, R. J. Payne, Org. Lett. **18**, 2788–2791 (2016)
58. K. Jin, I.H. Sam, K.H.L. Po, D. Lin, E.H. Ghazvini Zadeh, S. Chen, Y. Yuan, X. Li, Nat. Commun. **7**, 12394 (2016)
59. Y. Zhang, C. Xu, H.Y. Lam, C.L. Lee, X. Li, Proc. Nat. Acad. Sci. U.S.A. **110**, 6657–6662 (2013)
60. I. Wiegand, K. Hilpert, R.E.W. Hancock, Nat. Protoc. **3**, 163–175 (2008)
61. N.K. Taneja, J.S. Tyagi, J. Antimicrob. Chemother. **60**, 288–293 (2007)
62. S.A.H. Abdel Monaim, Y.E. Jad, E.J. Ramchuran, A. El-Faham, T. Govender, H.G. Kruger, B.G. de la Torre, F. Albericio, ACS Omega **1**, 1262–1265 (2016)
63. K. Jin, K.H.L. Po, S. Wang, J.A. Reuven, C.N. Wai, H.T. Lau, T.H. Chan, S. Chen, X. Li, Bioorg. Med. Chem. (2017) In Press
64. A. Parmar, S.H. Prior, A. Iyer, C.S. Vincent, D. Van Lysebetten, E. Breukink, A. Madder, E. J. Taylor, I. Singh, Chem. Commun. **53**, 2016–2019 (2017)
65. C. Wu, Z. Pan, G. Yao, W. Wang, L. Fang, W. Su, RSC Adv. **7**, 1923–1926 (2017)

66. H. Yang, D.R. Du Bois, J.W. Ziller, J.S. Nowick, Chem. Commun. **53**, 2772–2775 (2017)
67. A.T. Tran, E.E. Watson, V. Pujari, T. Conroy, L.J. Dowman, A.M. Giltrap, A. Pang, W.R. Wong, R.G. Linington, S. Mahapatra, J. Saunders, S.A. Charman, N.P. West, T.D.H. Bugg, J. Tod, C.G. Dowson, D.I. Roper, D.C. Crick, W.J. Britton, R.J. Payne, Nat. Commun. **8**, 14414 (2017)

Chapter 3
Synthesis of Deshydroxy Skyllamycins A–C

3.1 Isolation of Skyllamycins A–C

There is enormous structural diversity amongst nonribosomal peptide natural products that contributes to the broad spectrum of biological activity that NRPs exhibit. One important family of NRPs are the skyllamycins (**101**–**103**) which are highly functionalised cyclic peptides comprised of a number of non-proteinogenic amino acids (Fig. 3.1). Specifically, they contain an *N*-terminal aromatic cinnamoyl moiety (red). Related motifs have been shown to be present in other natural products including the pepticinnamins [1] and the recently isolated coprisamides [2]. They also contain a β-methylated aspartic acid (β-Me-Asp) residue (green) which is present in other NRPs such as the lipopeptide antibiotic friulimicin [3]. While these modifications are interesting, one of the most striking features of the skyllamycins is the high level of hydroxylation present. Three of the residues are hydroxylated at the β-position, namely β-hydroxy phenylalanine (β-OH-Phe), β-hydroxy leucine (β-OH-Leu) and β-hydroxy-*O*-methyl tyrosine (β-OH-*O*-Me-Tyr) (blue). It should be noted that β-hydroxylated amino acids are common in NRPs, and are present in a number of NRP antibiotics, including vancomycin (**8**) [4].

Finally, and most unusually, the skyllamycins contain an α-hydroxylated glycine (α-OH-Gly) residue (purple). This interesting modification is extremely rare and has to date only been found in one other natural product, the anti-tumour linear peptide natural product spergualin [5, 6]. Interestingly, α-OH-Gly is an important biosynthetic intermediate in the formation of *C*-terminal carboxamides in proteins, however it is rarely found in a mature product. Residues other than glycine have also been shown to be α-hydroxylated in a small number of natural products. Specifically, the natural product thanamycin [7] possesses α-hydroxylated ornithine residue, and the formation of α-OH-proline has been shown to be critical to the maturation of bacterial polysaccharide deacetylase enzymes [8].

A. Giltrap, *Total Synthesis of Natural Products with Antimicrobial Activity*, Springer Theses, https://doi.org/10.1007/978-981-10-8806-3_3

3.1.1 Isolation and Initial Biological Analysis

Skyllamycin A (**101**) was first isolated as RP-1776 in 2001 from a culture of *Streptomyces* sp. KY11784, a Japanese soil bacterium [9]. It was discovered during a screen for inhibitors of the platelet-derived growth factor (PDGF) signalling pathway. PDGF signalling plays a significant role in cell proliferation and movement, and is thought to be critical in the development and maintenance of cells and wound healing. However, incorrect signalling is implicated in the pathogenesis of a number of human disease states, including various types of cancer [10]. RP-1776 was found to selectively inhibit the binding of PDGF B-type dimer to the PDGF β-receptor.

In 2011, after screening for NRPSs in the genome of the bacteria *Streptomyces* sp. Acta 2897, RP-1776 and an analogue containing Asp rather than β-Me-Asp, were isolated by Süssmuth et al. and renamed skyllamycins A (**101**) and B (**102**), respectively [11] (Fig. 3.1). The authors also carried out in depth studies on the biosynthetic pathway and these are described below.

3.1.2 Skyllamycins A–C—Biofilm Inhibition Studies

One method by which bacteria evade eradication by antibiotics is via the formation of biofilms (Sect. 1.2.2). *Pseudomonas aeruginosa* is a Gram-negative bacterium which readily forms biofilms. These biofilms are particularly significant in patients with cystic fibrosis (CF), a genetic disease which results in an inability to clear mucus from the lungs. CF patients have a significantly decreased life expectancy, and a key cause of this is *P. aeruginosa* biofilm infections in the lungs, leading to respiratory failure and death [12].

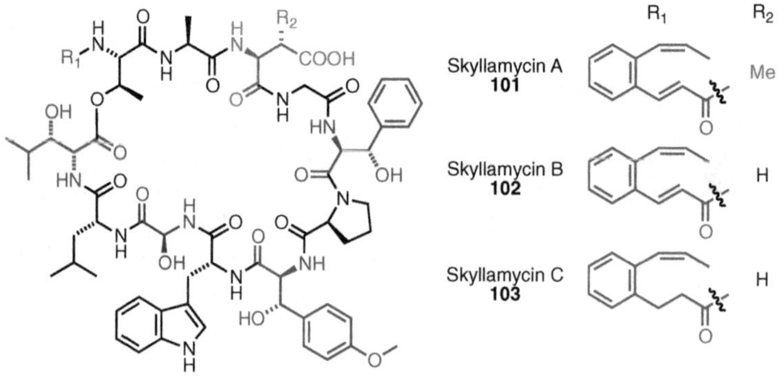

Fig. 3.1 Structure of skyllamycins A–C (**101–103**) with modified residues highlighted: cinnomyl moiety (red), β-Me-Asp (green), β-OH-residues (blue), α-OH-Gly (purple)

Recently, Linington and co-workers developed a high throughput screen to identify inhibitors and dispersers of *P. aeruginosa* biofilm formation, utilising fluorescence microscopy [13]. Biofilm inhibition is defined as the ability to prevent bacteria forming a biofilm, whereas biofilm dispersers can clear a pre-formed biofilm. To measure inhibition of biofilms, the compound of interest (analyte) and bacterial culture are added to the culture plate at the same time. However, to measure dispersion the bacterial culture was allowed to grow for 2 h prior to the addition of the analyte. A further important aspect of the screen is the ability to determine whether the analyte has the ability to inhibit/disperse the biofilm, or whether it acts as a traditional antibiotic and prevents bacterial growth. As such, the biofilm screen was coupled with a colorimetric assay to measure cellular activity.

During a screen of 312 natural product pre-fractions, Linington and co-workers found skyllamycin B (**102**), and a reduced analogue, skyllamycin C (**103**), to be inhibitors of biofilm formation with 50% effective concentrations (EC_{50}s) of 30 and 60 μM, respectively (Fig. 3.2) [13]. Skyllamycin B (**102**) was shown to be capable of clearing pre-attached biofilms in the dispersion assay with an EC_{50} of 60 μM. Interestingly, skyllamycin A (**101**) did not show any activity in either the inhibition or dispersion assays, which suggests the Asp residue plays an important role in the bioactivity. Skyllamycins A–C (**101–103**) did not inhibit bacterial cellular activity. Linington and co-workers subsequently showed that a co-dose treatment of skyllamycin B (**102**) and azithromycin, an antibiotic incapable of clearing biofilms, was able to disperse the *P. aeruginosa* biofilm and supress cellular activity, a result

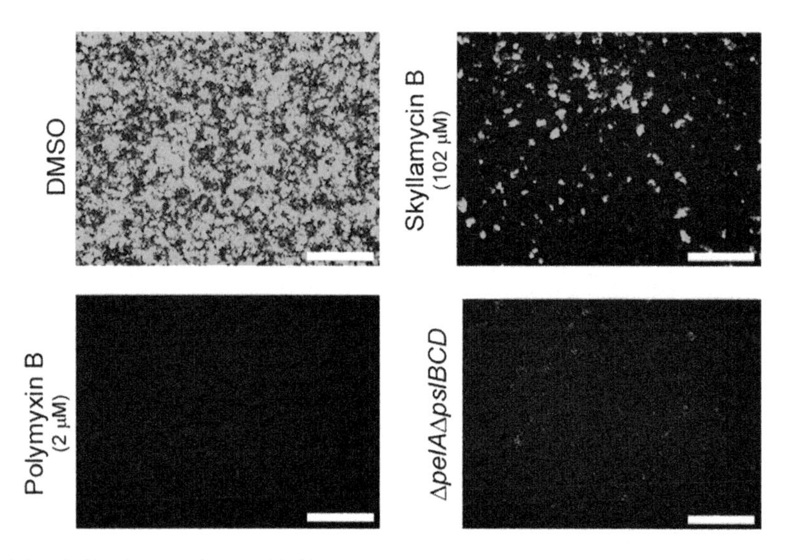

Fig. 3.2 Biofilm images from a biofilm inhibition assay. Negative control—top left—healthy biofilm with DMSO vehicle added. Chemical positive control—bottom left—no biofilm formation due to addition of antibiotic polymyxin B. Genetic positive control—bottom right—genes required for biofilm formation knocked out. Analyte—top right—addition of skyllamycin B (**102**) at 102 μM final concentration [13]

which neither molecule could achieve alone. Biofilm infections are normally treated with large doses of antibiotics, which can lead to resistance if the infection is not totally cleared. As such, co-dosing with a traditional antibiotic and a biofilm disperser is an attractive strategy for treating biofilm infections.

3.1.3 Full Stereochemical Assignment of Skyllamycins A–C

When the isolation of RP-1776 was reported in 200 [9], the stereochemistry of the amino acids was unassigned. During the subsequent isolation by Süssmuth and co-workers in 2011 [11], the stereochemistry of the amino acids was determined except for the three β-OH amino acids and configuration of the unusual α-OH-Gly residue. A follow up publication from the Süssmuth group determined the configuration of the three β-OH amino acids and postulated the configuration of the α-OH-Gly residue [14]. To carry this out, skyllamycin A (**101**) was hydrolysed into its constituent amino acids by treatment in refluxing 6 M HCl. The non-functionalised amino acids were subsequently compared to authentic standards of the L or D-amino acids utilising chiral gas chromatography mass spectrometry (GC-MS) or HPLC-MS. This revealed the presence of L-Thr, L-Ala, L-Pro, and D-Leu. The authors were unable to determine the configuration of the Trp residue using chiral GC/HPLC-MS. As such, this was analysed utilising Marfey's method [15], which enabled the Trp residue to be assigned the D-configuration.

In order to determine the stereochemistry of the β-substituted amino acids, authentic standards of all four stereoisomers [(2*S*,3*S*), (2*S*,3*R*), (2*R*,3*S*), (2*R*,3*S*)] of each amino acid were synthesised according to known literature procedures and compared to those in the natural product via chiral GC/HPLC-MS. This established the following configurations: (2*S*,3*S*)-β-Me-Asp, (2*S*,3*S*)-β-OH-Phe and (2*R*,3*S*)-β-OH-Leu. During the harsh acidic conditions required to hydrolyse the peptide the β-OH-*O*-Me-Tyr residue decomposed, meaning its configuration could not be assigned. The authors were able to assign the configuration following a similar procedure which was used to determine the configuration of β-*O*Me-Tyr in the NRP callipeltin A [16]. Specifically, the β-OH-*O*-Me-Tyr standards and natural product were subject to oxidative ozonolysis which produced the corresponding β-OH-Asp. This amino acid is stable to the conditions required to hydrolyse the natural product and, as such, could be compared to the authentic standards. Using this method, the configuration of the β-OMe-Tyr in the skyllamycins was assigned as (2*S*,3*S*)-β-OH-*O*-Me-Tyr.

The final residue requiring assignment was the α-OH-Gly. Due to its inherent instability, the configuration could not be determined by the usual methods which were utilised for the other amino acids. To begin, the authors calculated the interatomic distances between protons utilising Nuclear Overhauser Effect SpectroscopY (NOESY)-NMR. Using these distances as constraints, simulated annealing and free molecular dynamics simulations were carried out, when α-OH Gly was in both the (*S*) and (*R*) configuration. Ultimately the computational

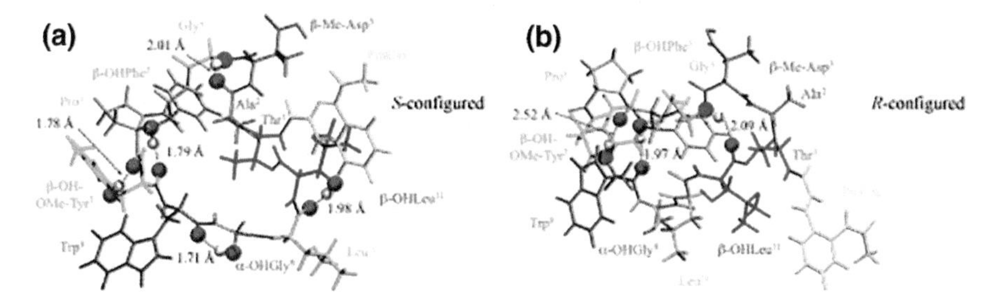

Fig. 3.3 Structures of skyllamycin A (**101**) generated from molecular dynamics simulations. **a** (*S*)-α-OH-Gly allows for five key hydrogen-bonding interactions (highlighted with ball and stick atoms). **b** In contrast, (*R*)-α-OH-Gly allows for only three key hydrogen bonding interactions [14]

experiments suggested that the α-OH-Gly present in the skyllamycins is (*S*)-configured. Interestingly, when in the (*S*) configuration, five strong intramolecular hydrogen bonds (H-bonds) are present, with the hydroxyl groups of the three β-OH residues, and α-OH-Gly all involved (Fig. 3.3). However, in the (*R*) configuration there are only three strong H-bonds and the hydroxyl of the α-OH-Gly residue was not involved. All the evidence considered, the authors suggest that the α-OH-Gly residue has the (*S*) configuration [14].

3.2 Biosynthesis of the Skyllamycins

During their isolation of skyllamycins A (**101**) and B (**102**) [11], Süssmuth and co-workers carried out extensive biosynthetic studies. To begin with, the researchers identified the gene cluster responsible for the synthesis of the skyllamycins by screening for known conserved motifs present in NRPSs. From the analysis, 49 genes were identified to be responsible for the biosynthesis of skyllamycins A (**101**) and B (**102**).

3.2.1 Biosynthesis of Building Blocks

β-Me-Aspartic Acid

Two genes in the biosynthetic gene cluster are responsible for the biosynthesis of the β-Me-Asp residue in the skyllamycins. The two proteins encoded have high homology to the glutamate mutase enzyme subunits present in the biosynthesis of β-Me-Asp in the friulimicins [3]. Glutamate mutase enzymes catalyse the conversion of (2*S*)-Glu (**104**) to (2*S*,3*S*)-β-Me-Asp (**105**) (Scheme 3.1). In order to verify this assignment, mutant strains of the *Streptomyces* sp. Acta 2897 with deletion of the genes *sky41* and *sky42* were generated. This resulted in the loss skyllamycin A (**101**), but the production of skyllamycin B (**102**) remained unaffected.

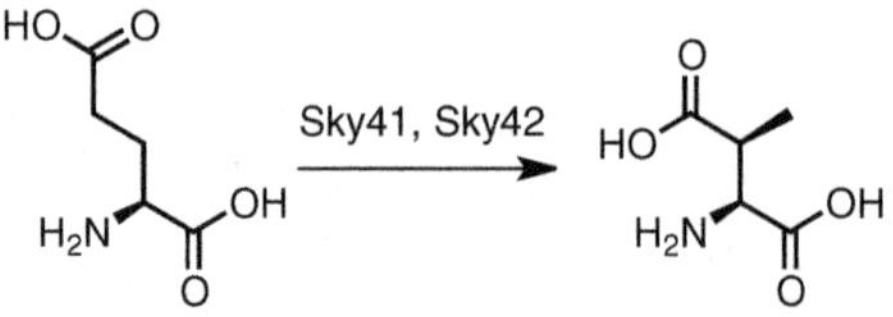

Scheme 3.1 Conversion of (2*S*)-Glu (**104**) to (2*S*,3*S*)-β-Me-Asp (**105**) catalysed by glutamate mutase

O-Me-Tyrosine

The gene product encoded by *sky37* carries out the *O*-methylation of tyrosine prior to incorporation into the NRPS. This was confirmed by the deletion of the *sky37* gene, and analysis of the products formed after culture. Interestingly, when analysed by mass spectrometry the product formed showed a loss of 30 Da compared to the natural skyllamycin, corresponding to loss of OCH_3, i.e. incorporation of phenylalanine as opposed to tyrosine. This was explained by analysis of the A-domain responsible for incorporation of the *O*-Me-Tyr. Surprisingly, it revealed specificity for phenylalanine rather than tyrosine. Spiking the culture broth of the *sky37* deletion mutant strain with *O*-Me-Tyr restored production of the native skyllamycins.

N-terminal Aromatic Cinnamoyl Residue

Aromatic moieties in natural products are often produced via polyketide synthase biosynthetic machinery. However, due to the precursors involved in this biosynthetic pathway the products are often polyhydroxylated aromatic rings. As the *N*-terminal aromatic cinnamoyl moiety of the skyllamycins is not hydroxylated, the

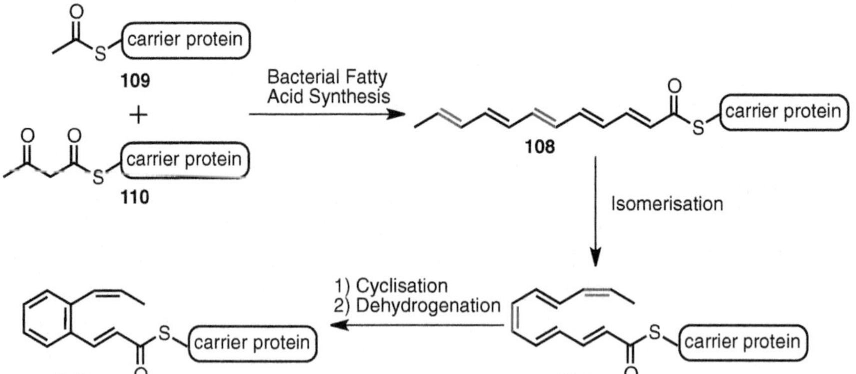

Scheme 3.2 Proposed biosynthetic assembly of the aromatic cinnamoyl residue through a bacterial fatty acid synthesis pathway

authors hypothesised that carrier protein-bound cinnamoyl moiety **106** is synthesised via a 6π-electrocyclisation reaction followed by dehydrogenation from the corresponding linear C_{12}-polyene **107** (Scheme 3.2). A bacterial fatty acid synthesis pathway could produce C_{12}-polyene **108** from the requisite building blocks, acetate and malonylate attached to carrier proteins **109** and **110** respectively. An important step is the isomerisation of two of the *trans* alkenes in **108** (green and red) to the *cis* alkenes in **107**, in order to prearrange the polyene for the electrocyclisation reaction. There are a number of genes in the skyllamycin gene cluster that support this hypothesis.

3.2.2 Assembly of the Peptide Backbone

In the skyllamycin biosynthetic gene cluster three genes (*sky29*, *sky30* and *sky31*) encode for NRPSs which contain a total of 11 modules consisting of at least a C, A and PCP domain (Fig. 3.4). Analysis of the residues present in the binding pockets of the A domains revealed that the first module of the NRPS activates the Thr residue, and the last (module 11) activates Leu. It is likely that module 1 couples the *N*–terminal aromatic cinnamoyl moiety to the Thr α-NH_2. Three modules (4, 8 and 10) contain E domains with modules 8 and 10 corresponding to D-Trp and D-Leu, respectively. Interestingly, module 4 corresponds to the Gly residue, which is achiral, and as such, the role of the E-domain is unknown. Furthermore, stereochemical analysis determined that the β-OH-Leu present in the skyllamycins has the

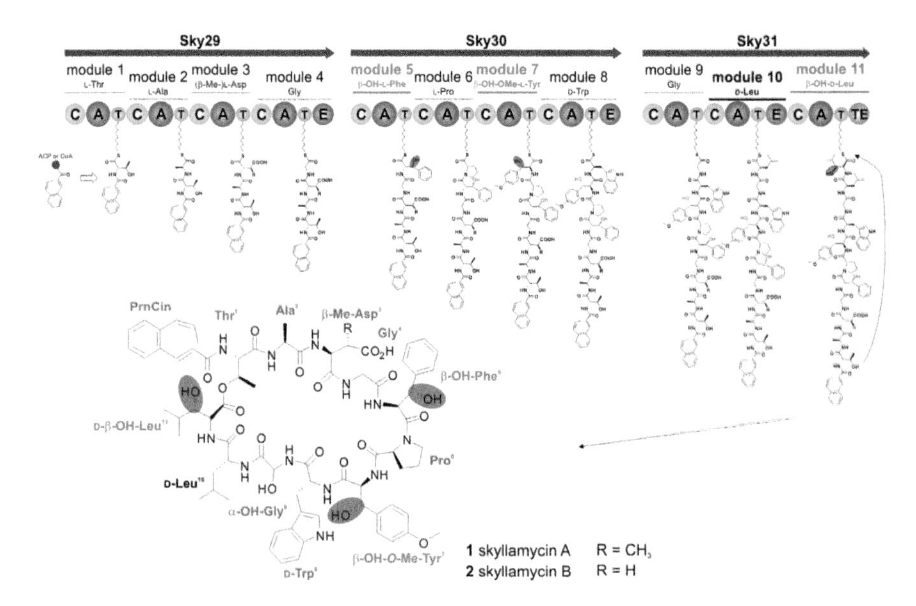

Fig. 3.4 Assembly of skyllamycins A (**101**) and B (**102**) on the NRPS [18]

D-configuration, however the corresponding NRPS module lacks an epimerase domain. It is possible that there is an 'in *trans* racemase enzyme' or that the adenylation domain activates D-Leu, however the biosynthetic mechanism by which this epimerisation takes place is currently unknown. The TE domain of module 11 catalyses the release of the peptide from the NRPS. This occurs via a macrolactonisation reaction between the Thr side chain OH and β-OH-Leu carboxylate.

3.2.3 β-Hydroxylation of Phe, O-Me-Tyr and Leu

During their initial studies Süssmuth and co-workers determined that the gene product encoded by *sky32* (Sky32) is responsible for the β-OH of Phe, *O*-Me-Tyr and Leu [11]. Indeed, the *sky*32 gene-deleted mutant produced a skyllamycin A analogue with a mass corresponding to the loss of three hydroxyl groups, consistent with a single enzyme being responsible for the hydroxylation of all three residues. Sky32 was found to be a cytochrome P450 monoxygenase with homology to a number of β-hydroxylating enzymes including OxyD, responsible for the formation of β-OH-Tyr in vancomycin biosynthesis (Fig. 1.14) [17].

Two further studies carried out by the Süssmuth group, in collaboration with the Cryle group, demonstrated that the Sky32 interacts specifically with the PCP domain of modules 5, 7 and 11 to hydroxylate the β position with (3*S*) stereochemistry [18]. Due to the transient nature of this interaction the authors were initially unable to obtain a crystal structure of the Sky32 in complex with the PCP. However, the synthesis of an analogue of the PCP domain of module 7, containing an imidazole tag capable of binding to the Fe(III) atom at the active site haem cofactor of the

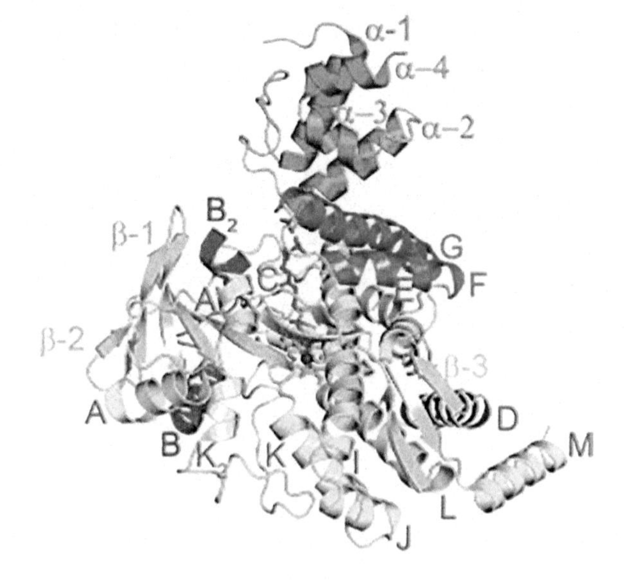

Fig. 3.5 The crystal structure of Sky32 (grey with interface forming helices G and F in blue) complexed with ^{7}PCP (gold and green) [19]

Sky32, resulted in a complex from which a crystal, suitable for an X-ray structure determination, was obtained (Fig. 3.5) [19]. Interestingly, the authors found that the main driver for the formation of the complex between the [7] PCP and Sky32 was hydrophobic forces. It seems that the selectivity for the PCP domains of modules 5 (Phe), 7 (Tyr) and 11 (Leu) is governed by small changes in the tertiary structure of two of the four helices making up the PCPs.

3.2.4 Formation of α-OH-Glycine

Nature has developed an interesting biosynthetic use for α-OH-Gly residues, namely the formation of a *C*-terminal carboxamide functionality on peptides. This is particularly important in peptide hormones, which often require a *C*-terminal carboxamide for full activity [20]. Two enzymes catalyse this reaction: a peptidylglycine α-hydroxylating monooxygenase (PHM) and a peptidyl-α-hydroxylglycine α-amidating lyase (PAL). In higher organisms these two enzymes make up one larger two-component enzyme (peptidylglycine α-amidating monooxygenase [PAM]) while in lower organisms these are discreet enzymes. Specifically, PHM, a copper-dependent monooxygenase catalyses the selective (*S*)-hydroxylation of the *C*-terminal Gly in **111**, forming an α-OH-Gly **112** (Scheme 3.3). The zinc dependent PAL enzyme then catalyses the cleavage reaction leading to the formation of a carboxamide **113** and glyoxylate **114** [20].

The formation of the most unusual structural feature of the skyllamycins, α-OH-Gly, occurs via a distinct mechanism, namely, catalysis by the gene product of *sky39*, a flavin-dependent monooxygenase (Sky39) [11]. Deletion of *sky39* resulted in the formation of a skyllamycin analogue with a mass of 16 Da less than skyllamycin A (**101**), corresponding to the loss of one hydroxyl group (and the incorporation of Gly). Interestingly, in the gene cluster *sky40* encodes for what is potentially a flavin mononucleotide reductase to regenerate the co-factor required by Sky39. The exact timing of this rare modification remains elusive. Due to its inherent instability, it seems unlikely that it would be formed by hydroxylation of a free or PCP-bound glycine. As such, the current hypothesis is that hydroxylation

Scheme 3.3 Formation of *C*-terminal carboxamides from α-OH-Gly catalysed by PHM and PAL

would occur either on a NRPS bound peptide, or on the cyclic peptide after cleavage from the NRPS. As discussed above, the extensive hydrogen bonding network which forms with (*S*)-α-OH-Gly most likely contributes to its stability in the mature natural product.

3.3 Analogue Design

Skyllamycins A–C (**101–103**) are highly complex cyclic depsipeptide NRPs. Due to their unusual structure and interesting bioactivity, in particular their biofilm inhibition and dispersal properties, skyllamycins A–C (**101–103**) served as exciting targets for a total synthesis program. Before embarking on the synthesis of the native natural products, a number of simplified analogues were designed. It was envisaged that this would enable the development of the synthetic methodology and synthetic route that could subsequently be applied to access the natural products, as well as to generate a small but targeted structural library. The first target chosen was simplified analogue **115** (Fig. 3.6) made up of unmodified amino acids, but containing the unsaturated cinnamoyl moiety present in skyllamycins A (**101**) and B (**102**). This analogue would help to determine the ease of cyclisation and also develop a robust synthesis of the cinnamoyl moiety.

The second generation of target analogues were dubbed deshydroxy skyllamycins A–C (**116–118**) (Fig. 3.7). These contain the modified amino acid residues, except the α-OH-Gly which was replaced by a native glycine. Due to the rarity and presumed instability of the α-OH-Gly residue in all but the final natural product, these analogues were chosen in order to develop robust synthetic routes to the modified amino acids, as well as probe the biological significance of the unusual α-OH-Gly modification. It was also envisaged that the generation of the three

Fig. 3.6 Structure of simplified analogue **115**

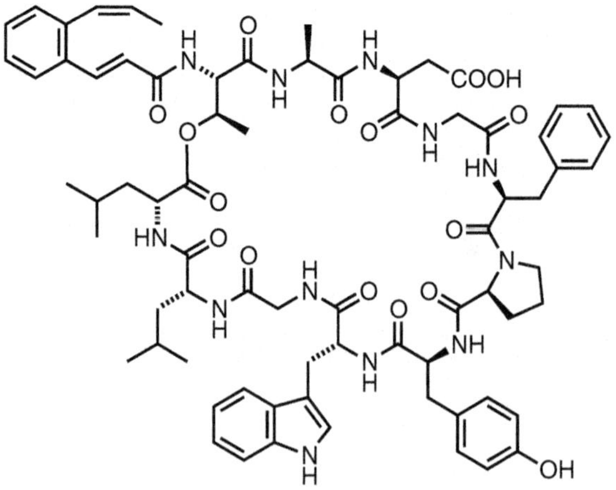

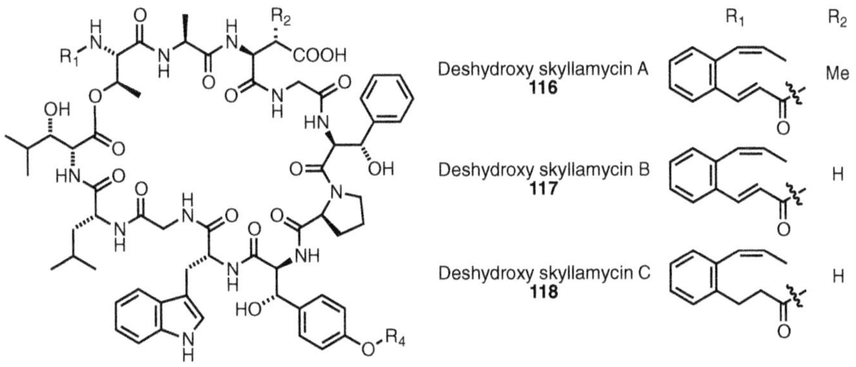

Fig. 3.7 Structure of second generation analogues—deshydroxy skyllamycins A–C (**116–118**)

deshydroxy analogues (**116–118**) could prove useful in biosynthetic studies. As discussed above, the mechanism by which the α-OH-Gly residue is installed in the skyllamycins is currently unknown. Exposure of deshydroxy skyllamycins A–C (**116–118**) to the α-hydroxylating enzyme Sky39 would shed light on whether the α-OH-Gly is installed in the free cyclic peptide, or on a peptide bound to the NRPS. Indeed, the methodology developed here could also be utilised to synthesise linear peptide intermediates bound to PCPs to further probe the biosynthetic installation of this rare modification. The four proposed structural analogues **115–118** of the skyllamycins would provide crucial synthetic insight into these unusual natural products, ultimately allowing for a total synthesis of the native natural products to be achieved.

3.4 Retrosynthetic Analysis of Skyllamycin Analogues

Before embarking on the synthesis of the aforementioned analogues a retrosynthetic analysis was carried out on skyllamycin analogues **115–118** (Scheme 3.4). In a similar approach to the synthesis of teixobactin (**28**), the cyclisation junction was chosen as an amide bond, rather than the lactone. Moreover, these peptides contain a number of favourable internal conformation elements including D-amino acids, and a proline residue, potentially improving the likelihood of cyclisation. The disconnection in the cyclic peptides was chosen to be the Phe/β-OH-Phe and Gly junction yielding the protected linear peptides **119**, **120**, **121**, and **122**. This disconnection point was chosen because the linear peptides **119**, **120**, **121**, and **122** contain a *C*-terminal glycine that cannot undergo epimerisation when activated during the cyclisation reaction. Glycine is also the smallest amino acid, which means there would be reduced steric hindrance during cyclisation.

Protected linear peptides **119–122** could be accessed from the linear resin bound intermediates **123**, **124**, **125**, and **126** (Scheme 3.4). It was envisioned that these

	⋯	R_2	R_3	R_4
115	⁄	H	H	H
Deshydroxy skyllamycin A 116	⁄	Me	OH	Me
Deshydroxy skyllamycin B 117	⁄	H	OH	Me
Deshydroxy skyllamycin C 118		H	OH	Me

1) Cyclisation
2) Deprotection

	⋯	R_2	R_3	R_4
119	⁄	H	H	H
120	⁄	Me	OPG	Me
121	⁄	H	OPG	Me
122		H	OPG	Me

	⋯	R_2	R_3	R_4
123	⁄	H	H	H
124	⁄	Me	OPG	Me
125	⁄	H	OPG	Me
126		H	OPG	Me

SPPS

2-CTC 127 128 129 105 130 131

Scheme 3.4 Retrosynthetic analysis of skyllamycin analogues **115–118**. PG: protecting group

could be accessed via SPPS beginning from the common 2-CTC resin. As discussed, the skyllamycins contain a number of modified amino acids, which are not commercially available and as such, the three β-OH residues **127**, **128** and **129**, β-Me-Asp **105** and both cinnamoyl moieties **130** and **131** required synthesis.

3.5 Synthesis of Simplified Skyllamycin Analogue 115

3.5.1 Synthesis of Cinnamoyl Moiety 130

Before simplified skyllamycin analogue **115** could be synthesised the cinnamoyl moiety **130** required synthesis. As discussed in Sect. 3.2.1, the biosynthetic formation of **130** is proposed to proceed via the formation of a linear C_{12} polyene followed by double bond isomerisation, electrocyclisation and dehydrogenation to yield the final aromatic cinnamoyl moiety (Scheme 3.3). Indeed, this hypothesis is supported by the biomimetic synthesis of a related natural product, psuedorubrenoic acid **132** [21] (Scheme 3.5). While an interesting approach to support the biosynthetic hypothesis, this methodology was plagued with a number of problems. The synthesis of linear precursor **133** via the alkyne **134** was lengthy, and yielded a mixture of isomers. This ultimately led to a poor yield (11%) for the cyclisation reaction to afford **135,** which was readily oxidised to **136**. The synthesis of the natural product was not completed due to a lack of material.

The synthesis of the aromatic cinnamoyl moiety **130** carried out in this work was adapted from the synthesis of the related pentenyl phenyl acrylic acid, a key component of the pepticinnamins [1], by Jiang and co-workers [22]. The synthesis began with the pyridinium chlorochromate mediated oxidation, of *o*-iodobenzyl alcohol (**137**) to the corresponding aldehyde **138** in good yield (Scheme 3.6). Subsequent Horner-Wadsworth-Emmons (HWE) [23, 24] olefination of **138** with triethyl phosphonoacetate (**139**) provided the (*E*)-alkene **140** in excellent yield. The

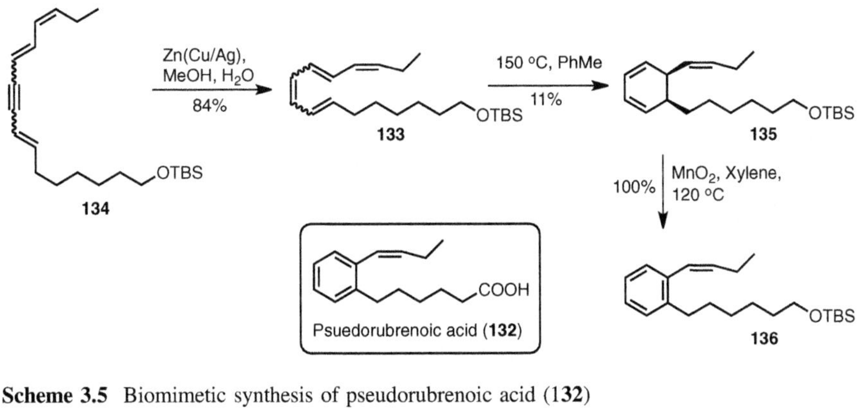

Scheme 3.5 Biomimetic synthesis of pseudorubrenoic acid (**132**)

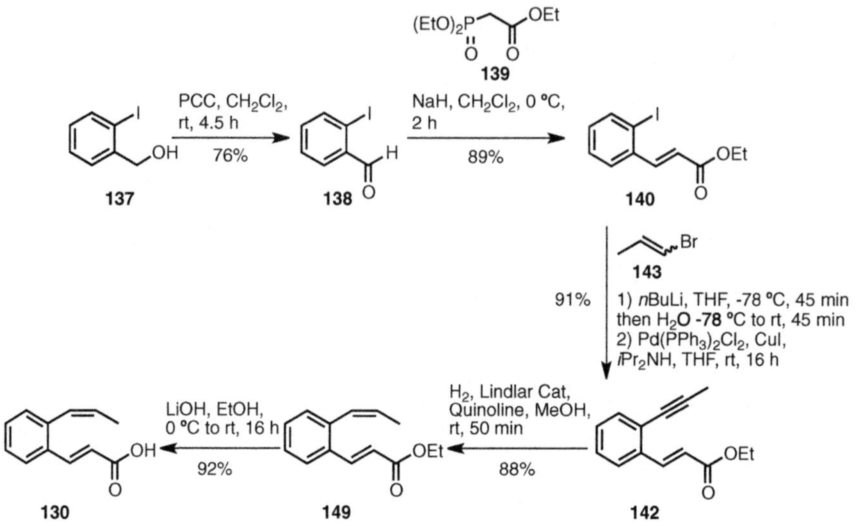

Scheme 3.6 Synthesis of cinnamoyl moiety **130**

stereoselectivity of the HWE reaction was confirmed by ^{1}H-NMR analysis. The observed large coupling constant (16 Hz) between the two olefinic protons provided convincing evidence for the presence of a *trans* double bond.

The key step in the synthesis of **130** was the Sonogashira cross coupling [25] between aryl iodide **140** and propyne (**141**) to form alkyne **142** (Scheme 3.7). Propyne is a gas, making its use in the laboratory inconvenient. Moreover, it is very expensive and in some cases is replaced by cheaper welding gas mixture which only contains a small percentage of propyne [26]. As such propyne (**141**) was generated in situ, by reaction of *Z*/*E*-1-bromopropene (**143**) with *n*-butyllithium, to form propynyllithium **144**, followed by quenching with water [27, 28]. This reaction is a variation of the Fritsch-Buttenberg-Wiechell rearrangement [29–31]. This solution of propyne was then utilised in the palladium-catalysed Sonogashira cross coupling reaction with aryl iodide **140**.

Palladium-catalysed cross-coupling reactions are extremely important in synthetic organic chemistry, so much so that the 2010 Nobel Prize in Chemistry was awarded to the inventors of three such reactions Heck, Suzuki and Negishi. A related Pd-catalysed cross-coupling is the Sonogashira reaction which was first reported in 1975 [25]. This coupling reaction proceeds with two interconnected catalytic cycles (Scheme 3.7), with the 'palladium cycle' similar to that present in other cross-couplings [32]. The first step is the oxidative addition of the active Pd(0) L_2 catalyst in the Aryl-X bond to form Pd(II) species **145**. This undergoes a transmetallation with the copper acetylide **146** leading to species **147**. Subsequent *trans-cis* isomerisation yields the final Pd(II) species **148** which undergoes reductive elimination to produce the coupled product **142** and regeneration of the active Pd(0)L_2 catalyst. The copper acetylide **146** is formed in the 'copper cycle' unique to

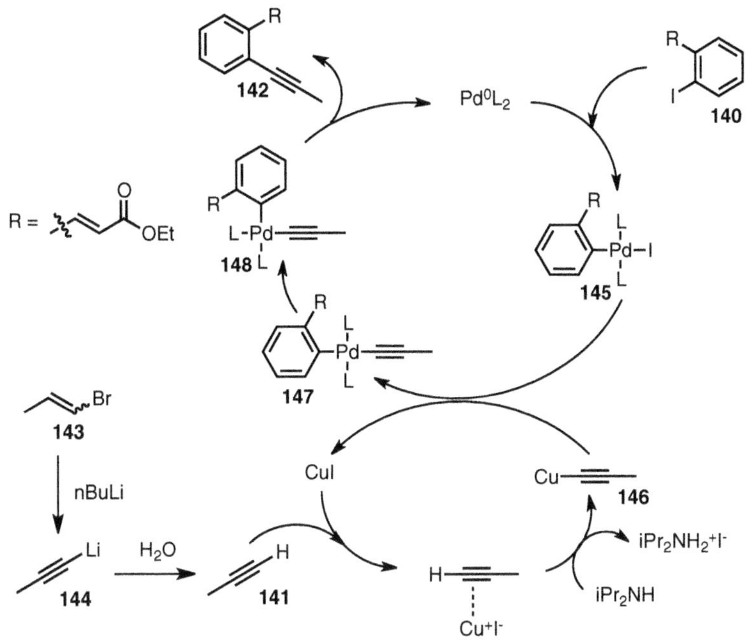

Scheme 3.7 Mechanism for the formation of propyne (**141**) and subsequent Sonogashira reaction with aryl iodide **140** to yield alkyne 142. L = PPh_3 ligand

the Sonogashira reaction. This cycle is not fully understood, however it is thought that the organic amine base (in this case *i*Pr_2NH) mediates the formation of **146** after a Cu-alkyne π-complex forms [32].

With alkyne **142** in hand, the triple bond was stereoselectively reduced to the (*Z*)-alkene **149**, employing hydrogen gas and Lindlar's catalyst [33], in excellent yield (Scheme 3.6). Lindlar's catalyst is a palladium catalyst poisoned with lead(II), which inhibits reduction of alkenes, but allows for the reduction of the alkyne. The selectivity was also improved by the addition of quinolone, commonly added to Lindlar reductions [34], which coordinates to the Pd active sites more strongly than the alkene, further preventing over-reduction. Finally, the ethyl ester of **149** was hydrolysed with lithium hydroxide, to yield the desired cinnamoyl moiety **130**. The synthesis was carried out in a highly efficient manner, requiring only five steps and proceeding in an overall yield of 50%.

3.5.2 *Solid Phase Assembly of 115*

With cinnamoyl moiety **130** in hand, the synthesis of simplified skyllamycin analogue **115** was carried out utilising an Fmoc-SPPS strategy. To begin, polystyrene resin functionalised with a 2-CTC linker was chosen for the synthesis [35]. The 2-CTC linker is extremely labile to acid, meaning the peptide can be cleaved from the resin selectively over the side chain protecting groups. This was an important consideration in the synthesis of **115** to ensure the cyclisation reaction would occur selectively at Gly carboxylate rather than the side chain Asp.

To begin Fmoc-Gly-OH was loaded to 2-CTC under mildly basic conditions to yield resin-bound amino acid **150** (Scheme 3.8). The Fmoc-group was subsequently removed utilising 10% v/v piperidine in DMF revealing the free amine **151**. Fmoc-Asp(*t*Bu)-OH and Fmoc-Ala-OH were then iteratively coupled using standard amino acid coupling conditions utilising PyBOP and NMM. To allow for the on-resin esterification reaction Fmoc-Thr-OH with an unprotected side chain was coupled, followed by Fmoc-deprotection to yield resin-bound tetrapeptide **152**. The cinnamoyl moiety **130** was then coupled utilising conditions employed for precious amino acids, utilising the more reactive coupling reagent HATU, the additive HOAt and Hünig's base. This yielded resin bound linear peptide **153**.

Resin-bound peptide **153** was exposed to on-resin esterification conditions, similar to those utilised in the total synthesis of teixobactin (Sect. 2.5.4), namely pre-formation of symmetric anhydride with Fmoc-D-Leu-OH and DIC followed by DMAP-catalysed esterification overnight. Gratifyingly, HPLC-MS analysis of an analytical scale resin cleavage reaction revealed this proceeded smoothly, yielding resin-bound branched peptide **154**. This was subjected to iterative SPPS utilising standard PyBOP coupling conditions with coupling of the appropriate L- and D-amino acids to yield the protected full-length linear peptide on resin **155**. The peptide was then cleaved from the resin with multiple treatments of the weakly acidic solution of 30% v/v hexafluoroisopropanol in dichloromethane. Removal of the solvent yielded protected linear peptide **156** (Fig. 3.8a).

The key reaction in the synthesis of skyllamycin analogue **115** was the final cyclisation. Initially two conditions were trialled: HATU-mediated cyclisation, and DMTMM.BF_4-mediated cyclisation. Both conditions were carried out at high dilution to prevent oligomerisation. Gratifyingly, analysis by HPLC-MS revealed that both conditions proceeded smoothly over 16 h to yield the desired protected cyclic peptide **157**. When the reaction was scaled up, the DMTMM.BF_4 conditions were utilised. The crude cyclisation mixture (Fig. 3.8b) was then subjected to an acidic deprotection cocktail containing TFA:*i*Pr_3SiH:H_2O (90:5:5 v/v/v) for 2 h to remove the side chain protecting groups (Scheme 3.9). After removal of the solvent, the crude mixture was subject to HPLC purification to yield the desired simplified skyllamycin analogue **115** (Fig. 3.8c) in 9.5% yield over 25 steps from initial resin loading.

3.5.3 Biological Evaluation

The simplified skyllamycin analogue **115** was then analysed in the *P. aeruginosa* biofilm inhibition assay at the University of California, Santa Cruz by Jake Haeckl and Roger Linington. Unfortunately, this analogue showed no biofilm inhibition activity, or reduction in bacterial cellular activity (Fig. 3.9). Whilst a disappointing

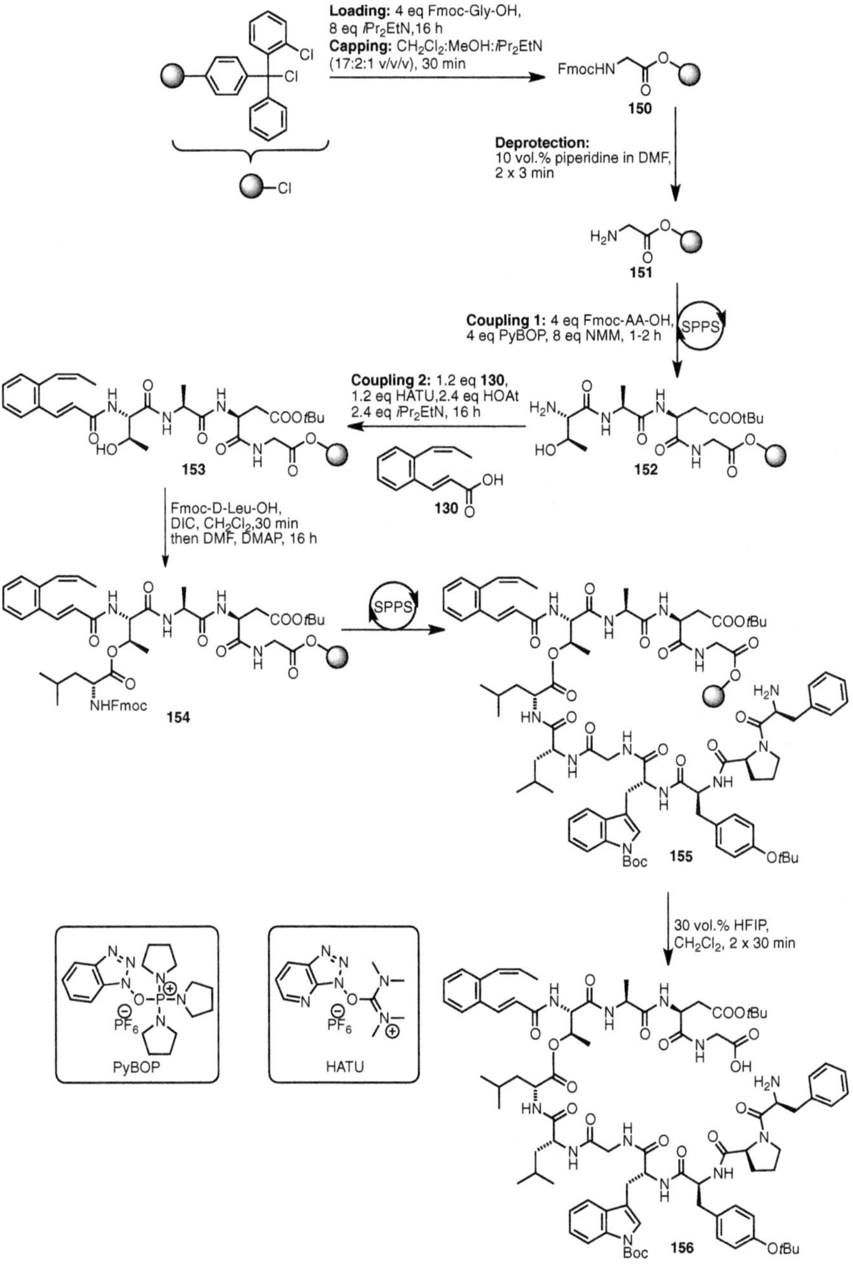

Loading: 4 eq Fmoc-Gly-OH, 8 eq iPr_2EtN, 16 h
Capping: CH_2Cl_2:MeOH:iPr_2EtN (17:2:1 v/v/v), 30 min
FmocHN 150
Deprotection: 10 vol.% piperidine in DMF, 2 x 3 min
H_2N 151
Coupling 1: 4 eq Fmoc-AA-OH, 4 eq PyBOP, 8 eq NMM, 1-2 h SPPS
152
Coupling 2: 1.2 eq 130, 1.2 eq HATU, 2.4 eq HOAt 2.4 eq iPr_2EtN, 16 h
130
153
Fmoc-D-Leu-OH, DIC, CH_2Cl_2, 30 min then DMF, DMAP, 16 h
154
SPPS
155
30 vol.% HFIP, CH_2Cl_2, 2 x 30 min
156
PyBOP
HATU

Scheme 3.8 Synthesis of linear peptide 156

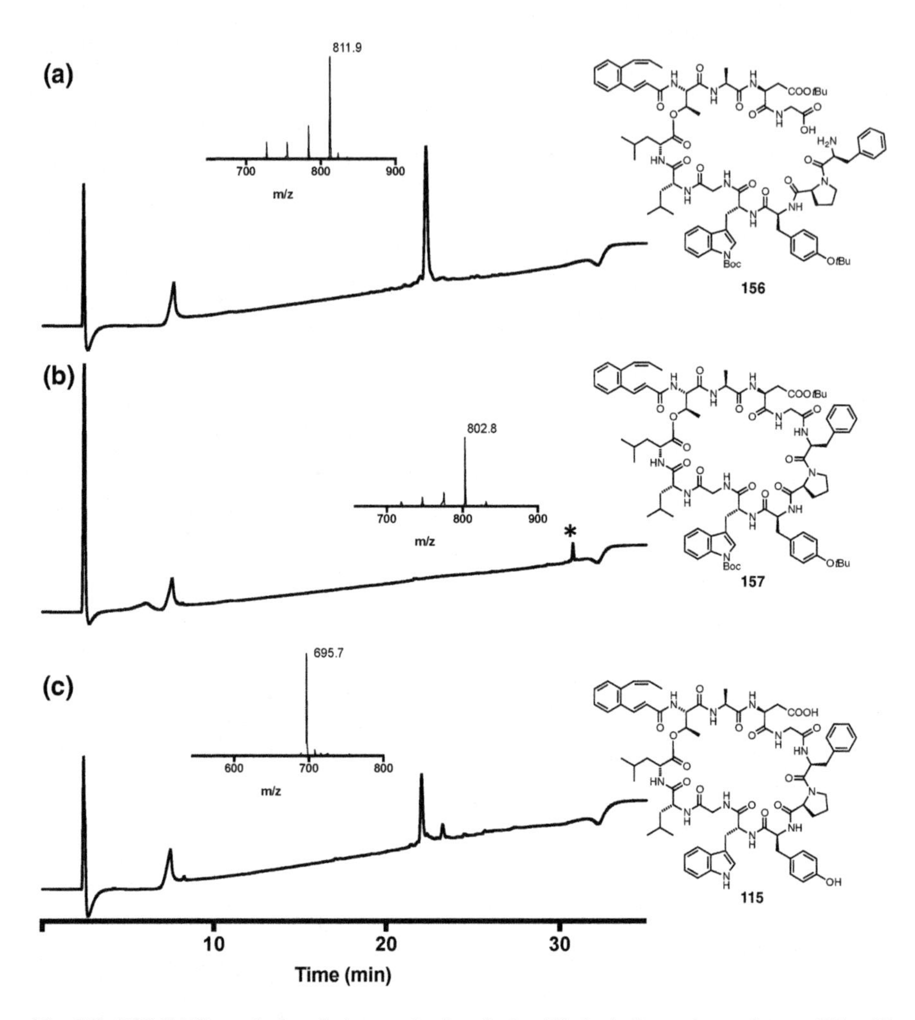

Fig. 3.8 HPLC-MS analysis of the synthesis of simplified skyllamycin analogue **115**. All chromatograms: 0–100% MeCN (0.1% formic acid) in H_2O (0.1% formic acid) over 30 min, $\lambda = 230$ nm. **a** Crude linear peptide **156**; *m/z* 811.4 $(M + 2H)^{2+}$; **b** crude cyclic peptide **157**; *m/z* 802.4 $(M + 2H)^{2+}$; **c** crude deprotected cyclic peptide **115**; *m/z* 696.3 $(M + 2H)^{2+}$

result, this demonstrates that the unusual modifications present in the native skyllamycins are vital for bioactivity. The hydroxyl groups present in the skyllamycin natural products form important internal hydrogen bonds clearly contributing to the three dimensional structure of the molecule. It is likely that **115**, lacking these H-bond donors, exists in a different three-dimensional conformation that cannot interact with the molecular target of the skyllamycins.

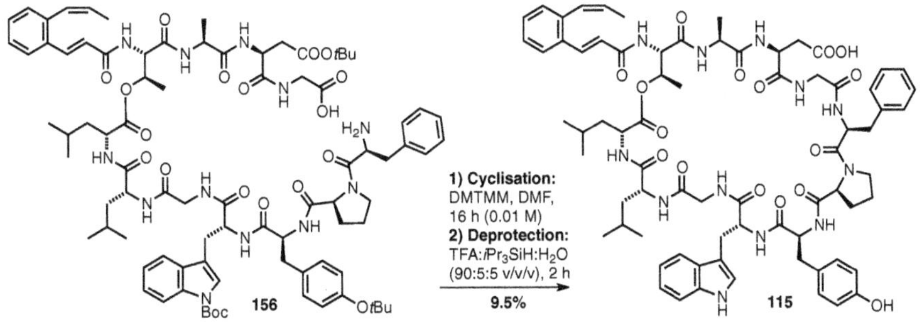

Scheme 3.9 Cyclisation of linear peptide **156** followed by acid deprotection to yield skyllamycin analogue **115**

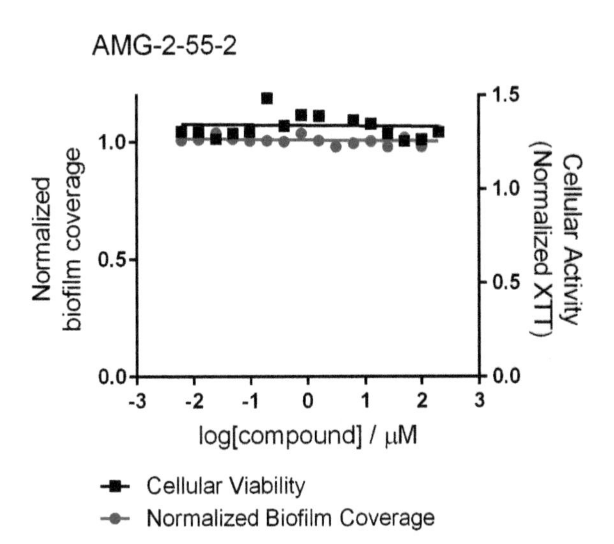

Fig. 3.9 *P. aeruginosa* biofilm inhibitory activity of simplified skyllamycin analogue **115**

3.6 Attempted Synthesis of Deshydroxy Skyllamycin B

With simplified analogue **115** in hand, the synthesis of deshydroxy skyllamycins A–C (**116–118**) could commence. To begin, deshydroxy skyllamycin B (**117**) was chosen as the initial target for a number of reasons. Firstly, the required cinnamoyl moiety **130** was already in hand, secondly, this analogue does not contain β-Me-Asp, which would require synthesis. Finally, skyllamycin B (**102**) is the most active of the three natural products in the biofilm inhibition/dispersion assay, so is of greatest interest from a biological activity standpoint. Before the assembly of deshydroxy skyllamycin B (**117**) could commence, the three non-commercially available β-OH amino acids required synthesis. β-OH amino acids are prevalent in a wide range of NRP natural products and, as such, there have been numerous synthetic routes developed to access them. One such method starts from (*R*)-

Garner's aldehyde (**158**) [36], a synthetically useful chiral aldehyde. Due to ongoing studies within our laboratory involving the synthesis of β-thiolated amino acids for use in native chemical ligation, there was considerable in house expertise in the use of (*R*)-Garner's aldehyde (**158**).

3.6.1 Garner's Aldehyde—A Useful Starting Point for the Synthesis of β-OH Amino Acids

(*S*)-Garner's aldehyde (**159**) was initially developed by Phillip Garner in 1984 as a method for synthesising β-OH amino acids present in biologically active natural products [37]. This chiral building block has been widely adopted in the synthetic community. Indeed, a more thorough follow up study by Garner and Park [36], published in 1987, had been cited more than 400 times at the time of writing. Both enantiomers of Garner's aldehyde (**158** and **159**) can be readily accessed from the cheap and available starting materials L-Ser and D-Ser (**160**). Garner's aldehyde has predominantly been used as an electrophile in a reaction with an organometallic reagent, to form alcohol **161**, which is varied depending on the R group utilised [38] (Scheme 3.10). The stereoselectivity of the reaction can be tuned by changing the metal, solvent and other reaction conditions. Importantly, after deprotection and oxidation of the primary alcohol, to form the β-OH amino acid **162**, the formal configuration of the α-centre changes. For example, to produce an L amino acid, (*R*)-Garner's aldehyde (**158**), produced from D-Ser (**160**), is utilised.

Garner's aldehyde was chosen as the common starting material for the synthesis of the three β-OH amino acids present in the skyllamycins. While both enantiomers of Garner's aldehyde (**158** and **159**) are commercially available they remain very expensive (>$250 per gram, Sigma-Aldrich). As such, both (*R*)-Garner's aldehyde (**158**) and (*S*)-Garner's aldehyde (**159**) were synthesised (Scheme 3.11). The synthesis of (*R*)-Garner's aldehyde (**158**) began with the Boc-protection of D-serine methylester hydrochloride (**163**), followed by protection of the carbamate nitrogen and primary alcohol with 2, 2-dimethoxy propane (2,2-DMP), promoted by the Lewis acid BF_3•OEt_2, to yield oxazolidine **164** in good yield over 2 steps. This was subjected to di*iso*butylaluminium hydride (DIBAL-H) reduction to yield (*R*)-Garner's aldehyde (**158**) in excellent yield. To ensure there was no over reduction to the alcohol, a solution of DIBAL-H was added slowly over 2 h to methyl ester **164** at −78 °C, ensuring the reaction temperature did not rise above −70 °C [36].

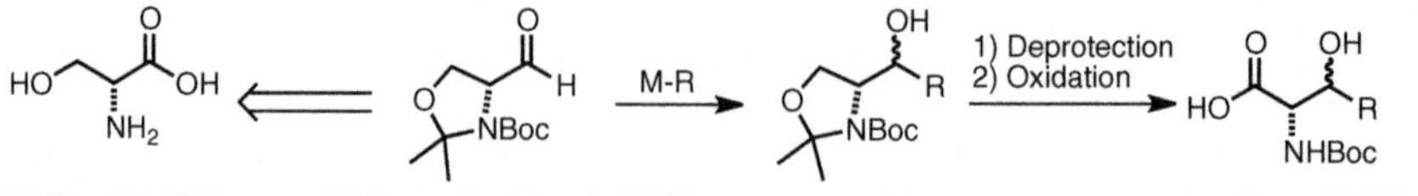

Scheme 3.10 Retrosynthesis and synthetic utility of (*R*)-Garner's aldehyde (**158**). M = metal

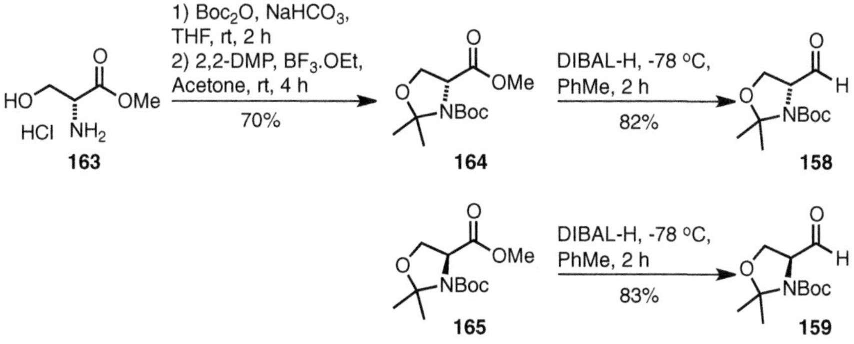

Scheme 3.11 Synthesis of both enantiomers of Garner's aldehyde (**158**) and (**159**)

Importantly, during their seminal studies Garner and Park determined that Garner's aldehyde is configurationally stable to silica gel chromatography, which is not always the case with *N*-protected α-amino aldehydes [36]. To expedite the synthesis of (*S*)-Garner's aldehyde (**159**), a cheap commercial source of (*S*)-methyl ester **165** (~ $13 per gram) was obtained, and this was also subjected to DIBAL-H reduction to yield (*S*)-Garner's aldehyde (**159**) in excellent yield.

3.6.2 Synthesis of Suitably Protected β-OH-Leu 166

With both enantiomers of Garner's aldehyde in hand, the synthesis of (2*R*,3*S*)-β-OH-Leu (**129**) could begin. For successful incorporation into the Fmoc-SPPS protocol the α-amine required protection with an Fmoc group. This amino acid forms the critical ester linkage with the side chain hydroxyl of the Thr residue in the skyllamycins. As such, to ensure that oligomerisation would not take place during the esterification reaction, the β-OH required protection. With this in mind, a novel pseudo-proline strategy was designed, in which the carbamate nitrogen and β-hydroxyl are protected as an oxazolidine, yielding (2*R*,3*S*)-β-OH-Leu **166**—suitably protected for incorporation in Fmoc-SPPS (Fig. 3.10).

The pseudo-proline protecting group was first developed as a method for the protection of Ser, Thr and Cys during SPPS [39]. The secondary role of the pseudo-proline is to disrupt the formation of secondary structure in the growing peptide chain during SPPS, which can significantly decrease yield [39]. They are

Fig. 3.10 (2*R*,3*S*)-β-OH-Leu (**129**) and the structure of its suitably protected variant **166**

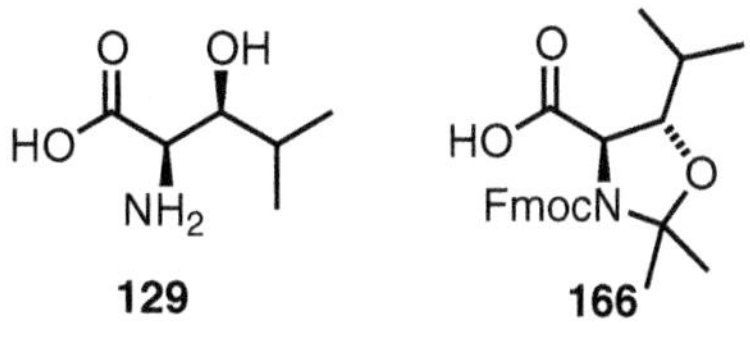

powerful turn inducers, i.e. they put a kink in the otherwise linear peptide chain, and this property has been exploited to synthesise otherwise synthetically intractable cyclic peptides [40]. Importantly they can be deprotected under standard acidic deprotection conditions used in SPPS chemistry. It was reasoned that pseudo-proline moiety in **166** would prevent oligomerisation during esterification, and also provide a turn inducer, further aiding cyclisation.

As discussed in Sect. 3.6.1, because the β-OH-Leu present in the skyllamycins has the (2*R*,3*S*) stereochemistry the synthesis began with (*S*)-Garner's aldehyde (**159**). To this end (*S*)-Garner's aldehyde (**159**) was reacted with the Grignard reagent, *iso*propylmagnesium bromide according to procedure by Joullié and coworkers [41] (Scheme 3.12). Unfortunately the yield for this reaction was poor. Efforts were undertaken to improve the yield of the reaction, including using a new bottle of *iso*propylmagnesium bromide, or by making the Grignard reagent fresh from *iso*propyl bromide and magnesium turnings, however the yield remained consistently low. Gratifyingly, the alcohol **167** was isolated exclusively as the desired *syn* diastereomer, consistent with the literature [41].

In order to explain the high *syn* stereoselectivity of the Grignard addition, a Cram chelation model can be used (Scheme 3.13). This model was developed to explain the observed stereochemical outcomes for the addition of an organometallic species to a chiral aldehyde bearing an atom capable of coordinating the metal at the α-position [42, 43]. It proposes that the Lewis acidic metal atom and the aldehyde form a short-lived chelated intermediate **168**. In this case the nitrogen is presumed to be too electron-deficient [41], and as such the chelate forms with the carbamate carbonyl. The nucleophile then adds to aldehyde from the less hindered *si*-face leading to the observed *syn* diastereomer.

With secondary alcohol **167** in hand, selective deprotection of the oxazolidine to yield diol **169**, in the presence of the Boc group, was next examined (Scheme 3.12). Fortunately, in this instance, the oxazolidine proved to be very labile to weak acid. Indeed, if alcohol **167** was heated, or submitted to $CDCl_3$ for NMR analysis it

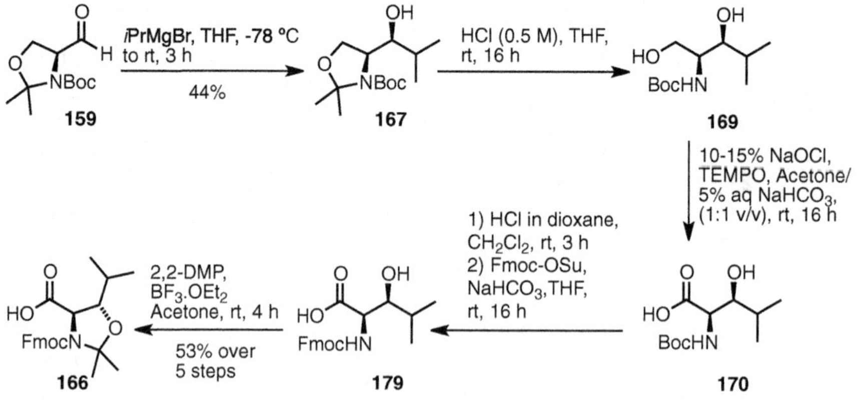

Scheme 3.12 Synthesis of suitably protected (2*R*,3*S*)-β-OH-Leu **166**

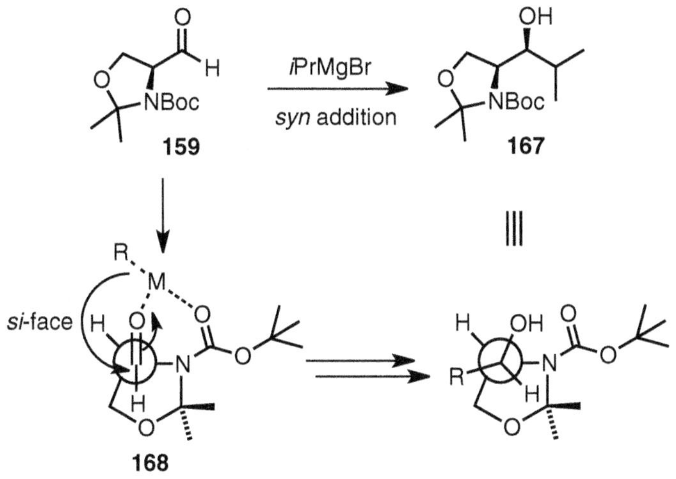

Scheme 3.13 Cram chelation model to explain the high *syn* selectivity observed in the Grignard addition of (*S*)-Garner's aldehyde (**159**)

would begin to decompose. This instability to acid may have contributed to the poor isolated yield of alcohol **167**. Nonetheless, following the procedure of Joullié and co-workers [41], alcohol **167** was treated with dilute aqueous HCl in THF, to afford diol **169** in near quantitative yield. Indeed, crude diol **169** was judged of sufficient purity by ^{1}H-NMR analysis and submitted to the next step without further purification.

The next step involved the key selective oxidation of the primary alcohol to the carboxylic acid, in the presence of the secondary alcohol, to yield acid **170**. Joullié and co-workers [41] attempted this transformation, utilising PtO_2 (Adam's catalyst) in air, however the authors isolated a mixture of product, and Boc-deprotected product in overall low yield. In order to complete this important transformation, a TEMPO-mediated oxidation was carried out, utilising sodium hypochlorite as the stoichiometric oxidant [44] (Scheme 3.12). The reaction proceeded smoothly to yield carboxylic acid **170**. Bisacetoxyiodobezene (BAIB) was initially utilised as the stoichiometric oxidant, however greater reproducibility was observed with NaOCl. In order to reduce the number of silica gel chromatography steps in the synthesis, the workup was optimised so that the product could be isolated with sufficient purity for the next reaction. To this end, the reaction mixture was diluted with water carefully ensuring the pH remained basic before it was washed with diethyl ether. This removed TEMPO, and other organic soluble by-products, whilst leaving the newly formed carboxylate in the aqueous layer. Subsequent acidification, and extraction with ethyl acetate yielded the desired product **170**.

The selectivity of the TEMPO-mediated oxidation was critical in the overall efficiency of the synthetic route towards suitably protected (2*R*,3*S*)-β-OH-Leu (**166**). 2,2,6,6-tetramethylpiperidinyloxy (TEMPO) (**171**) is a stable nitroxyl radical. It reacts with the stoichiometric oxidant NaOCl to form an oxoammonium salt **172** that acts as the primary oxidant (Scheme 3.14). This reacts with the primary alcohol **173** and proceeds through a compact five membered transition state **174**, liberating aldehyde **175** and hydroxylamine **176**. The compact nature of the transition state **174** imparts selectivity for primary alcohols over secondary alcohols. Hydroxylamine **176** is then reoxidised to TEMPO **171** completing the catalytic cycle [45]. By controlling the reaction conditions, aldehyde **175** can be isolated, however in this case it reacts with water to form hydrate **177** which undergoes a second reaction with oxoammonium salt **172** to yield the desired carboxylic acid **178**.

In order to complete the synthesis of suitably protected (2*R*,3*S*)-β-OH-Leu **166** a number of protecting group manipulations were carried out (Scheme 3.12). To begin, carboxylic acid **170** was treated with hydrochloric acid in dioxane to remove the Boc protecting group. Following removal of the solvent, the crude reaction mixture was treated with Fmoc succinimide under basic conditions to selectively protect the primary amine to yield Fmoc amino acid **179**. The workup was carried out in a similar fashion to that used after the TEMPO oxidation, which resulted in the isolation of Fmoc amino acid **179** with high purity. The final step in the synthesis was the formation of the oxazolidine, which was carried out using 2,2-DMP and $BF_3{\bullet}OEt_2$ as a Lewis acid catalyst. The crude product was then purified by silica gel chromatography to yield suitably protected (2*R*,3*S*)-β-OH-Leu **166** in 53% yield over 5 steps which equates to an average of 88% yield per

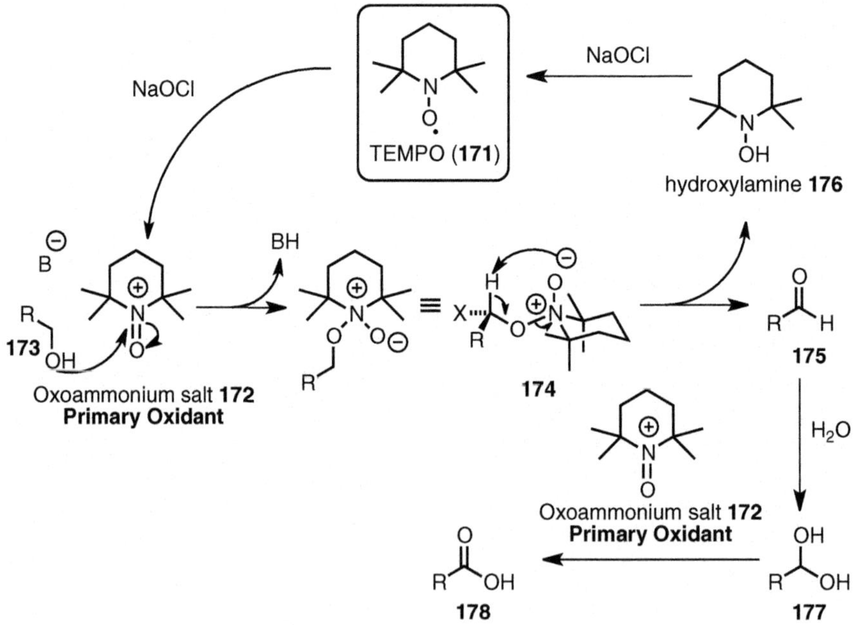

Scheme 3.14 Mechanism of the TEMPO (**171**) mediated oxidation

step. While unfortunately the first step was low yielding, the desired product was isolated as a single diastereomer. Overall the synthesis was highly efficient and scalable (2.7 g of final product produced), and was optimised to include only two silica gel chromatography steps.

3.6.3 Synthesis of Suitably Protected β-OH-Phe 180

With suitably protected (2*R*,3*S*)-β-OH-Leu **166** in hand we turned our attention to the synthesis of (2*S*,3*S*)-β-OH-Phe (**127**), suitably protected for the incorporation into Fmoc-SPPS. This residue is the *N*-terminal residue in the linear peptide **121**, which would take part in the cyclisation reaction to form deshydroxy skyllamycin B (**117**). As such, it was decided to leave the alcohol unprotected to reduce steric bulk at the cyclisation junction, thus yielding suitably protected Fmoc-(2*S*,3*S*)-β-OH-Phe (**180**) which contains only an Fmoc group (Fig. 3.11).

In order to synthesise **180** (*R*)-Garner's aldehyde (**158**) was utilised. Due to the stereoselectivity of Sky32 during the biosynthesis of the skyllamycins, the β-centre is (*S*)-configured in all three β-OH amino acids. This results in an *anti* relationship between the amine and hydroxyl groups in **180**, and as such the Cram chelation-controlled addition to Garner's aldehyde could not be utilised. Interestingly, Joullié and co-workers demonstrated that modest *anti*-selectivity (5:1) could be obtained when Garner's aldehyde is treated with the phenyl Grignard reagent (PhMgBr) in THF [41] producing alcohol **181** (Scheme 3.15). This is in contrast to the complete *syn*-selectivity observed for *i*PrMgBr as discussed in the previous section.

The authors suggest this is due to the reactivity of the Grignard reagent. The highly reactive Grignard, PhMgBr reacts with Garner's aldehyde before the chelation pathway can take place, leading to the observed *anti*-selectivity. The *anti*-selectivity can be explained by the Felkin-Anh model [46, 47] which takes into account steric and electronic factors (Scheme 3.15). This model requires that the lowest energy conformation of the aldehyde will have the largest substituent perpendicular to the aldehyde, in this case the *N*-Boc moiety (**182** in Scheme 3.15). Electronic factors also contribute to the stereoselectivity. Placing the electronegative atom (in this case also the *N* atom), perpendicular to the aldehyde, allows for the greatest orbital overlap between the C-N σ* and C-O π* orbitals, lowering the energy of the C-O π* orbital and thus increasing its electrophilicity. The nucleophile then approaches from the least hindered side, at the Burgi-Dunitz angle [48],

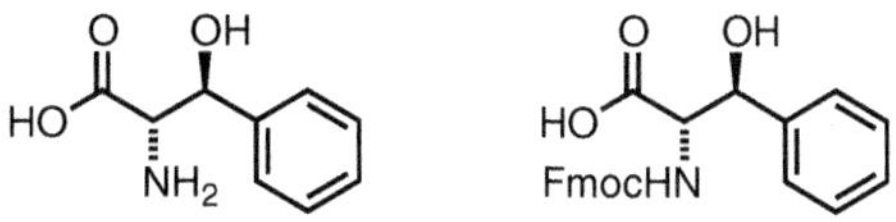

Fig. 3.11 (2*S*,3*S*)-β-OH-Phe (**127**) and suitably protected Fmoc-(2*S*,3*S*)-β-OH-Phe (**180**)

Scheme 3.15 Felkin-Anh model to explain the *anti* addition of PhMgBr to (*R*)-Garner's aldehyde (**158**). M = magnesium

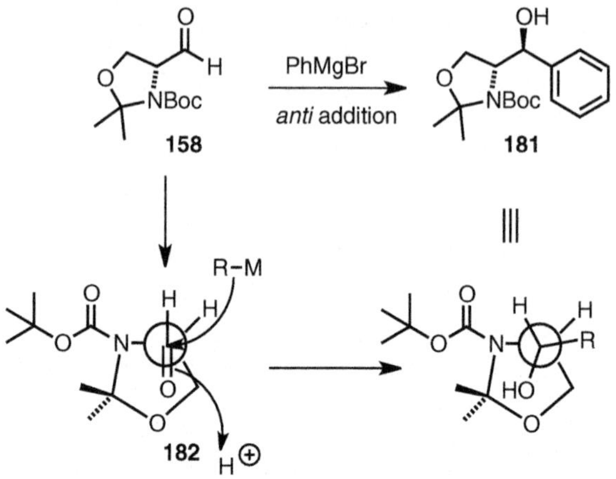

leading to the observed *anti*-stereoselectivity. The same stereochemical outcome can also be rationalised by invoking the Cornforth model [49]. Changing the solvent from the strongly coordinating THF to the weakly coordinating diethylether resulted in a reversal of *anti*-selectivity (2:3 *anti*:*syn*) [41]. This is presumably because THF has a greater ability to coordinate to the Mg, reducing the ability for chelation control.

Due to the modest *anti*-selectivity observed in the reaction between Garner's aldehyde and PhMgBr, a more stereoselective method was required. Previous work in our lab utilised a simple method for accessing the *anti* Grignard product during the synthesis of β-thiolated [50] and β-selenated [51] phenylalanine, and this method was adapted in this synthesis (Scheme 3.16 and Scheme 3.17). To begin, the Grignard reaction between PhMgBr and (*R*)-Garner's aldehyde (**158**) was carried out in Et_2O, in this case yielding alcohol **183** as a mixture of diastereomers

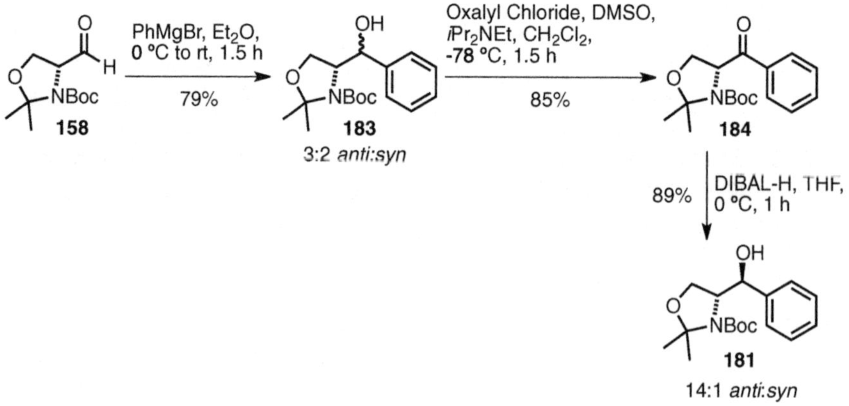

Scheme 3.16 Synthesis of *anti*-secondary alcohol **181**

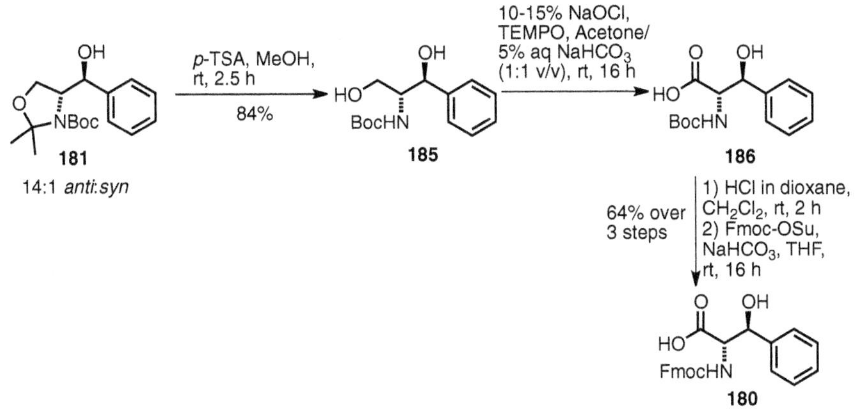

Scheme 3.17 Completion of the synthesis of Fmoc-β-OH-Phe-OH (**180**)

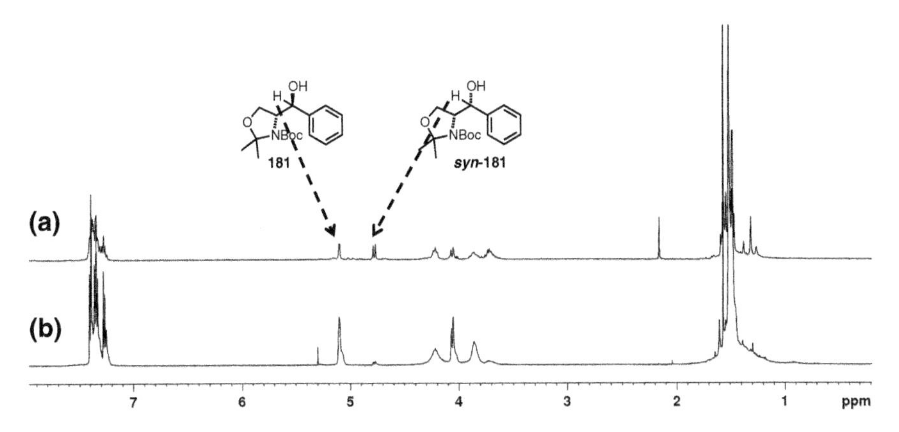

Fig. 3.12 ^{1}H-NMR analysis ($CDCl_3$, A-400 MHz, B-500 MHz, 328 K) of **a** alcohol **183** from Grignard addition and **b** alcohol **181** after reduction of ketone **184** with DIBAL-H

in a ratio of 3:2 *anti:syn* (Fig. 3.12a). Secondary alcohol **183** was then oxidised using Swern oxidation [52] conditions to afford ketone **184** in high yield.

The next step was the most crucial in the synthesis, namely the stereoselective reduction of ketone **184** to anti-alcohol **181**. An extensive study on the reduction of ketone **184** was carried out by Nishida et al. [53]. in which the authors screened a number of reducing agents, and found that DIBAL-H at 0 °C was highly *anti*-selective. As such ketone **184** was treated with DIBAL-H at 0 °C to forming *anti*-enriched alcohol **181** (14:1, *anti:syn*) (Fig. 3.12b) in excellent yield. The stereoselectivity of the reaction was determined by ^{1}H-NMR spectroscopy. The hindered nature of the molecule results in rotamers on the NMR time scale and as such the spectra were recorded at elevated temperature (328 K). The chemical shifts of the newly formed benzylic proton were easily identified *anti* (5.10 ppm) and *syn*

(4.78 ppm) and the ratio of the two products was determined by the integration (Fig. 3.12).

With the desired *anti* relationship now set, the oxazolidine protecting group was selectively removed in the presence of the Boc group using *para*-toluenesulfonic acid (*p*-TSA) in methanol to yield diol **185** in excellent yield (Scheme 3.17). Diol **185** was then subjected to the same reaction conditions described for the synthesis of β-OH-Leu, namely selective oxidation of the primary alcohol to form acid **186**, followed by the removal of the Boc protecting group and installation of the Fmoc group to yield the desired Fmoc-(2*S*,3*S*)-β-OH-Phe (**180**) in 64% yield over the three steps. Although two extra steps were required to install the desired *anti*-stereochemistry, the synthesis remained highly efficient due to the high yields of the Swern oxidation and DIBAL-H reduction steps.

3.6.4 Synthesis of Suitably Protected β-OH-O-Me-Tyr 187

In order to begin the assembly of deshydroxy skyllamycin B (**117**), (2*S*,3*S*)-β-OH-*O*-Me-Tyr (**128**) suitably protected for incorporation into Fmoc-SPPS also required synthesis (Fig. 3.13). Secondary alcohols are relatively unreactive functional groups to conditions employed during SPPS, however they are protected in routine SPPS protocols. Unlike β-OH-Phe, which is present at the *N*-terminus, the β-OH-*O*-Me-Tyr residue is present within the growing peptide chain. As such the oxazolidine protecting group utilised for β-OH-Leu was also chosen as the protecting group strategy for (2*S*,3*S*)-β-OH-*O*-Me-Tyr resulting in **187** as the target suitably protected variant.

It was envisioned that this could be carried out in a similar fashion to the synthesis of (2*S*,3*S*)-β-OH-Phe as described above. As such, to synthesise *anti* alcohol **188**, (*R*)-Garner's aldehyde (**158**) was reacted with the Grignard reagent generated from *para*-bromoanisole (**189**) affording alcohol **190** as a mixture of diastereomers in moderate yield (Scheme 3.18). This could be explained by the electron rich nature of the carbon bromine bond, making it more difficult for the magnesium to insert, and thus poor formation of the Grignard reagent. Nevertheless, alcohol **190** was then oxidised to the ketone **191** using Dess-Martin

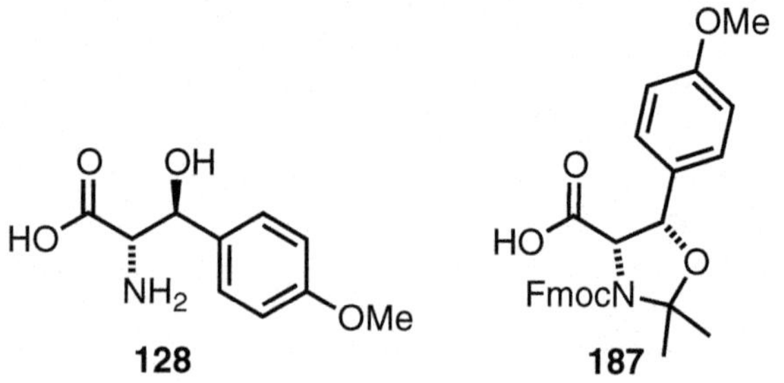

Fig. 3.13 (2*S*,3*S*)-β-OH-*O*-Me-Tyr (**128**) and the structure of its suitably protected variant **187**

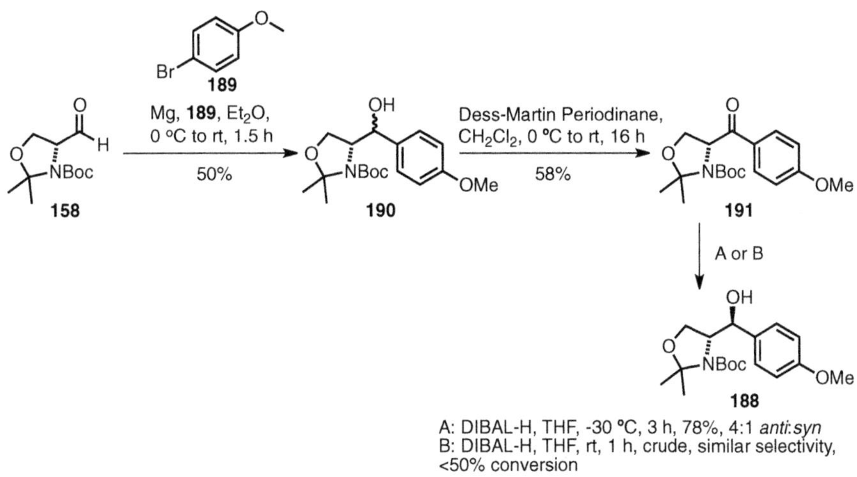

Scheme 3.18 Attempted synthesis of *anti* alcohol **188**

periodinane [54] in moderate yield. In an analogous fashion to the method described above for β-OH-Phe, ketone **191** was treated with DIBAL-H under two different reaction conditions, −30 °C or room temperature, however unfortunately neither reaction proceeded in high yield or high *anti* diastereoselectivity. Due to the electron rich aromatic ring, the ketone is presumably less electrophilic, impeding the stereoselective reduction.

An alternate method for the formation of alcohol **188** was investigated. This was inspired by Hamada and co-workers who carried out a synthesis of all four diastereomers of β-*O*Me-Tyr-OH in their synthetic efforts towards the papuamide natural products [55]. To this end, *para*-bromoanisole (**189**) was reacted with *n*-BuLi, undergoing lithium halogen exchange to form the lithiated species. Subsequent reaction with (*R*)-Garner's aldehyde (**158**) produced alcohol **188** in good yield (Scheme 3.19). Pleasingly, after analysis of the ^{1}H-NMR spectrum, the reaction was shown to have proceeded with high *anti* selectivity (5:1) (Fig. 3.14). This selectivity can be explained in a similar manner to the synthesis of β-OH-Phe

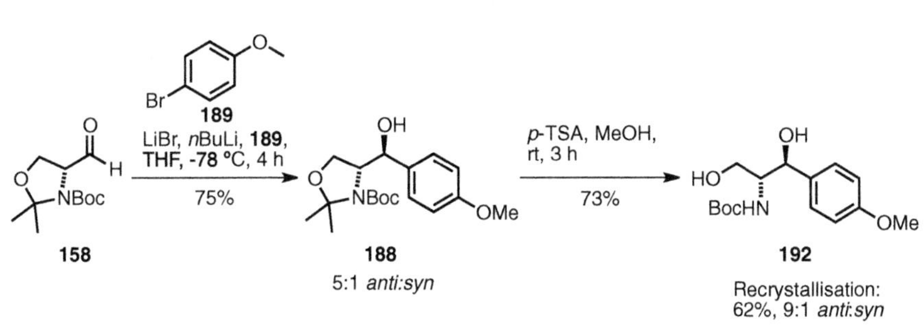

Scheme 3.19 Synthesis of *anti* diol **192**

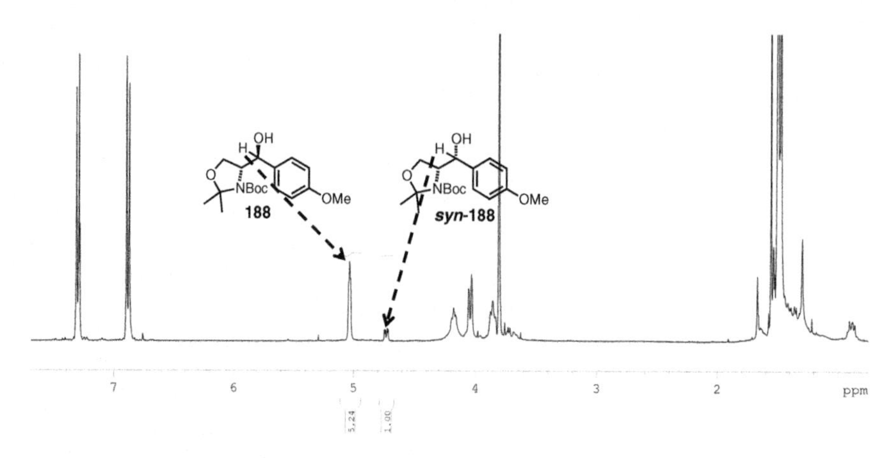

Fig. 3.14 ^{1}H-NMR analysis ($CDCl_3$, 400 MHz, 328 K) of alcohol **188**. The key protons used to determine the *anti*:*syn* ratio are highlighted

(Sect. 3.6.3), in which the organometallic addition occurs rapidly under Felkin-Anh control, before chelation can occur.

With alcohol **188** in hand, the oxazolidine protecting group was removed using *p*-TSA in methanol to yield diol **192** (Scheme 3.19). At this point a number of recrystallisation trials were carried out in order to improve the diastereomeric purity of diol **192**. After a number of trials, a solvent system (EtOAc/pentane) was developed in which crystals formed however upon analysis of the ^{1}H-NMR it was revealed that the diastereomeric ratio had only slightly improved to 9:1, however the overall yield of crystallisation was poor. In any case, diol **192** was carried through the remainder of the synthetic sequence (Scheme 3.20). The first step was

Scheme 3.20 Completion of the synthesis of oxazolidine protected Fmoc-β-OH-*O*-Me-Tyr-OH **187**

again selective oxidation, with TEMPO and NaOCl, of the primary alcohol to yield the carboxylic acid **193**. There have been some reports of chlorinated side products when this reaction is carried out in the presence of electron rich aromatics, pleasingly in this case they were not observed [56].

Carboxylic acid **193** was then subject to protecting group manipulations, namely removal of the Boc group and installation of an Fmoc functionality yielded amino acid **194**. The final step in the synthesis was the formation of oxazolidine **187** which proceeded smoothly (Scheme 3.20). Thin layer chromatography (TLC) analysis revealed the presence of two spots and partial separation of these could be achieved by silica gel chromatography. Unsurprisingly, the fractions containing both spots (Fraction B, Fig. 3.15) showed two peaks by HPLC analysis, however both contained the mass of the desired product **187**, suggesting that partial separation of the desired *anti* and undesired *syn* diastereomers could be achieved at this stage. Ultimately, separation proved successful, and the desired major product, oxazolidine protected Fmoc-β-OH-*O*-Me-Tyr-OH **187**, could be isolated as a single diastereomer in high yield. As such, the oxazolidine protecting group proved to be an extremely convenient tool in enabling the separation of the diastereomers. To this end, when the synthesis was scaled up, the recrystallisation step was omitted to avoid unnecessary loss of yield, and the final product **187** was isolated as a single diastereomer after silica gel chromatography.

3.6.5 Attempted Assembly of Deshydroxy Skyllamycin B

With all three β-OH amino acids in hand, the synthesis of deshydroxy skyllamycin B (**117**) could commence. Starting from common resin-bound intermediate **153**, which was formed in the synthesis of simplified analogue **115**, the first step in the synthesis was the formation of the ester bond between the Thr residue and β-OH-Leu **166**. This was initially carried out by preformation of the symmetric anhydride using DIC and subsequent DMAP-catalysed esterification. In forming the

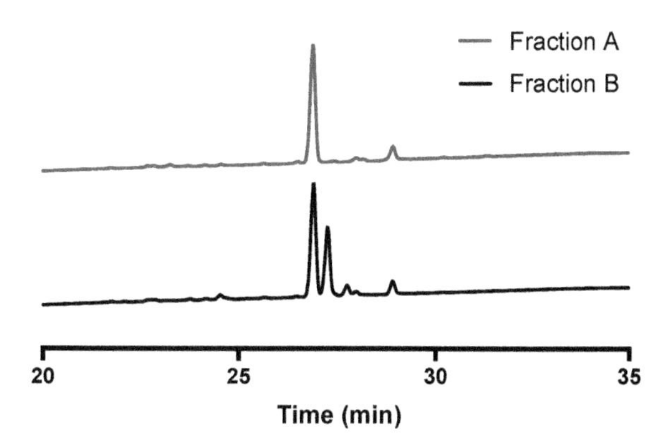

Fig. 3.15 HPLC analysis of fractions A and B collected during the purification of **187**. Chromatogram: 0–100% MeCN (0.1% formic acid) in H_2O (0.1% formic acid) over 30 min, λ = 230 nm

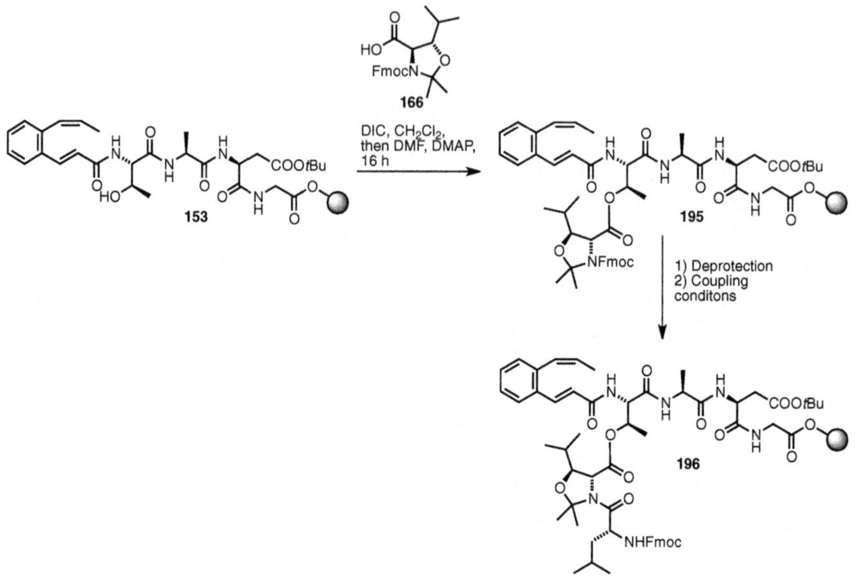

Scheme 3.21 On-resin esterification reaction and optimisation of the Fmoc-D-Leu-OH coupling

symmetric anhydride, an equivalent of the precious protected β-OH-Leu **166** is wasted. In order to reduce this waste, the conditions were optimised such that equimolar amounts of DIC and β-OH-Leu were reacted, preforming the DIC active ester. This was then added to the resin in the presence of DMAP. Pleasingly, the reaction proceeded smoothly to generate resin bound ester **195** (Scheme 3.21).

Following Fmoc-deprotection of **195**, the next step in the synthesis was the coupling of Fmoc-D-Leu-OH to generate resin-bound **196**. To begin, standard coupling conditions were trialled, and the reactions analysed by HPLC-MS, however unfortunately the reaction did not proceed to completion. This is presumably due to the increased steric bulk at the amine due to the oxazolidine protecting group. Indeed, in standard Fmoc-SPPS protocols, pseudo-prolines are coupled as the preformed dipeptide to avoid this problem. A number of other conditions were trailed, including the more active coupling reagents HATU and COMU, however quantitative product formation was never observed (Table 3.1). Eventually, the reaction was trialled with microwave irradiation at 50 °C for 2 h. Pleasingly, quantitative product formation was observed. After further investigations it was found quantitative coupling could be achieved with microwave irradiation at 50 °C for 1 h.

Having successfully formed this key peptide bond, the synthesis was continued. Specifically, Gly and D-Trp residues were installed under standard conditions to yield resin-bound peptide **197** (Scheme 3.22). Coupling of the precious protected β-OH-*O*-Me-Tyr-OH **187** was carried out using HATU, HOAt and Hünig's base and proceeded smoothly yielding resin-bound peptide **198**. HPLC-MS analysis of

Table 3.1 Conditions trialled for hindered coupling to resin bound peptide **195**

Conditions	Result
Fmoc-D-Leu-OH, (4 eq), PyBOP (4 eq), NMM (8 eq), DMF, rt, 2 h	20% product
Fmoc-D-Leu-OH, (2 eq), HATU (2 eq), *i*Pr_2NEt (4 eq), DMF, rt, 2 h	30% product
Fmoc-D-Leu-OH, (2 eq), COMU (2 eq), *i*Pr_2NEt (4 eq), DMF, rt, 2 h	50% product
Fmoc-D-Leu-OH, (4 eq), HATU (4 eq), HOAt (8 eq), *i*Pr_2NEt (8 eq), DMF, rt, 2 h Microwave, 50 °C, 2 h	100% product

Scheme 3.22 SPPS of resin-bound intermediate **199**

an analytical scale resin cleavage under standard acidic cleavage cocktail conditions revealed the desired product (Fig. 3.16a). Coupling of the Fmoc-Pro-OH was next carried out using the optimised microwave conditions described above for coupling to an oxazolidine-protected amino acid. However when an analytical amount of resin-bound peptide **199** was cleaved and analysed by HPLC-MS the expected product was not observed, instead a complex trace was observed (Fig. 3.16b) presumably due to degradation of the peptide. It was proposed that this was most likely caused via dehydration of the β-OH-*O*-Me-Tyr-OH residue resulting in the formation of a complex mixture of (*E*) and (*Z*) dehydrated *O*-Me-Tyr as well the reduction of this double bond to form a mixture of diastereomers of *O*-Me-Tyr.

Interestingly, this elimination does not occur when the β-OH-*O*-Me-Tyr residue is at the *N*-terminus, as in **198**, presumably because the amine is protonated, reducing the ability for dehydration to occur. Efforts were made to optimise the cleavage and deprotection, by reducing the amount of acid used, however significant elimination still occurred even with 50% v/v TFA in dichloromethane. β-OH-*O*-Me-Tyr was found to be stable at lower concentrations of acid, namely 1% v/v TFA in dichloromethane, but this was not sufficient to remove the Boc and *t*-butyl protecting groups. An alternate route to synthesise deshydroxy skyllamycin B (**117**) was therefore sought.

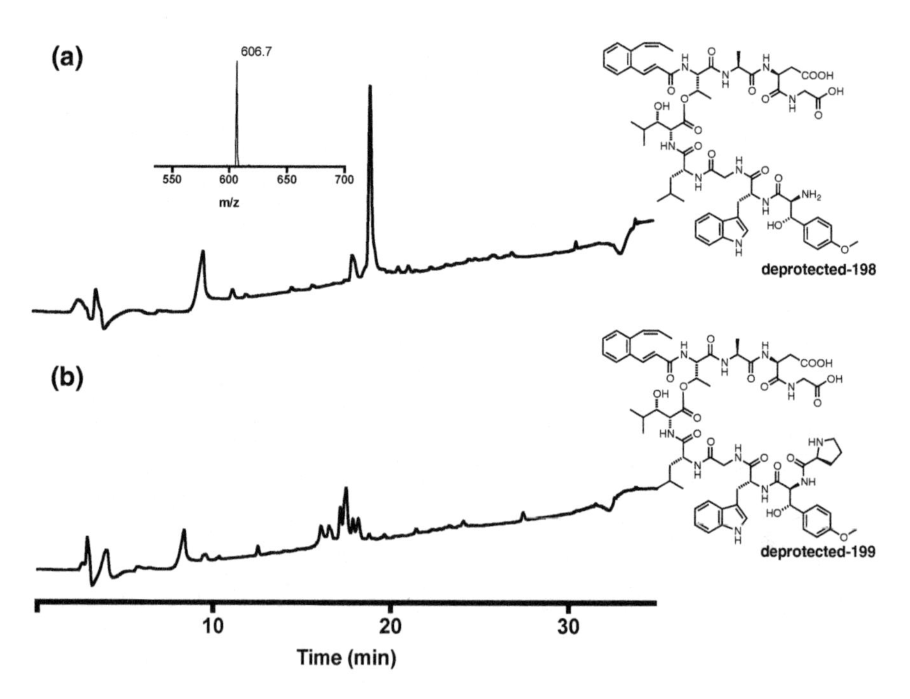

Fig. 3.16 HPLC-MS analyses of analytical scale cleave of resin-bound **198** and **198**. All chromatograms: 0–100% MeCN (0.1% formic acid) in H_2O (0.1% formic acid) over 30 min, λ = 230 nm. **a** expected peptide product **deprotected-198**; *m/z* 606.3 $(M + 2H)^{2+}$; **b** expected peptide product **deprotected-199**

Whilst it was disappointing that a new synthetic strategy needed to be investigated, a number of key steps were optimised during this initial synthetic endeavour, including the coupling of Fmoc-D-Leu-OH to the hindered oxazolidine under microwave conditions. Importantly, during these initial studies it was observed that the oxazolidine protecting groups were highly acid labile. Indeed, they were never intact on the peptide by HPLC-MS analysis, even when the mildly HFIP was used to cleave the peptide from resin. While this hyper-acid lability is not fully understood, it may be that the bulkiness of the amino acid side chains increases the acid lability of this protecting group.

3.7 Second Generation Synthesis of Deshydroxy Skyllamycin B

3.7.1 Revised Synthetic Strategy

Due to the instability of the β-OH-*O*-Me-Tyr residue to acid, a new protecting group strategy for the synthesis of the deshydroxy skyllamycin B (**117**) was required. It was envisioned that hyper acid-labile protecting groups, which could be cleaved with 1% v/v TFA in dichloromethane, would be suitable. To this end, analysis of deshydroxy skyllamycin B (**117**) was carried out, and functionalities requiring protection were identified. Specifically, the key functionality that required protection was the side-chain carboxylic acid of the Asp residue; to ensure the final cyclisation reaction occurred specifically at the Gly *C*-terminus. As such, the phenyl *iso*propyl ester (Ph*i*Pr) group was chosen (red, Scheme 3.23), as this can be readily cleaved with 1% v/v TFA/CH_2Cl_2 [57]. During the initial synthesis, the indole nitrogen of the D-Trp residue was protected with a Boc group. No alternative hyper acid-labile protecting group for the Trp indole was identified in the literature. As such, it was envisioned that the indole could remain unprotected during the

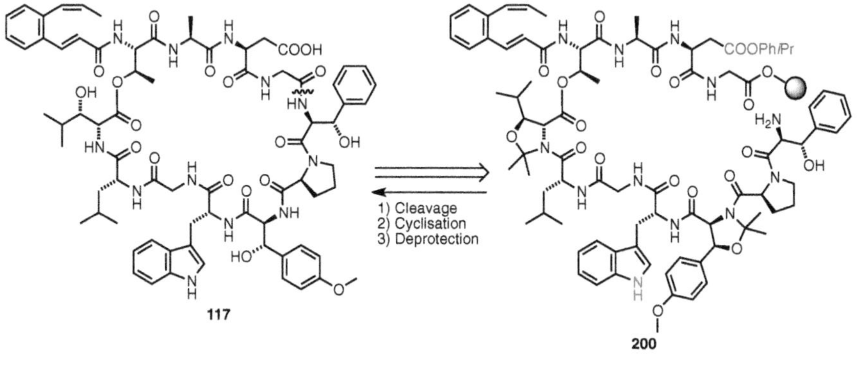

Scheme 3.23 Revised synthetic strategy for the assembly of deshydroxy skyllamycin B (**117**)

synthesis (green, Scheme 3.23) which is not uncommon in Fmoc-SPPS [58]. The oxazolidine protecting groups utilised for β-OH-*O*-Me-Tyr and β-OH-Leu proved to be highly acid labile, and as such, they would be retained in the synthesis, resulting in target resin-bound protected linear peptide **200**.

3.7.2 Completion of the Synthesis of Deshydroxy Skyllamycin B

Having carefully planned a revised strategy to prevent dehydration of the β-OH-*O*-Me-Tyr residue, the synthesis of deshydroxy skyllamycin B (**117**) was undertaken (Scheme 3.24). Beginning with resin-bound Fmoc-Gly **150**, the starting point in the synthesis of simplified analogue **115**, Fmoc-Asp(Ph*i*Pr)-OH was coupled using

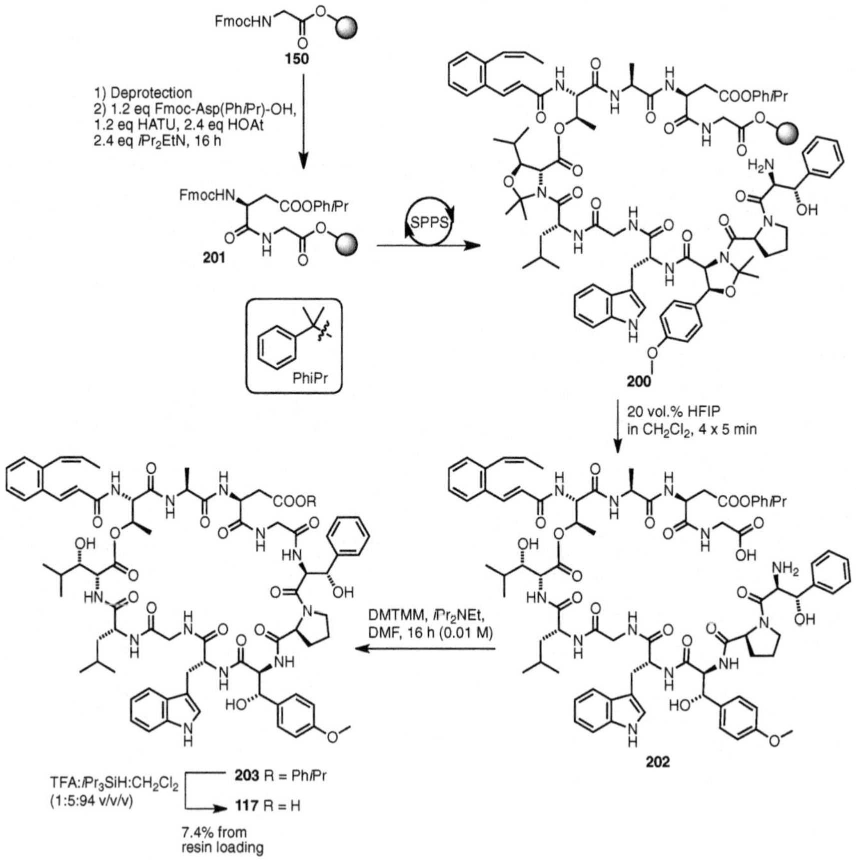

FmocHN **150**

1) Deprotection
2) 1.2 eq Fmoc-Asp(Ph*i*Pr)-OH, 1.2 eq HATU, 2.4 eq HOAt 2.4 eq iPr_2EtN, 16 h

FmocHN COOPh*i*Pr **201**

PhiPr

SPPS

COOPh*i*Pr **200**

20 vol.% HFIP in CH_2Cl_2, 4 x 5 min

COOPh*i*Pr **202**

DMTMM, iPr_2NEt, DMF, 16 h (0.01 M)

COOR

TFA:iPr_3SiH:CH_2Cl_2 (1:5:94 v/v/v)

203 R = Ph*i*Pr

117 R = H

7.4% from resin loading

Scheme 3.24 Synthesis of deshydroxy skyllamycin B (**117**)

HATU conditions yielding resin-bound dipeptide **201**. Subsequent Fmoc-SPPS was carried out using the conditions optimised in the above Sect. 3.6.5, with the only difference being use of side chain unprotected D-Trp. The final coupling of Fmoc-β-OH-Phe-OH (**180**) was carried out using HATU coupling conditions and this yielded resin-bound linear peptide **200**. Pleasingly, when an analytical scale cleavage with 1% v/v TFA/CH_2Cl_2 was performed and analysed by HPLC-MS, only one peak was observed with the mass corresponding to that of the desired peptide—on this occasion no degradation of the peptide occurred.

With resin-bound linear peptide **200** in hand the key final steps of the synthesis were next attempted, namely cleavage from the resin and cyclisation. To this end, **200** was treated with 20% v/v HFIP/CH_2Cl_2 (4 × 5 min) and the solvent concentrated to yield protected linear peptide **202** (Fig. 3.17a). It should be noted that owing to the hyper acid-lability of the Ph*i*Pr group, the HFIP treatments were kept short as prolonged treatment was found to result in some hydrolysis of the Ph*i*Pr protecting group.

Protected linear peptide **202** was next exposed to the cyclisation conditions developed in Sect. 3.5.2, namely exposure to the coupling reagent DMTMM.BF_4 at dilute concentration for 16 h. Pleasingly, LC-MS analysis of the crude mixture showed complete conversion to the protected cyclic peptide **203** (Fig. 3.17b). After removal of the solvent, and extensive drying, the crude protected cyclic peptide **203** was treated with 1% v/v TFA/CH_2Cl_2 containing 5% TIS for 30 min (Fig. 3.17c). Unfortunately, initial efforts yielded product, together with a new peak with the same mass as the starting material. This was presumed to be due to the formation of a covalent adduct between the liberated Ph*i*Pr carbocation and the Trp indole moiety. By running the reaction at high dilution, the formation of this adduct was suppressed. Purification of the crude reaction mixture by HPLC yielded the pure deshydroxy skyllamycin B (**117**) in 7.4% overall yield based on the initial resin loading (Fig. 3.17d). While this yield may seem low, it corresponds to an average of 90% per step over the course of the SPPS as well as the final cleavage, cyclisation and deprotection steps (25 steps in total). Overall, this strategy represents an efficient method for the synthesis of the deshydroxy skyllamycin analogues and should be amenable to application to the synthesis of deshydroxy skyllamycins A (**116**) and C (**118**).

3.8 Synthesis of Deshydroxy Skyllamycin A and C

With deshydroxy skyllamycin B (**117**) successfully synthesised, the synthesis of deshydroxy skyllamycins A (**116**) and C (**118**) was next commenced. Given that the three β-OH amino acids had already been prepared, only the reduced cinnamoyl moiety **131** and β-Me-Asp-OH **105** required synthesis.

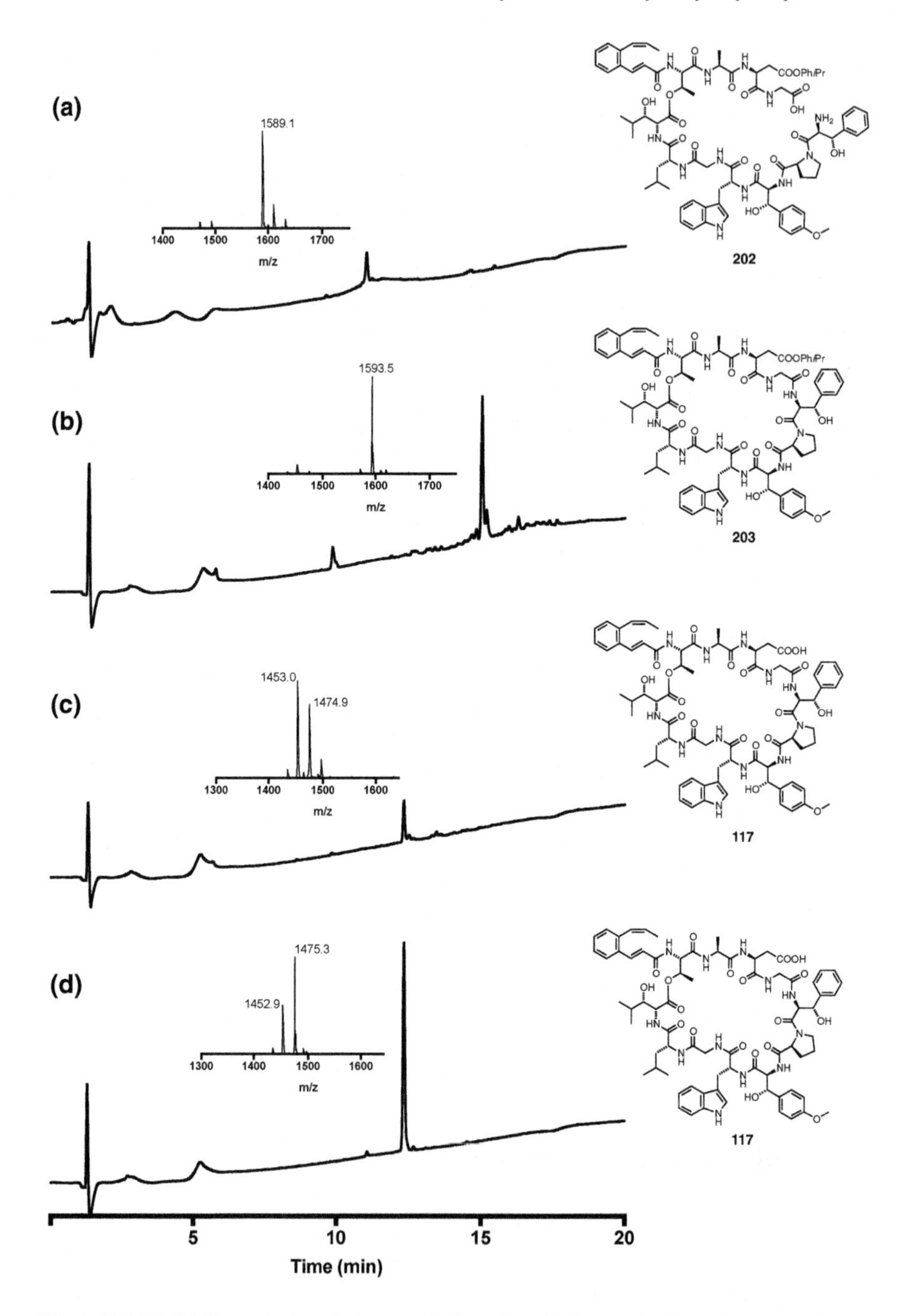

Fig. 3.17 HPLC-MS analysis of the synthesis of deshydroxy skyllamycin B (**117**). All chromatograms: 0–100% MeCN (0.1% formic acid) in H_2O (0.1% formic acid) over 15 min, $\lambda = 230$ nm. **a** Crude linear peptide **202**; *m/z* 1589.8 $(M + H)^+$; **b** crude cyclic peptide **203**; *m/z* 1593.7 $(M + Na)^+$; **c** crude deprotected cyclic peptide **117**; *m/z* 1453.7 $(M + H)^+$, 1475.7 $(M + Na)^+$; **d** purified deshydroxy skyllamycin B (**117**); *m/z* 1453.7 $(M + H)^+$, 1475.7 $(M + Na)^+$

3.8.1 Synthesis of Reduced Cinnamoyl Moiety 131

Reduced cinnamoyl moiety **131** was synthesised using a similar route as that described for **130** described in Sect. 3.5.1. The first step was the generation of reduced iodide **204** from HWE product **140** synthesised previously (Scheme 3.25). Due to the presence of the aryl iodide moiety, the double bond could not simply be reduced using a Pd-catalysed hydrogenation. The first step was the hydrolysis of the ethyl ester with LiOH. Free carboxylic acid **205** was then treated with hydrazine and guanidine nitrate to reduce the double bond affording reduced acid **206** [59]. Initially this reaction was attempted on ethyl ester **140**, and reduction of the double bond was observed along with a by-product, presumably the acyl hydrazide, caused by unwanted hydrazinolysis. As such, the ethyl ester was hydrolysed prior to the reduction reaction. With the double bond successfully reduced, the ethyl ester was reinstalled to yield reduced iodide **204** ready for the ensuing Sonogashira cross coupling reaction. These transformations proceeded extremely efficiently and were carried out with just one final silica gel chromatography step, producing the desired product **131** in 94% yield over the three steps.

Whilst reduction of alkenes was recently reported to proceed in the presence of hydrazine (8 equivalents) in refluxing ethanol [60], due to the high temperature, and large excess of the highly toxic hydrazine required, the guanidine nitrate-catalysed reaction conditions were selected here [59]. This reaction is proposed to proceed by guanidine-catalysed oxidation of hydrazine (**207**) to diimide (**208**) (Scheme 3.26). Diimide (**208**), a well known olefin reducing agent [61], is able to effect the reduction of the double bond, producing nitrogen gas as a by-product.

With reduced iodide **204** in hand, the synthesis could then follow the same route used for cinnamoyl moiety **130** (Scheme 3.27). Specifically, reduced iodide **204** was reacted in a Sonogashira cross-coupling with in situ generated propyne to yield alkyne **209** in good yield. Alkyne **209** was selectively reduced to (*Z*)-alkene **210**

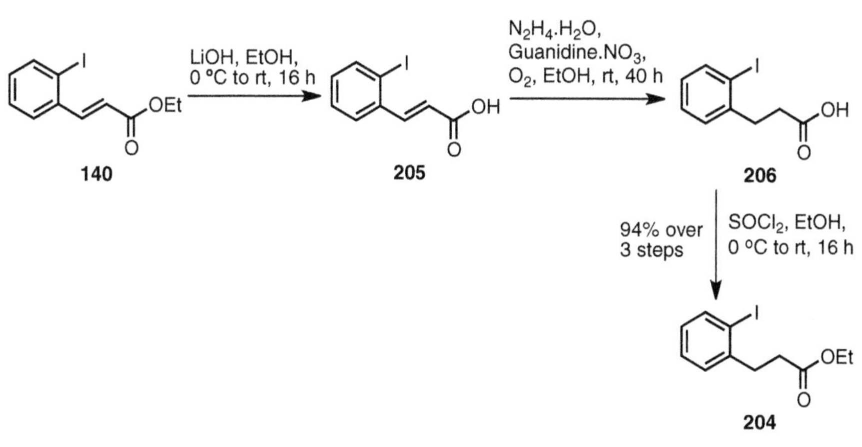

LiOH, EtOH, 0 °C to rt, 16 h
$N_2H_4.H_2O$, Guanidine.NO_3, O_2, EtOH, rt, 40 h
94% over 3 steps
$SOCl_2$, EtOH, 0 °C to rt, 16 h
140 205 206 204

Scheme 3.25 Synthesis of reduced iodide **204**

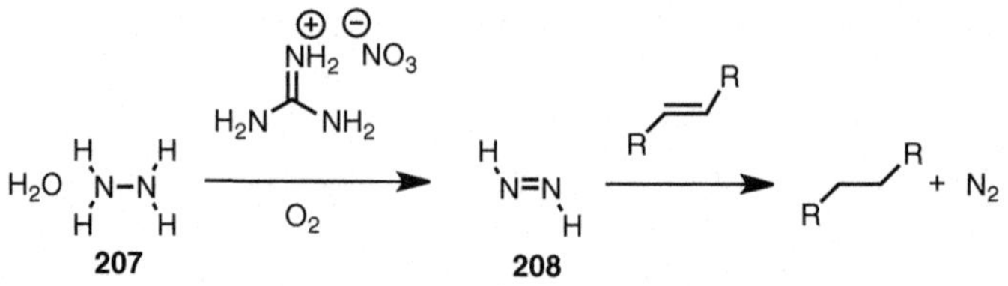

Scheme 3.26 Proposed mechanism for the guanidine nitrate-catalysed reduction of olefins with hydrazine (**207**)

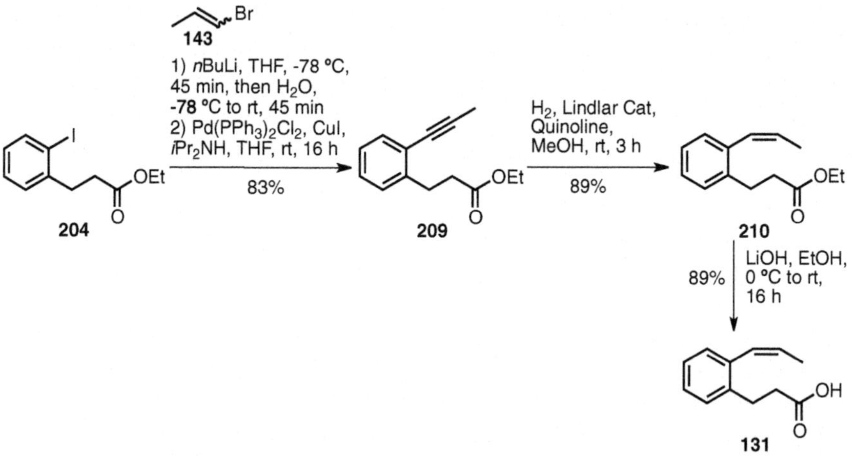

Scheme 3.27 Completion of the synthesis of reduced cinnamoyl moiety **131**

using Lindlar's catalyst and hydrogen gas in excellent yield. This reaction did not proceed as readily as the reduction of alkyne **142**. The reaction was carefully monitored via TLC and was dosed with Lindlar's catalyst a number of times until the starting material was fully consumed. It should be noted that if the reaction mixture was left too long in the presence of the catalyst (16 h), the (*E*)-alkene was observed, presumably due to the Pd-catalysed isomerisation of the double bond. The final reaction was hydrolysis of ethyl ester **210** to produce reduced cinnamoyl moiety **131** in excellent yield.

3.8.2 Synthesis of Suitably Protected β-Me-Asp 212

The final amino acid requiring synthesis was (2*S*,3*S*)-β-Me-Asp and this was carried out based on work by Goodman and co-workers [62]. In their study, the authors successfully synthesised the protected analogue **211** with a side-chain *t*-butyl protecting group. Amino acid **211** was not suitable for incorporation into the synthetic route to deshydroxy skyllamycin A (**116**) and so the suitably protected analogue, Fmoc-β-Me-Asp(Ph*i*Pr)-OH (**212**) instead was targeted (Fig. 3.18). It was envisioned that **212** could be accessed via simple manipulations from **211**, generated through the method of Goodman and co-workers.

To this end, side chain *t*-butyl-protected Asp **213** was protected in a two-step sequence (Scheme 3.28). The first step was the protection of the α-amine with the 9-phenylfluorene (PhF) protecting group. This was carried out via temporary formation of the trimethyl silyl ester, followed by PhF protection, mediated by lead (II) nitrate. Prior studies from Rapoport and co-workers [63] showed that in the absence of the bromide scavenger, lead (II) nitrate, the reaction was sluggish and could not be pushed to completion. The TMS ether was then rapidly cleaved in the acidic work up yielding intermediate **214**. Due to the steric bulk of the PhF group, the α-proton of the amino acid is shielded from deprotonation by the strong base required in the next step of the synthesis [64]. Protection of the carboxylic acid as a benzyl ester was then carried out, to yield protected Asp **215** in 64% yield over two steps. Protected Asp **215** was then selectively enolised at the side chain ester using the strong base lithium hexamethyldisilazide (LiHMDS). Subsequent reaction with methyl iodide yielded protected β-Me-Asp **216** in excellent yield as a mixture of diastereomers (3:1, *syn:anti*), with the desired (*syn*) diastereomer in excess, in agreement with the work from Goodman and co-workers [62].

At this stage, the diastereomers could not be separated by silica gel chromatography. However, after removal of the PhF and benzyl ester protecting groups

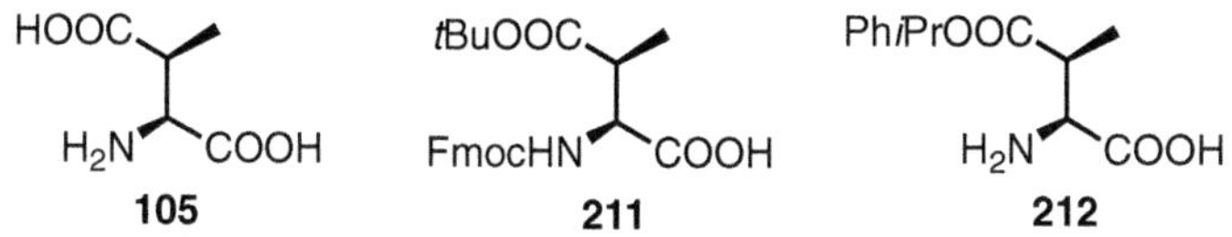

Fig. 3.18 Structure of (2*S*,3*S*)-β-Me-Asp (**105**) and protected variants

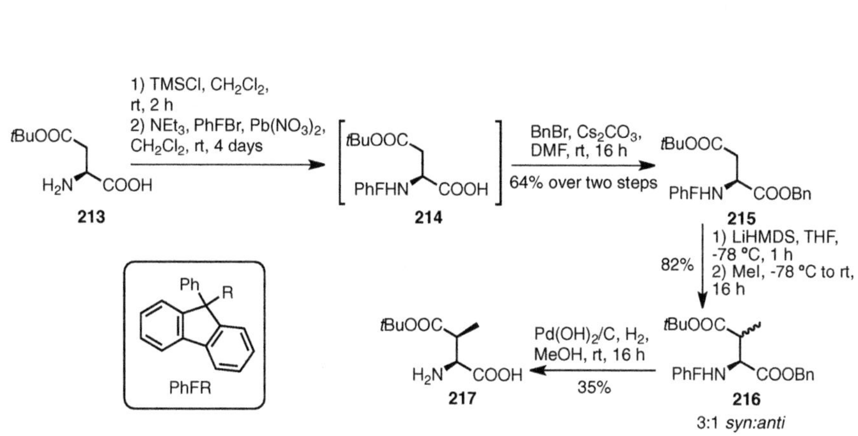

Scheme 3.28 Synthesis of β-Me-Asp **217**

by hydrogenolysis, Goodman and coworkers [62] showed that the diastereomers could be separated. Surprisingly, considering the polarity of the zwitterion **217**, this was achieved using normal phase silica chromatography with the unusual solvent system of EtOAc:*i*PrOH:H_2O (8:2:1, v/v/v). Using this purification method the desired *syn* amino acid **217** was isolated in 35% overall yield after hydrogenation and chromatographic separation (Scheme 3.28). Pleasingly, the ^{1}H-NMR and optical rotation matched those reported by Goodman and coworkers [62].

To complete the synthesis of the desired Fmoc-β-Me-Asp(Ph*i*Pr)-OH (**212**) a number of protecting group manipulations were performed in order to install the Ph*i*Pr protecting group (Scheme 3.29). Specifically, free amino acid **217** was protected with an Fmoc group, followed by formation of an allyl ester at the α-carboxylic acid to yield **218**. Removal of the *t*-butyl ester and treatment with phenyl*iso*propyl trichloroacetimidate (**219**) yielded β-Me-Asp **220** decorated with the desired orthogonal protecting groups. These protecting group manipulations were accomplished in an excellent yield of 49% over the four steps. The final step was the removal of the allyl ester, with catalytic $Pd(PPh_3)_4$ and $PhSiH_3$ as an allyl scavenger, which proceeded quantitatively, yielding the desired Fmoc-β-Me-Asp (Ph*i*Pr)-OH (**212**) suitable for incorporation into deshydroxy skyllamycin A (**116**).

3.8.3 Synthesis of Deshydroxy Skyllamycins A and C

With the final two modified amino acids in hand, the synthesis of deshydroxy skyllamycins A (**116**) and C (**118**) was commenced (Scheme 3.30). This followed the same strategy utilised to synthesise deshydroxy skyllamycin B (**117**) detailed in Sect. 3.7. To begin with recently synthesised Fmoc-β-Me-Asp(Ph*i*Pr)-OH (**212**) or Fmoc-Asp-(Ph*i*Pr)-OH were coupled to resin-bound Gly **150** yielding resin-bound dipeptides **221** and **201**, respectively. Standard Fmoc-SPPS yielded **222** and **223**,

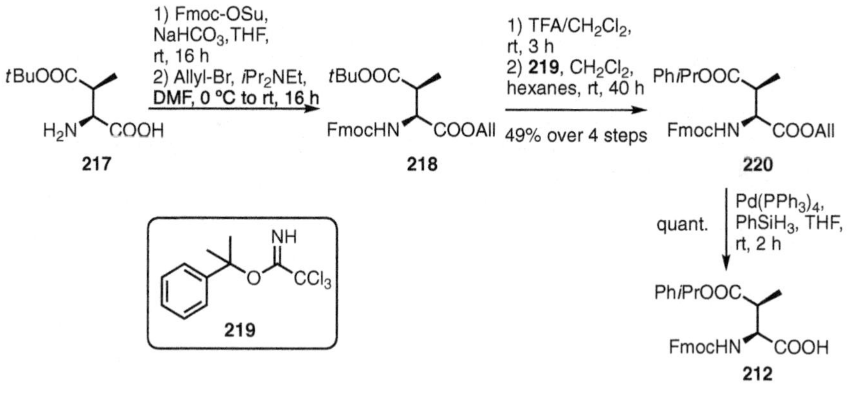

*t*BuOOC
H_2N COOH
217
1) Fmoc-OSu, $NaHCO_3$,THF, rt, 16 h
2) Allyl-Br, *i*Pr_2NEt, DMF, 0 °C to rt, 16 h
*t*BuOOC
FmocHN COOAll
218
1) TFA/CH_2Cl_2, rt, 3 h
2) **219**, CH_2Cl_2, hexanes, rt, 40 h
49% over 4 steps
Ph*i*PrOOC
FmocHN COOAll
220
quant.
$Pd(PPh_3)_4$, $PhSiH_3$, THF, rt, 2 h
Ph*i*PrOOC
FmocHN COOH
212
NH
O
CCl_3
219

Scheme 3.29 Completion of the synthesis of Fmoc-β-Me-Asp(Ph*i*Pr)-OH (**212**)

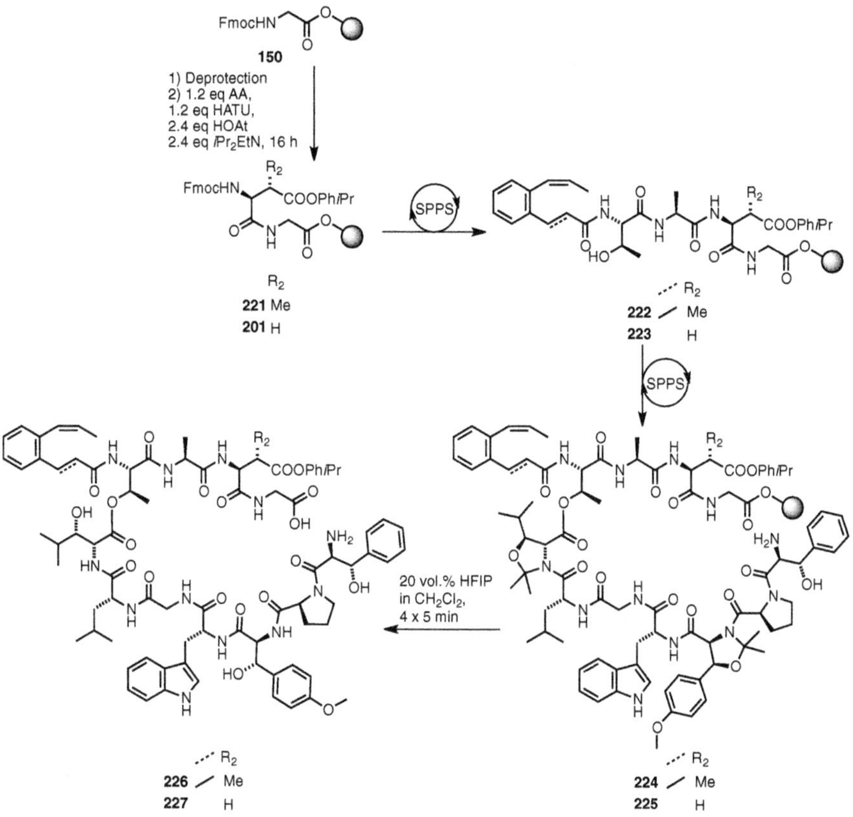

Scheme 3.30 SPPS of protected linear peptides **226** and **227**

bearing cinnamoyl moiety **130** and reduced cinnamoyl moiety **131**, respectively. With these two peptides in hand, the remaining amino acids that required installation were the same for both deshydroxy skyllamycins A (**116**) and C (**118**). As such, synthesis of the two peptides proceeded under identical conditions to yield resin-bound linear peptides **224** and **225**.

Completed linear resin-bound peptides **224** and **225** were next cleaved with HFIP to yield linear protected peptides **226** and **227** (Scheme 3.30). Linear peptides **226** and **227** were then cyclised using DMTMM.BF_4 in DMF to yield the protected cyclic peptides (Scheme 3.31). Gratifyingly, deprotection of the Ph*i*Pr group with mild acidic treatment provided the target deshydroxy skyllamycins A (**116**) and C (**117**) in 5.2% and 3.9% yield, respectively from the initial resin loading. These yields correspond to an average of between 88–89% per step for the entire synthesis (25 steps from resin loading).

1) DMTMM, *i*Pr_2NEt, DMF, 16 h (0.01 M)
2) TFA:*i*Pr_3SiH:CH_2Cl_2 (1:5:94 v/v/v)

116 5.2% over 25 steps
118 3.9% over 25 steps

226 R_2 = Me
227 R_2 = H

deshydroxy skyllamycin A **116** R_2 = Me
deshydroxy skyllamycin C **118** R_2 = H

Scheme 3.31 Cyclisation and deprotection to yield deshydroxy skyllamycins A (**116**) and C (**118**)

3.9 Biological Evaluation

Having successfully prepared deshydroxy skyllamycins A–C (**116–118**), the synthetic analogues were next analysed in the *P. aeruginosa* biofilm inhibition assay at the University of California, Santa Cruz by Jake Haeckl and Roger Linington. Unfortunately, the compounds displayed poor anti biofilm activity, in which they only inhibited biofilm formation at the highest concentration tested. As such, an accurate MIC could not be determined. However, this result suggests that the unusual α-OH-Gly residue in the skyllamycins is important for optimal anti-biofilm activity. Based on the computational work carried out by Süssmuth and co-workers [14], the stereochemistry of the α-OH-Gly plays an important role in the three dimensional conformation of the skyllamycins. Presumably, the deshydroxy skyllamycin analogues are unable to adopt the correct confirmation critical for the anti biofilm activity.

3.10 Conclusions and Future Directions

This chapter describes the design and efficient synthesis of deshydroxy skyllamycins A–C (**116–118**) and simplified skyllamycin analogue **115**. The synthesis was enabled by the preparation of protected variants of the six modified amino acids present in the skyllamycin family. Suitably protected variants of the three key β-OH amino acids (**166**, **180** and **187**) were prepared beginning from the chiral building block, Garner's aldehyde (**158** and **159**), through a unified strategy. Differences in the reactivity of the chosen organometallic reagents, allowed for the key stereochemical relationship to be established. Cinnamoyl moieties **130** and **131** were synthesised utilising a Sonogashira cross coupling step as the key transformation.

Suitably protected β-Me-Asp **212** was synthesised from Asp following a modified literature procedure.

Utilising the synthesised modified amino acids, an efficient SPPS protocol, followed by a key cyclisation reaction and acidolytic deprotection, facilitated the synthesis of the four skyllamycin analogues **115–118**. During the course of the work, the instability of the β-OH-Tyr residue to acidic deprotection conditions was discovered. In order to circumvent this issue, the highly acid labile Ph*i*Pr protecting group was utilised. Unfortunately, all four analogues **115–118** proved to have poor activity in the *P. aeruginosa* biofilm inhibition assay, at least an order of magnitude less than the natural products. This work highlights the critical nature of the α-OH-Gly residue to the skyllamycins bioactivity. The work also lays the foundation for the synthesis of the native skyllamycin natural products which is described in the following chapter.

References

1. K. Shiomi, H. Yang, J. Inokoshi, D.V.D. Pyl, A. Nakagawa, H. Takeshima, S. Omura, J. Antibiot. **46**, 229–234 (1993)
2. S. Um, S.H. Park, J. Kim, H.J. Park, K. Ko, H.-S. Bang, S.K. Lee, J. Shin, D.-C. Oh, Org. Lett. **17**, 1272–1275 (2015)
3. E. Heinzelmann, S. Berger, O. Puk, B. Reichenstein, W. Wohlleben, D. Schwartz, Antimicrob. Agents Chemother. **47**, 447–457 (2003)
4. C.M. Harris, T.M. Harris, J. Am. Chem. Soc. **104**, 4293–4295 (1982)
5. T. Takeuchi, H. Iinuma, S. Kunimoto, T. Masuda, M. Ishizuka, M. Takeuchi, M. Hamada, H. Naganawa, S. Kondo, H. Umezawa, J. Antibiot. **34**, 1619–1621 (1981)
6. H. Umezawa, S. Kondo, H. Iinuma, S. Kunimoto, Y. Ikeda, H. Iwasawa, D. Ikeda, T. Takeuchi, J. Antibiot. **34**, 1622–1624 (1981)
7. C.W. Johnston, M.A. Skinnider, M.A. Wyatt, X. Li, M.R.M. Ranieri, L. Yang, D.L. Zechel, B. Ma, N.A. Magarvey, Nat. Commun. **6**, 8421 (2015)
8. V.E. Fadouloglou, S. Balomenou, M. Aivaliotis, D. Kotsifaki, S. Arnaouteli, A. Tomatsidou, G. Efstathiou, N. Kountourakis, S. Miliara, M. Griniezaki, A. Tsalafouta, S.A. Pergantis, I.G. Boneca, N.M. Glykos, V. Bouriotis, M. Kokkinidis, J. Am. Chem. Soc. (2017)
9. S. Toki, T. Agatsuma, K. Ochiai, Y. Saitoh, K. Ando, S. Nakanishi, N.A. Lokker, N.A. Giese, Y. Matsuda, J. Antibiot. **54**, 405–414 (2001)
10. J. Andrae, R. Gallini, C. Betsholtz, Genes Dev. **22**, 1276–1312 (2008)
11. S. Pohle, C. Appelt, M. Roux, H.P. Fiedler, R.D. Süssmuth, J. Am. Chem. Soc. **133**, 6194–6205 (2011)
12. R.M. Donlan, J.W. Costerton, Clin. Microbiol. Rev. **15**, 167–193 (2002)
13. G. Navarro, A.T. Cheng, K.C. Peach, W.M. Bray, V.S. Bernan, F.H. Yildiz, R.G. Linington, Antimicrob. Agents Chemother. **58**, 1092–1099 (2014)
14. V. Schubert, F. Di Meo, P.L. Saaidi, S. Bartoschek, H.P. Fiedler, P. Trouillas, R.D. Süssmuth, Chem. Eur. J. **20**, 4948–4955 (2014)
15. R. Bhushan, H. Brückner, Amino Acids **27**, 231–247 (2004)
16. A. Zampella, R. D'Orsi, V. Sepe, A. Casapullo, M.C. Monti, M.V. D'Auria, Org. Lett. **7**, 3585–3588 (2005)
17. O. Puk, D. Bischoff, C. Kittel, S. Pelzer, S. Weist, E. Stegmann, R.D. Süssmuth, W. Wohlleben, J. Bacteriol. **186**, 6093–6100 (2004)

18. S. Uhlmann, R.D. Süssmuth, M.J. Cryle, ACS Chem. Biol. **8**, 2586–2596 (2013)
19. K. Haslinger, C. Brieke, S. Uhlmann, L. Sieverling, R.D. Süssmuth, M.J. Cryle, Angew. Chem. Int. Ed. **53**, 8518–8522 (2014)
20. S.T. Prigge, R.E. Mains, B.A. Eipper, L.M. Amzel, Cell. Mol. Life Sci. **57**, 1236–1259 (2000)
21. R.W. Rickards, D. Skropeta, Tetrahedron **58**, 3793–3800 (2002)
22. D. Sun, P. Lai, W. Xie, J. Deng, Y. Jiang, Synth. Commun. **37**, 2989–2994 (2007)
23. L. Horner, H. Hoffmann, H.G. Wippel, Chem. Ber. **91**, 61–63 (1958)
24. W.S. Wadsworth, W.D. Emmons, J. Am. Chem. Soc. **83**, 1733–1738 (1961)
25. K. Sonogashira, Y. Tohda, N. Hagihara, Tetrahedron Lett. **16**, 4467–4470 (1975)
26. J. Jauch, D. Schmalzing, V. Schurig, R. Emberger, R. Hopp, M. Köpsel, W. Silberzahn, P. Werkhoff, Angew. Chem. Int. Ed. **28**, 1022–1023 (1989)
27. E. Abraham, Suffert, J. Synlett, 328–330 (2002)
28. J. Suffert, D. Toussaint, J. Org. Chem. **60**, 3550–3553 (1995)
29. W.P. Buttenberg, Justus Liebigs Ann. Chem. **279**, 324–337 (1894)
30. P. Fritsch, Justus Liebigs Ann. Chem. **279**, 319–323 (1894)
31. H. Wiechell, Justus Liebigs Ann. Chem. **279**, 337–344 (1894)
32. R. Chinchilla, C. Najera, Chem. Soc. Rev. **40**, 5084–5121 (2011)
33. H. Lindlar, Helv. Chim. Acta **35**, 446–450 (1952)
34. A.B. McEwen, M.J. Guttieri, W.F. Maier, R.M. Laine, Y. Shvo, J. Org. Chem. **48**, 4436–4438 (1983)
35. K. Barlos, D. Gatos, J. Kallitsis, G. Papaphotiu, P. Sotiriu, Y. Wenqing, W. Schäfer, Tetrahedron Lett. **30**, 3943–3946 (1989)
36. P. Garner, J.M. Park, J. Org. Chem. **52**, 2361–2364 (1987)
37. P. Garner, Tetrahedron Lett. **25**, 5855–5858 (1984)
38. X. Liang, J. Andersch, M. Bols, J. Chem. Soc. Perkin Trans. **1**, 2136–2157 (2001)
39. T. Wöhr, F. Wahl, A. Nefzi, B. Rohwedder, T. Sato, X. Sun, M. Mutter, J. Am. Chem. Soc. **118**, 9218–9227 (1996)
40. D. Skropeta, K.A. Jolliffe, P. Turner, J. Org. Chem. **69**, 8804–8809 (2004)
41. L. Williams, Z.D. Zhang, F. Shao, P.J. Carroll, M.M. Joullié, Tetrahedron **52**, 11673–11694 (1996)
42. D.J. Cram, F.A.A. Elhafez, J. Am. Chem. Soc. **74**, 5828–5835 (1952)
43. M.T. Reetz, M. Hüllmann, T. Seitz, Angew. Chem. Int. Ed. **26**, 477–479 (1987)
44. P. Lucio Anelli, C. Biffi, F. Montanari, S. Quici, J. Org. Chem. **52**, 2559–2562 (1987)
45. G. Tojo, M. Fernández, In *Oxidation of Primary Alcohols to Carboxylic Acids: A Guide to Current Common Practice* (Springer New York: New York, NY, 2007), pp. 79–103
46. M. Chérest, H. Felkin, N. Prudent, Tetrahedron Lett. **9**, 2199–2204 (1968)
47. A. Nguyen Trong, O. Eisenstein, J.M. Lefour, Tran Huu Dau. M. E. J. Am. Chem. Soc. **95**, 6146–6147 (1973)
48. H.B. Bürgi, J.D. Dunitz, J.M. Lehn, G. Wipff, Tetrahedron **30**, 1563–1572 (1974)
49. J.W. Cornforth, R.H. Cornforth, K.K. Mathew, *J. Chem. Soc.* 112–127 (1959)
50. L.R. Malins, A.M. Giltrap, L.J. Dowman, R.J. Payne, Org. Lett. **17**, 2070–2073 (2015)
51. L.R. Malins, R.J. Payne, Org. Lett. **14**, 3142–3145 (2012)
52. A.J. Mancuso, S.-L. Huang, D. Swern, J. Org. Chem. **43**, 2480–2482 (1978)
53. A. Nishida, H. Sorimachi, M. Iwaida, M. Matsumizu, T. Kawate, M. Nakagawa, Synlett, 389–390 (1998)
54. D.B. Dess, J.C. Martin, J. Org. Chem. **48**, 4155–4156 (1983)
55. N. Okamoto, O. Hara, K. Makino, Y. Hamada, J. Org. Chem. **67**, 9210–9215 (2002)
56. C.S. Rye, S.G. Withers, J. Am. Chem. Soc. **124**, 9756–9767 (2002)
57. C. Yue, J. Thierry, P. Potier, Tetrahedron Lett. **34**, 323–326 (1993)
58. A. Isidro-Llobet, M. Álvarez, F. Albericio, Chem. Rev. **109**, 2455–2504 (2009)
59. M. Lamani, R.S. Guralamata, K.R. Prabhu, Chem. Commun. **48**, 6583–6585 (2012)
60. H. Chen, J.M. Wang, X.C. Hong, H.B. Zhou, C.N. Dong, Can. J. Chem. **90**, 758–761 (2012)

61. S. Hünig, H.R. Müller, W. Thier, Angew. Chem. Int. Ed. **4**, 271–280 (1965)
62. S. Schabbert, M.D. Pierschbacher, R.H. Mattern, M. Goodman, Bioorg. Med. Chem. **10**, 3331–3337 (2002)
63. B.D. Christie, H. Rapoport, J. Org. Chem. **50**, 1239–1246 (1985)
64. J.P. Wolf, H. Rapoport, J. Org. Chem. **54**, 3164–3173 (1989)

Chapter 4
Total Synthesis of Skyllamycins A–C

The synthesis of deshydroxy skyllamycins A–C (**116–118**) described in the previous chapter laid the foundation for the synthesis of the native natural products, skyllamycins A–C (**101–103**). While a number of synthetic challenges were overcome during the synthesis of deshydroxy skyllamycins A–C (**116–118**), the installation of the unusual α-OH-Gly residue was not investigated. As such, before beginning the synthesis of skyllamycins A–C (**101–103**) previous methods for investigating the installation of the unusual α-OH-Gly motif were reviewed.

4.1 Previous Approaches to α-OH-Gly Moieties

While the α-OH-Gly residue is an important intermediate in the biosynthetic formation of *C*-terminal carboxamides in proteins (Sect. 3.2.4), it is extremely rare in mature natural products, and to date has only been found in the skyllamycins and the linear peptide anti-tumour natural product spergualin (**228**) [1, 2]. The α-OH-Gly residue and its derivatives have also found use in the synthesis of unnatural amino acids.

4.1.1 Methods for the Synthesis of α-Functionalised-Gly Moieties

There have been a number of studies carried out on the synthesis and properties of α-heteroatom-functionalised glycine moieties. Steglich and co-workers have carried out a significant body of work on the development of methods to install unusual amino acid side chains via an electrophilic glycine intermediate (Scheme 4.1a) [3–5]. This methodology involves the mild selective cleavage of a serine or

A. Giltrap, *Total Synthesis of Natural Products with Antimicrobial Activity*,
Springer Theses, https://doi.org/10.1007/978-981-10-8806-3_4

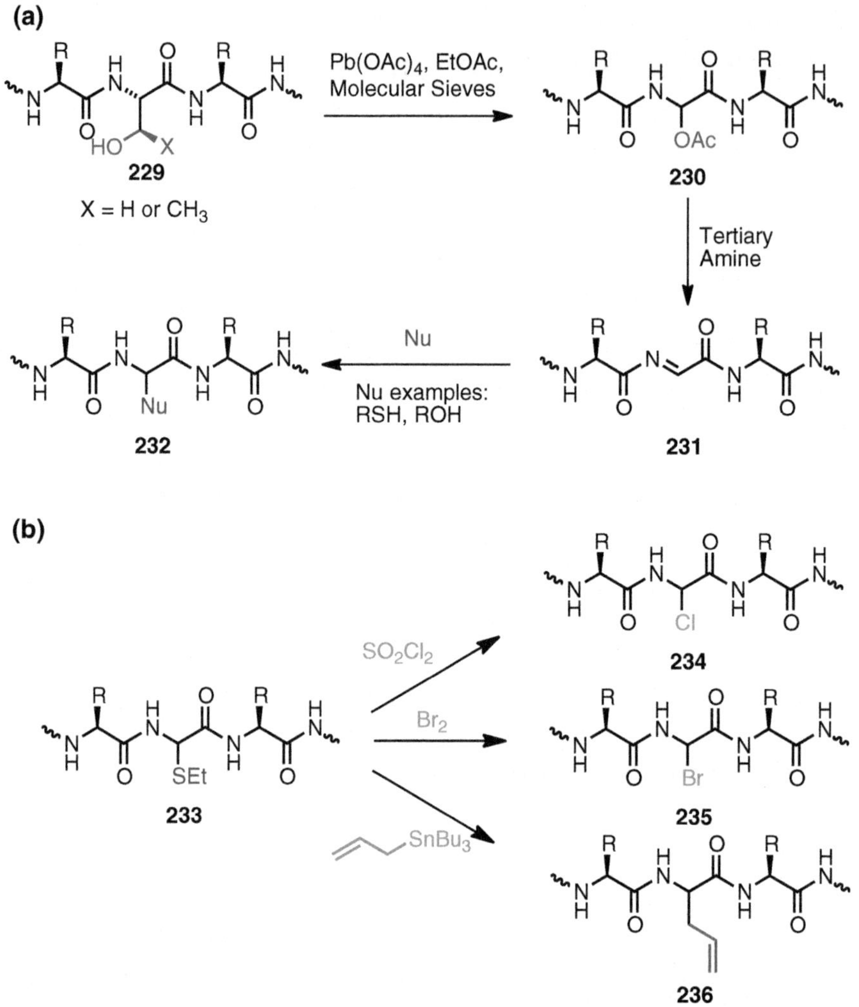

Scheme 4.1 **a** Formation of α-substituted glycine derivatives. **b** Transformations of α-thiol-glycine derivative **233**

threonine residue in a peptide **229** to the corresponding α-OAc-Gly derivative **230** by treatment of the peptide with lead (IV) acetate. Subsequent treatment with a tertiary amine forms the dehydroglycine intermediate **231**, which can react with a nucleophile to form peptide **232**. A large variety of nucleophiles have been utilised, including oxygen, sulfur and carbon nucleophiles. Steglich and co-workers have also demonstrated that the α-thiol-glycine derivative **233** can also be manipulated in a number of ways to form the α-chloro **234**, bromo **235** and even allyl **236** derivatives (Scheme 4.1b). The α-chloro **234** and bromo **235** derivatives can be further functionalised [5].

4.1.2 Studies Towards the Synthesis of Spergualin and 15-Deoxyspergualin

A number of total syntheses of spergualin (**228**) [6], and its deoxy analogue, 15-deoxyspergualin (**237**) [7] have been completed (Fig. 4.1). Renaut and co-workers carried out an elegant total synthesis of 15-deoxyspergualin (**237**) in 1998 [8]. Importantly, during the synthesis the authors were able to definitively confirm the (*S*)-configuration at the α-OH-Gly residue. To install the key α-OH-Gly moiety, amide **238** was reacted with methyl 2-hydroxy-2-methoxyacetate (**239**) in refluxing dichloromethane to yield the desired α-OH-Gly methyl ester **240** as a racemic mixture (Scheme 4.2). Chlorination, to yield the α-Cl-Gly **241**, similar to that described in the above section, and subsequent substitution with (*S*)-α-methyl-2-naphthenemethanol (**242**) yielded **243**. Hydrolysis of the methyl ester gave **244** and subsequent coupling of amine **245** afforded naphtyl ether **246**.

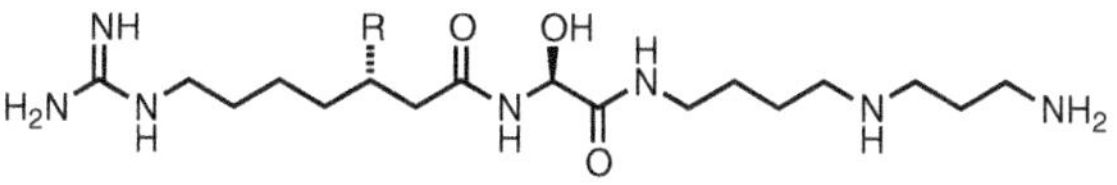

R = OH Spergualin **228**
R = H 15-Deoxyspergualin **237**

Fig. 4.1 Structure of spergualin (**228**) and 15-deoxyspergualin (**237**)

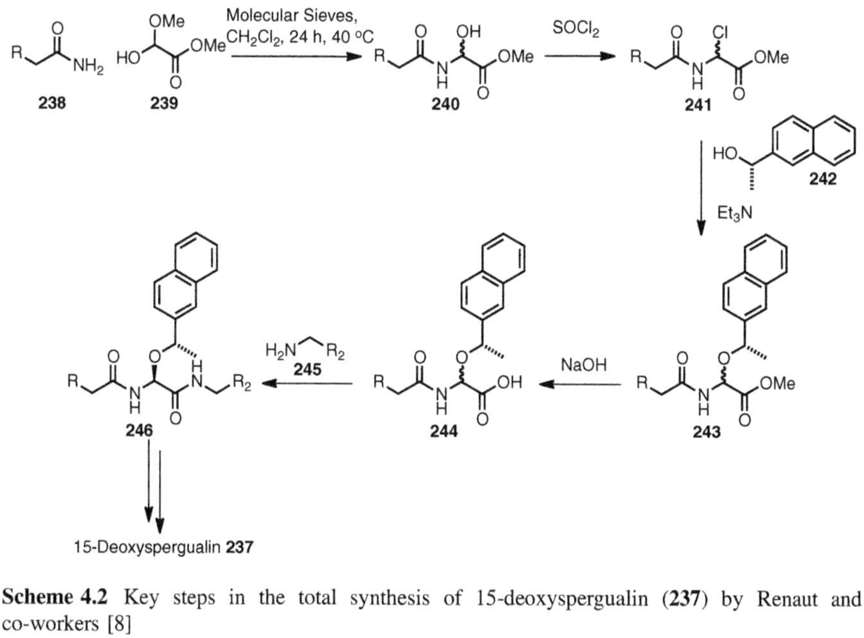

Scheme 4.2 Key steps in the total synthesis of 15-deoxyspergualin (**237**) by Renaut and co-workers [8]

Importantly, at this stage, the mixture of diastereomers could be separated and the compound with the desired (*S*) configuration at the α-OH-Gly centre was carried on to complete the total synthesis of 15-deoxyspergualin (**237**).

4.2 Retrosynthetic Analysis of Skyllamycins A–C

Before the total synthesis of skyllamycins A–C (**101–103**) was embarked upon, a retrosynthetic analysis was carried out. As discussed above, there has been a large body of work concerning the addition of nucleophiles into dehydroglycine equivalents. This methodology was employed in the total synthesis of 15-deoxysperagulin. However, one of the limitations of this methodology is the difficulty in the controlling the stereochemistry at the α-OH-Gly position. This could be overcome in a very simple system such as 15-deoxyspergualin (**237**), by using as chiral auxiliary, however it would be difficult to utilise this type of methodology in the synthesis of more structurally complex natural products, such as skyllamycins A–C (**101–103**). As well as this, the known instability of the α-OH-Gly residue in spergualin (**228**) and 15-deoxyspergualin (**237**) [9] is evidence that the late stage formation of this moiety in the skyllamycins would be the most appropriate strategy.

Considering the above, it was envisioned that a final cyclisation and concurrent formation of the α-OH-Gly residue could overcome these issues. The skyllamycins possess a strong internal hydrogen bond network in which the α-OH-Gly residue plays an important role [10]. It was hypothesised that the formation of the desired (*S*)-configuration of the α-OH-Gly residue would be favoured during the final cyclisation step of the linear peptide driven by this hydrogen bond network. Furthermore, formation of this unusual residue in the last step would obviate the requirement for any final deprotection of the peptide, which could result in the decomposition of this unique functional group. With this in mind, the disconnection of the skyllamycins (**101–103**) was made at α-OH-Gly residue, which would yield linear peptides **247–249** bearing an *N*-terminal aldehyde moiety (red, Scheme 4.3), and a *C*-terminal carboxamide. It was envisaged that formation of the aldehyde in **247–249** could be achieved by oxidative cleavage of linear peptides bearing an *N*-terminal serine residue **250–252**. In turn, linear peptides **250–252** could be accessed from the protected resin-bound linear peptides **253–255**, which could be synthesised via Fmoc-SPPS starting from Sieber amide resin [11]. Importantly, all the non-commercially available amino acids necessary for the synthesis of skyllamycins A–C (**101–103**) were accessed during the synthesis of deshydroxy skyllamycins A–C (**116–118**) described in the previous chapter and were therefore ready for direct installation into the peptide by Fmoc-SPPS.

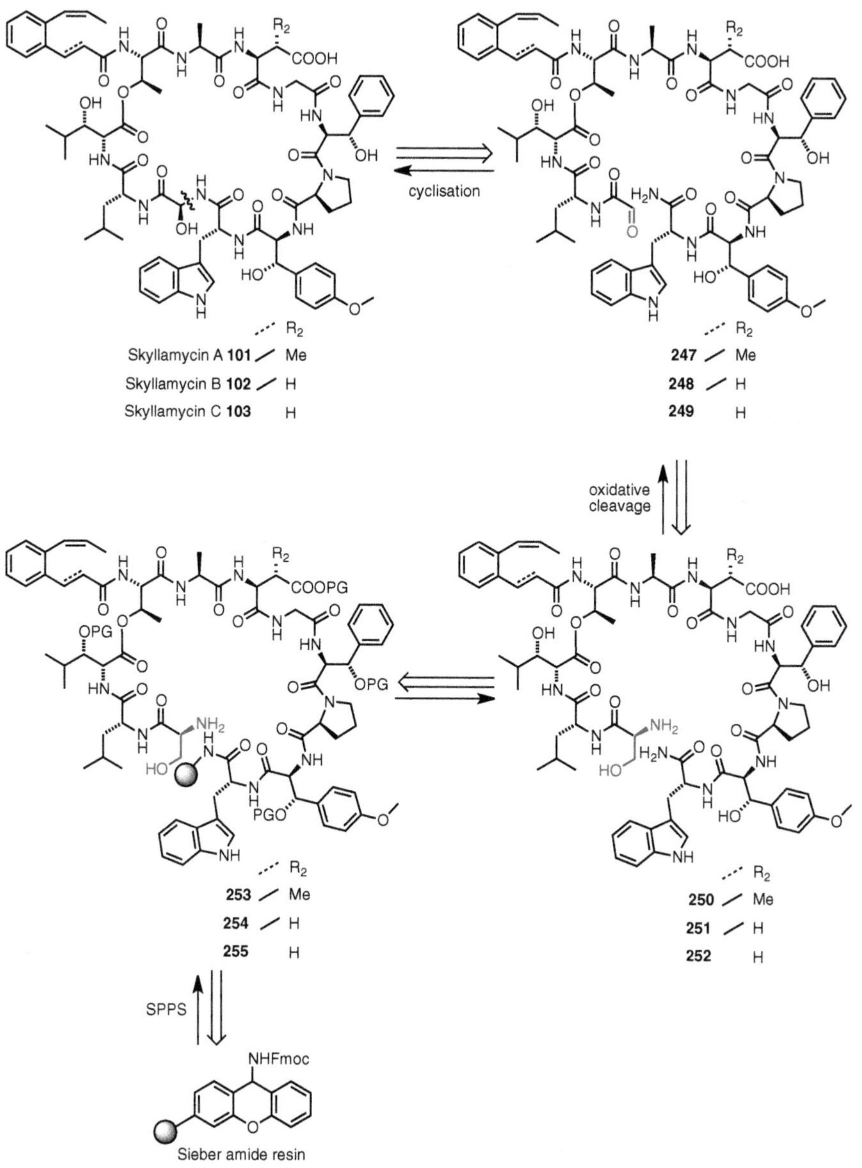

Scheme 4.3 Retrosynthetic analysis of skyllamycins A–C (**101–103**)

4.3 Synthesis of Skyllamycin B

In order to verify the synthetic strategy described above, skyllamycin B (**102**) was chosen as the initial target for synthesis. This was due to the presence of an unmodified Asp rather than the precious β-Me-Asp residue, as well as the synthetic accessibility of the cinnamoyl moiety **130**. It was envisioned that once the synthetic strategy was optimised for skyllamycin B (**102**) it could be readily applied to the synthesis of skyllamycin A (**101**) and C (**103**).

4.3.1 Synthesis of Appropriately Protected β-OH-Phe 256

While the modified amino acids were synthesised in the previous chapter, the β-OH group of Fmoc-β-OH-Phe **180** was utilised unprotected in the synthesis of deshydroxy skyllamycins A–C (**116–118**). Due to the change in the cyclisation junction in the proposed synthetic strategy for skyllamycins A–C (**101–103**) (Scheme 4.3), the key on-resin esterification of the β-OH-Leu residue would be carried out after installation of the β-OH-*O*-Me-Tyr and β-OH-Phe residues. As such, protection of the side chain of the β-OH-Phe residue would be necessary to prevent unwanted esterification. Due to the convenience and acid lability of the oxazolidine protecting group employed for both β-OH-Leu and β-OH-*O*-Me-Tyr, this strategy was also chosen for β-OH-Phe. To this end, Fmoc-β-OH-Phe **180** was reacted with 2,2-DMP with BF_3•OEt_2 as a Lewis acid catalyst to yield the desired oxazolidine protected β-OH-Phe **256** in excellent yield (Scheme 4.4).

4.3.2 Synthesis of Linear Peptide 251

With all the suitably protected modified amino acids in hand, the synthesis of skyllamycin B (**102**) was commenced. This began with the loading of Sieber amide resin with Fmoc-D-Trp-OH generating resin-bound **257** (Scheme 4.5). This reaction was carried out using 7-Azabenzotriazol-1-yloxy tripyrrolidinophosphonium hexafluorophosphate (PyAOP), the HOAt containing derivative of PyBOP. The resin was reacted with acetic anhydride/pyridine to cap any unreacted amine

2,2-DMP, Acetone, BF_3.OEt_2, rt, 4 h

94%

Scheme 4.4 Synthesis of oxazolidine protected β-OH-Phe **256**

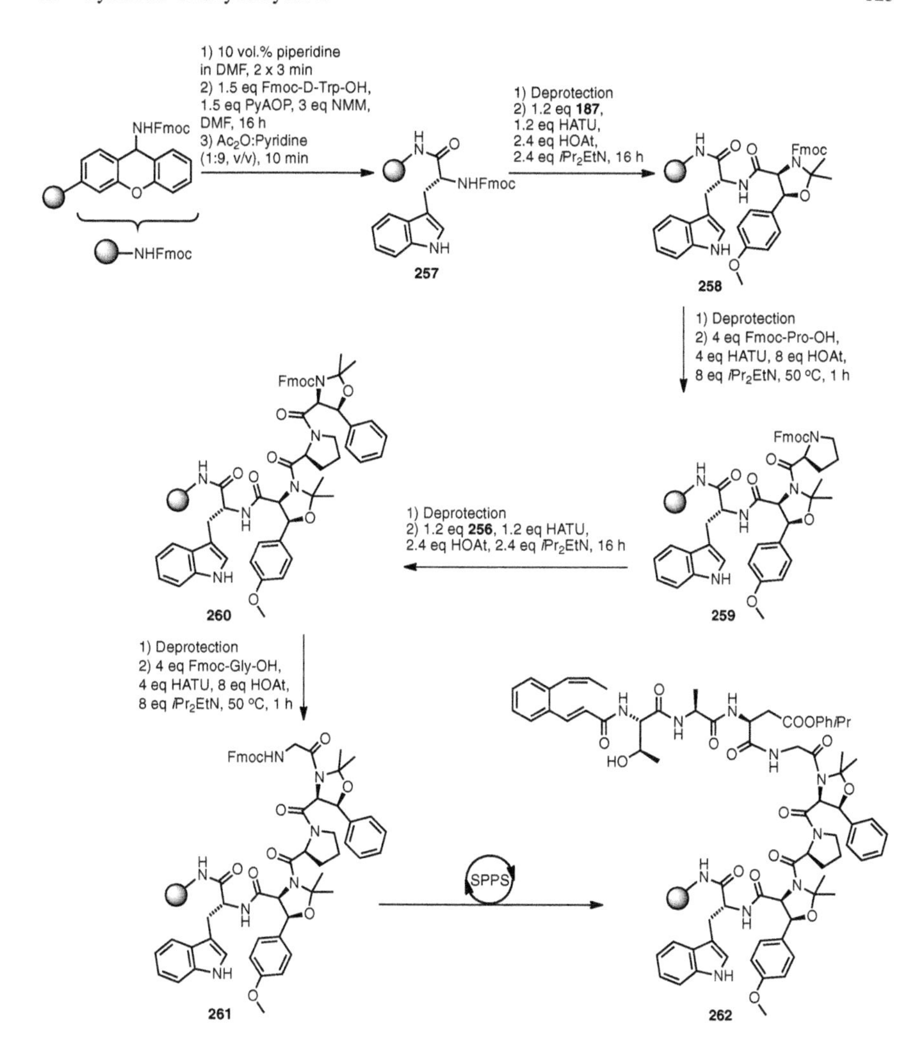

Scheme 4.5 SPPS of resin-bound peptide **262**

functionalities. In 1987 Peter Sieber developed this functionalised resin suitable for synthesising protected peptide fragments containing a *C*-terminal carboxamide [11]. This resin is highly acid labile and can be cleaved with as little as 1% TFA/CH_2Cl_2. As such, it was an ideal choice for the synthesis of the skyllamycins as cleavage from resin could be carried out with concomitant cleavage of the Asp(Ph*i*Pr) group to yield the linear peptides **250–252**.

With resin-bound **257** in hand, following Fmoc-deprotection, suitably protected β-OH-*O*-Me-Tyr-OH **187** was coupled using HATU coupling conditions to yield resin-bound dipeptide **258**. Subsequent coupling of Fmoc-Pro-OH under optimised microwave coupling conditions yielded tripeptide **259** to which suitably protected β-OH-Phe-OH **256** was coupled under HATU conditions producing resin-bound peptide **260**. Again, microwave-assisted coupling of Fmoc-Gly-OH yielded resin-bound pentapeptide **261**, which was a key divergence point for the synthesis of the three skyllamycin natural products. From here, the synthesis continued using the SPPS conditions developed in the previous chapter, namely coupling of Fmoc-Asp(Ph*i*Pr)-OH using HATU, followed by PyBOP coupling of the Fmoc-Ala-OH and Fmoc-Thr-OH residues. Subsequent coupling of the precious cinnamoyl moiety **130** was carried out using HATU conditions to yield the resin-bound peptide **262**.

With resin-bound **262** in hand, the on-resin esterification reaction was carried out using 4 equivalents of suitably protected β-OH-Leu-OH **166** to yield the desired branched peptide bound to the resin **263** (Scheme 4.6). At this point HPLC-MS analysis of an analytical scale resin cleavage confirmed the presence of the desired peptide. However, unfortunately it was determined that there were also peaks corresponding to peptides containing two or three β-OH-Leu moieties.

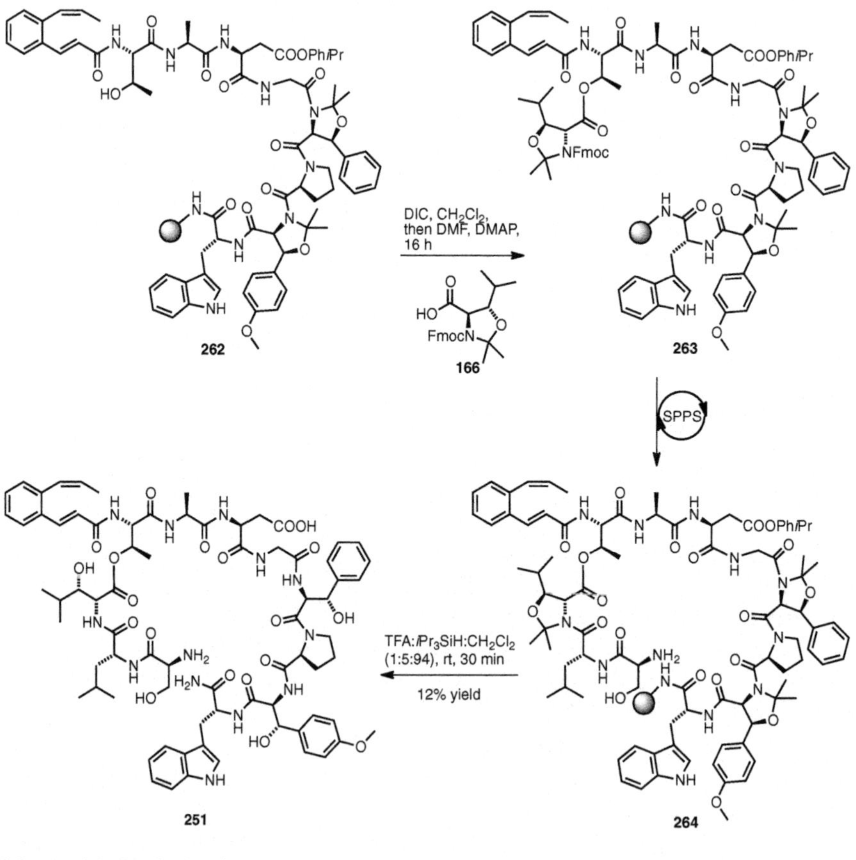

Scheme 4.6 Synthesis of unprotected linear peptide **251**

This suggested that the oxazolidine protecting groups on the β-OH-Phe and β-OH-*O*-Me-Tyr residues were removed at some point during the elongation of the peptide by SPPS. To test the acid stability of this protecting group, oxazolidine protected Fmoc-β-OH-Phe-OH (**256**) was treated with acid at various concentrations. Surprisingly, a solution of 1% TFA in CH_2Cl_2 was unable to remove the protecting group, indeed it was determined a 2 h treatment with 50% TFA in CH_2Cl_2 was necessary. Presumably, the formation of an amide bond, as opposed to the Fmoc-carbamate, plays a role in the acid stability of this protecting group. It is also possible that conformational effects exerted by the peptide backbone also lead to increased lability.

While this result was disappointing, efforts were carried out to optimise this on-resin esterification reaction, by reducing the number of equivalents of suitably protected β-OH-Leu-OH **166**. It was determined that an initial treatment of the resin with 3 equivalents of protected β-OH-Leu-OH **166**, followed by a further treatment with 0.5 equivalents was sufficient to produce quantitative esterification, with minimal over esterification. Resin-bound **263** was then reacted under microwave conditions with Fmoc-D-Leu-OH followed by the final coupling in the synthesis to install a moiety suitable for oxidative cleavage to yield a *C*-terminal aldehyde. To this end, side chain unprotected Fmoc-Ser-OH was coupled to yield the desired resin bound peptide **264**. This was then cleaved from resin, with concomitant removal of the Asp(Ph*i*Pr) protecting group, utilising a dilute solution of 1% TFA in CH_2Cl_2 with *i*Pr_3SiH as a scavenger, to yield the desired unprotected linear peptide **251**. Preparative HPLC purification yielded the desired peptide **251** in 12% yield from the initial resin loading, corresponding to an average of 92% per step over the 26 steps of the synthesis (Fig. 4.2a).

4.3.3 Oxidative Cleavage Reaction to Aldehyde 248

With linear peptide **251** in hand, the key oxidative cleavage reaction was carried out. A significant body of work has been carried out concerning this type of reaction. Indeed, in 1939 it was reported that α-aminoalcohols, such as serine and threonine, can be oxidatively cleaved to the corresponding aldehyde by treatment with periodate [12]. This reactivity has been used to selectively modify the *N*-terminus of a number of proteins [13, 14] and subsequently carry out selective bio-conjugation reactions [15, 16]. To this end, linear peptide **251** was dissolved in a 1:1 mixture of aqueous Na_2HPO_4 buffer (pH = 9) and acetonitrile (Scheme 4.7). Subsequently, a solution of sodium *meta*periodate in water was added to the solution of peptide and allowed to react at room temperature. The oxidative cleavage is known to proceed rapidly at basic pH and therefore the reaction was quenched after 10 min with an excess of ethylene glycol to avoid any over oxidation of the tryptophan or tyrosine side chains. Pleasingly, HPLC-MS analysis showed complete consumption of starting material and formation of the product **248**. The product was characterised by the distinctive signal in the mass-spectrum

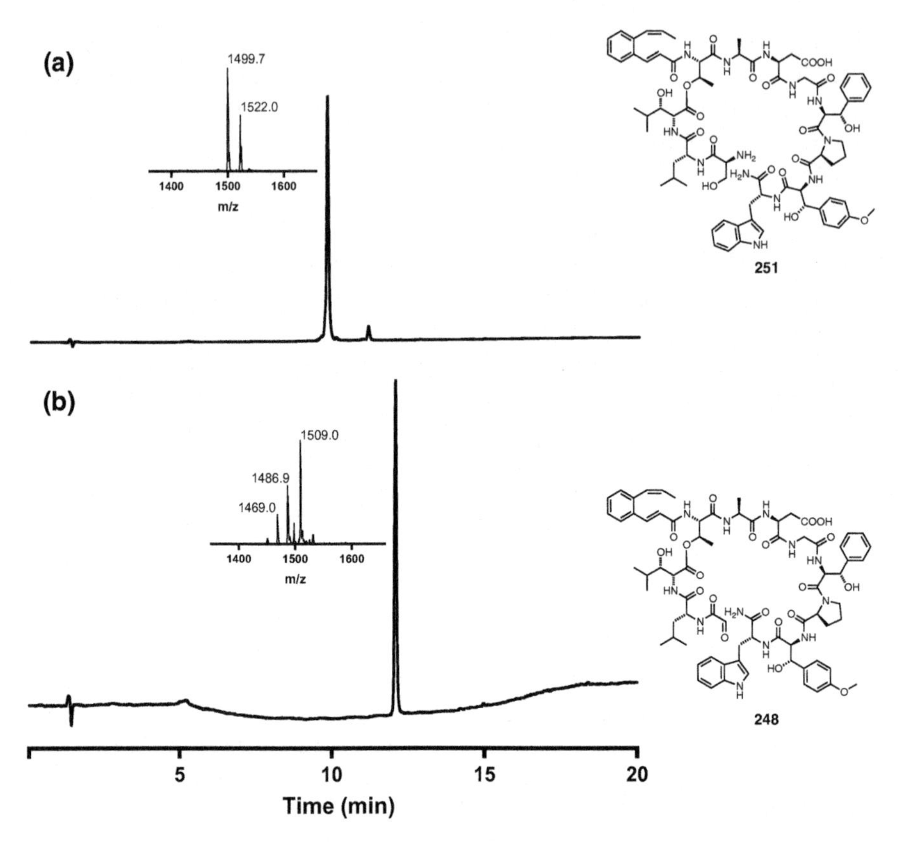

Fig. 4.2 HPLC-MS analysis of peptide **251** and aldehyde **148**. All chromatograms: 0–100% MeCN (0.1% formic acid) in H_2O (0.1% formic acid) over 15 min, λ = 280 nm. **a** Purified linear peptide **251**; *m/z* 1500.7 $(M+H)^+$, 1522.7 $(M+Na)^+$; **b** purified aldehyde **248**; *m/z* 1469.7 $(M+H)^+$, 1487.7 $(M+H_2O+H)^+$, 1509.7 $(M+H_2O+Na)^+$

NaIO$_4$, Na$_2$HPO$_4$ buffer, 10 min, then (CH$_2$OH)$_2$

63%

251

248

Scheme 4.7 Oxidative cleavage of **251** to form aldehyde **248**

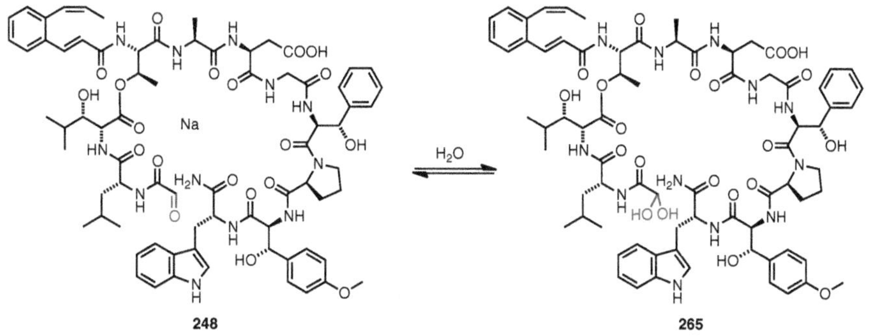

Scheme 4.8 Equilibrium between peptide aldehyde **248** and hydrate **265**

corresponding to the hydrate **265** (Scheme 4.8). It is well established that peptide aldehydes exist in this form [15]. The reaction was immediately purified by HPLC to yield the desired aldehyde **248** in 63% yield (Fig. 4.2b). Rapid purification was important, as over incubation of the peptide with $NaIO_4$ led to the formation of unwanted by-products.

4.3.4 Initial Cyclisation Trial

With aldehyde **248** in hand, the key cyclisation reaction was attempted (Scheme 4.9). To begin with a number of reaction conditions were trialled (Table 4.1) and analysed by HPLC-MS. Incubation of aldehyde **248** in a mixture of MeCN and H_2O with 0.1% TFA did not result in any reaction taking place. It was

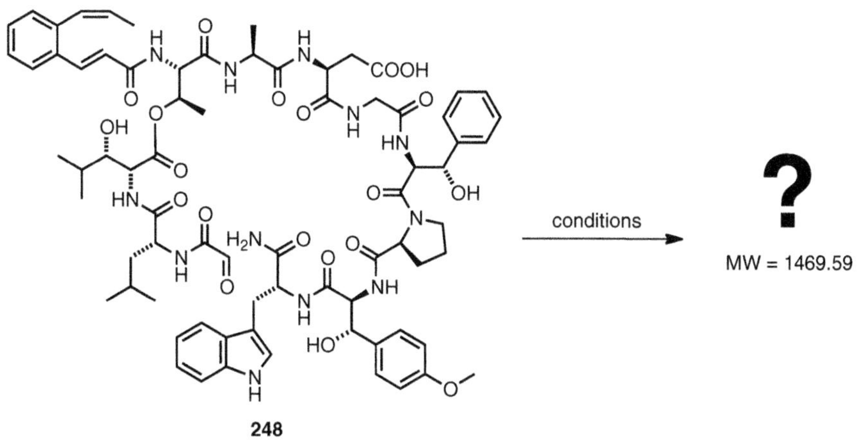

Scheme 4.9 Initial cyclisation studies

Table 4.1 Initial cyclisation reaction conditions trialled on aldehyde **248**

Conditions	Result
1:1 H_2O: MeCN + 0.1% TFA, rt, 16 h	No change
MeCN, 30 °C, 16 h	No change
MeCN + 0.1% TFA, rt, 20 h	Formation of 2 new peaks, <5% starting material
MeCN + 1% TFA, rt, 16 h	Formation of 2 new peaks, no starting material

hypothesised that the reaction would proceed slowly in the presence of water as the cyclisation of the hydrate form of aldehyde **265** would require the release of one water molecule. As such, anhydrous conditions were trialled. To this end, **248** was incubated at slightly elevated temperature (30 °C) in neat MeCN, however again no reaction was observed. It was reasoned that the cyclisation reaction might be acid catalysed and, as such, two methods were trialled namely incubation in MeCN with the addition of 0.1% or 1% TFA. Gratifyingly, HPLC-MS analysis of both reactions revealed the formation of two news peaks with the desired mass (1469.25 Da), and the complete consumption of the starting aldehyde **248**. The reaction with 1% TFA added gave the same result, but the reaction proceeded slightly faster.

Encouraged by these results the reaction was scaled up, and aldehyde **248** was incubated in MeCN with the milder acidic conditions, namely 0.1% TFA. After 20 h, the reaction was purified by HPLC and the two new peaks (Peak A [blue], Peak B [orange], Fig. 4.3) were isolated and analysed by ^{1}H-NMR. The spectra obtained were then compared to the ^{1}H-NMR of the authentic skyllamycin B (**102**) obtained by A/Prof Roger Linington during the isolation of the skyllamycins [17]. As shown in Fig. 4.4, while the spectra were similar, there are a number of significant differences suggesting that neither of the compounds isolated were the skyllamycin B (**102**) natural product. In particular, the chemical shift of the α-proton of the α-OH-Gly residue was significantly different (Fig. 4.4, denoted by *) in the isolated sample compared to the authentic material. It was hypothesised that these

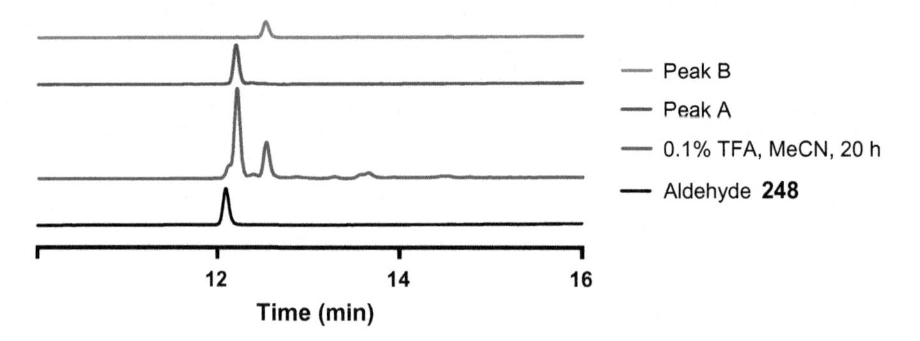

Fig. 4.3 HPLC analysis of the initial cyclisation reaction. All chromatograms: 0–100% MeCN (0.1% formic acid) in H_2O (0.1% formic acid) over 15 min, λ = 280 nm

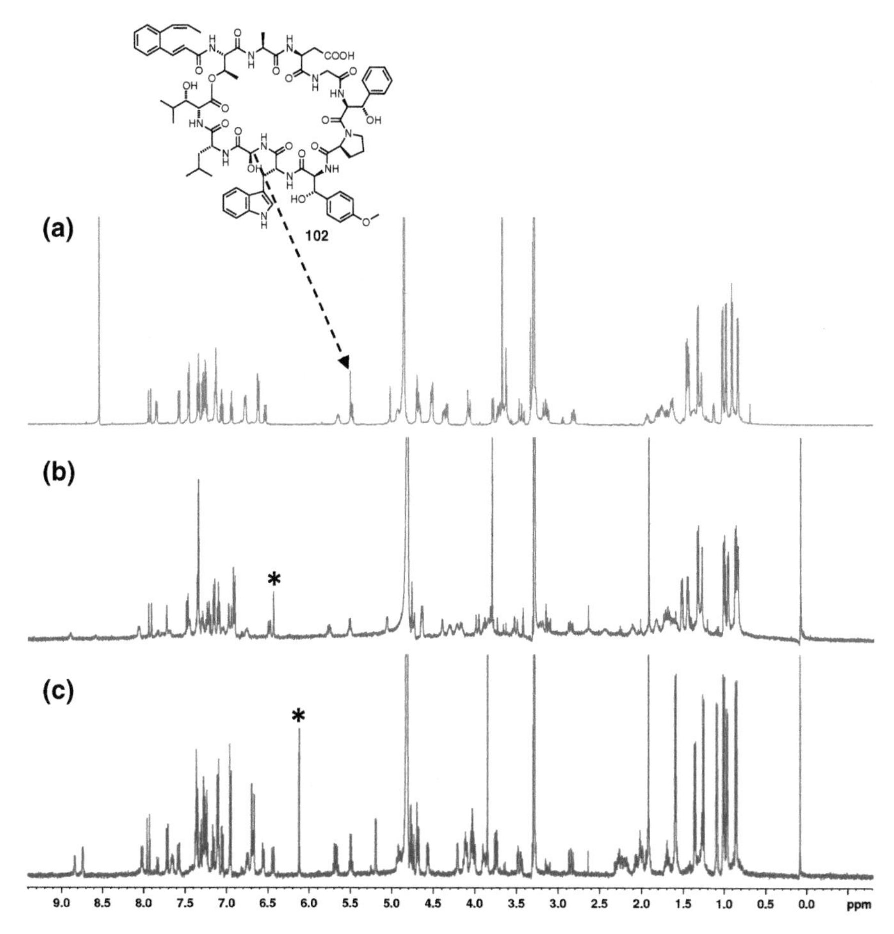

Fig. 4.4 ^{1}H-NMR comparison of authentic skyllamycin B (**102**) **a** (600 MHz, CD_3OD) with Peak B **b** (500 MHz, CD_3OD) and Peak A **c** (500 MHz, CD_3OD) formed in the cyclisation reaction. The α-proton of the α-OH-Gly residue is highlighted in the authentic sample (arrow). * corresponds to the presumed hemiacetal proton in Peaks A and B

compounds may be cyclic hemiacetals, formed by reaction of one of the β-OH groups and the aldehyde, however this was not confirmed. While the result was disappointing, the fact that a cyclisation reaction took place was very encouraging, and so new reaction conditions were interrogated.

4.3.5 Synthesis of Skyllamycin B

To this end, further reaction conditions were trialled, namely incubation of aldehyde **248** in MeCN at 60 °C overnight (Scheme 4.10). Gratifyingly, this resulted in

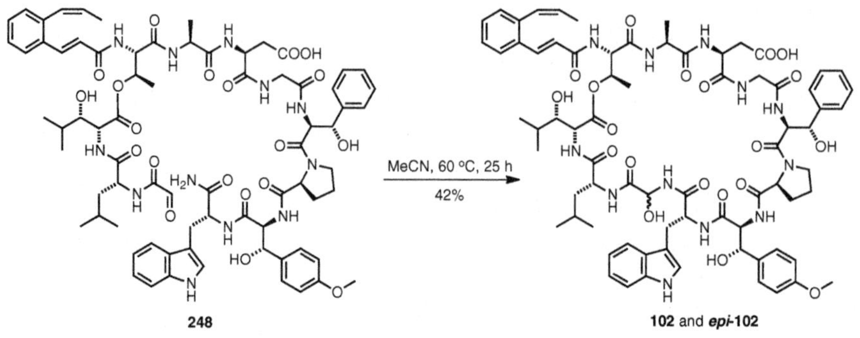

Scheme 4.10 Synthesis of skyllamycin B (**102**) and *epi*-skyllamycin B (***epi*-102**)

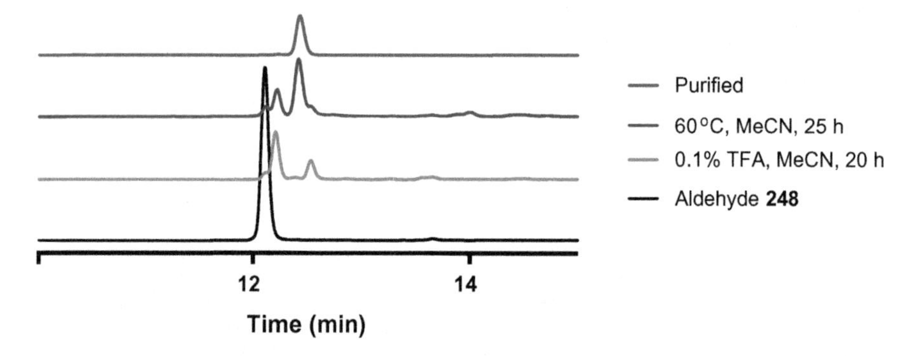

Fig. 4.5 HPLC analysis of the cyclisation reaction of aldehyde **248**. All chromatograms: 0–100% MeCN (0.1% formic acid) in H_2O (0.1% formic acid) over 15 min, λ = 280 nm

the consumption of the starting material and formation of two peaks with the desired mass as observed by HPLC-MS analysis. While one of these peaks was found to be the same as the major compound isolated in the acid-catalysed cyclisation experiment described above (Peak A, Fig. 4.5), the other peak had a distinct retention time. HPLC purification of the reaction mixture resulted in the isolation of the new peak (red, Fig. 4.5) in 42% yield from the cyclisation reaction. Interestingly, the ^{1}H-NMR spectrum of the synthetic material (Fig. 4.6b) revealed the presence of two compounds in a 1:1 ratio, however upon comparison with the NMR of authentic skyllamycin B (**102**) (Fig. 4.6a) significant similarities were observed, particularly in the region containing the key α-proton signals (3.0–6.0 ppm). It was hypothesised that the two compounds present in the sample could be a mixture of skyllamycin B (**102**) and the epimer at the newly formed α-OH-Gly residue.

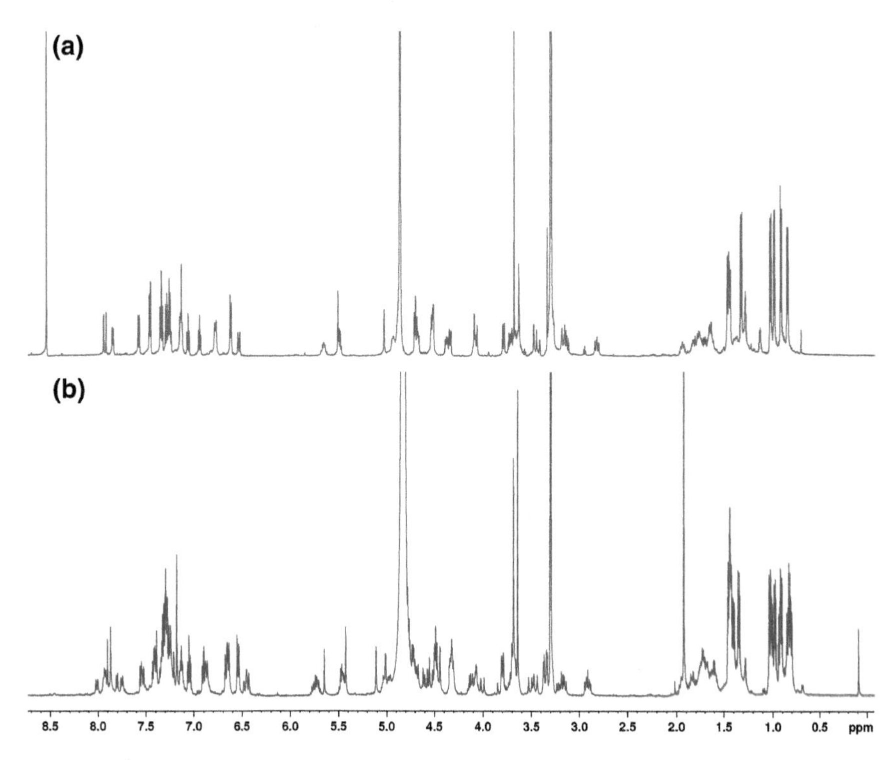

Fig. 4.6 **a** ^{1}H-NMR spectrum of authentic skyllamycin B (**102**) (600 MHz, CD_3OD) **b** ^{1}H-NMR spectrum of synthetic skyllamycin B (**102**)—a 1:1 mixture of two compounds (500 MHz, CD_3OD)

Excited by these results, significant efforts were made to separate the two compounds by HPLC. It was found that some separation could be achieved by using a 300 Å wide pore C18 column. However, unfortunately this only ever achieved partial separation on an analytical scale, and was not amenable to scale up. However, we were able to work with our collaborator, A/Prof Roger Linington at Simon Fraser University (SFU), who has significant expertise in the separation of complex natural products and was responsible for the isolation of the skyllamycins [17]. Fortunately, the group at SFU were able to separate the two compounds by employing an optimised three component solvent system involving a shallow gradient of MeOH in H_2O with constant 20% MeCN as a third solvent and 0.02% formic acid in each buffer. UHPLC-MS co-injection experiments revealed that the compound with the longer retention time (Peak 2) was likely the desired skyllamycin B (**102**). This was isolated in 46% yield from the separation, which translates to an overall yield of 19% yield of skyllamycin B (**102**) from the initial cyclisation reaction. Further structural confirmation and characterisation of the putative synthetic skyllamycin B (**102**) was carried out and is discussed in detail later in this chapter.

4.4 Synthesis of Skyllamycin A and C

With the synthesis of skyllamycin B (**102**) complete, attention turned to the synthesis of the remaining two natural products, skyllamycin A (**101**) and skyllamycin C (**103**). To access these two natural products, the same synthetic approach used in the synthesis of skyllamycin B (**102**) was employed.

4.4.1 Synthesis of Linear Peptides 250 and 252

The synthesis of skyllamycins A (**101**) and C (**103**) began with the construction of the linear peptides **250** and **252** by SPPS. Starting from key resin-bound intermediate **261**, constructed during the synthesis of skyllamycin B (**101**), either the commercially available Fmoc-Asp(Ph*i*Pr)-OH, or the synthesised Fmoc-β-Me-Asp (Ph*i*Pr)-OH (**212**) were coupled under HATU conditions to yield resin-bound intermediates **266** and **267** respectively (Scheme 4.11). Coupling of Fmoc-Ala-OH and Fmoc-Thr-OH under standard PyBOP coupling conditions followed by reaction with either cinnamoyl **130**, or reduced cinnamoyl moieties **131** afforded resin-bound peptides **268** and **269**. Both **268** and **269** were then subjected to on-resin esterification conditions with **166** to yield the branched peptides **270** and **271**.

The next step in the synthesis was the coupling of Fmoc-D-Leu-OH to **271** under the optimised microwave conditions (Scheme 4.12). Pleasingly, HPLC-MS analysis of an analytical scale resin cleavage revealed that this reaction proceeded smoothly. As such, in order to complete the synthesis of the skyllamycin C peptide, Fmoc-Ser-OH was coupled, which afforded the complete resin-bound peptide **272** (Scheme 4.12).

Fmoc-D-Leu-OH then also coupled to resin-bound peptide **270** under optimised microwave conditions (Scheme 4.13). Surprisingly however, when an analytical scale resin cleavage of resin-bound peptide **273** was analysed by HPLC-MS a significant amount of a by-product **274**, with a mass of -18 Da compared to the desired product **275** was observed (Fig. 4.7). The identity of the by-product was hypothesised to be peptide **274** caused by aspartimide formation during peptide elongation. Aspartimides are a well known by-product in SPPS, and they form through repeated Fmoc-deprotection steps [18]. As aspartimide formation had not been observed during the synthesis of skyllamycin B (**101**) it was presumed that the presence of the β-Me group modified the conformation of the peptide backbone and promoted formation of by-product **274** through intramolecular cyclisation.

Indeed it seemed likely that the increased temperature in the microwave coupling, as well as the addition of 8 equivalents of Hünig's base, led to the formation of the aspartimide **274**. In order to overcome this issue, coupling conditions that omitted an additional external base were trialled, namely use of DIC and HOAt as the coupling reagents with microwave heating at 50 °C. Gratifyingly, this resulted

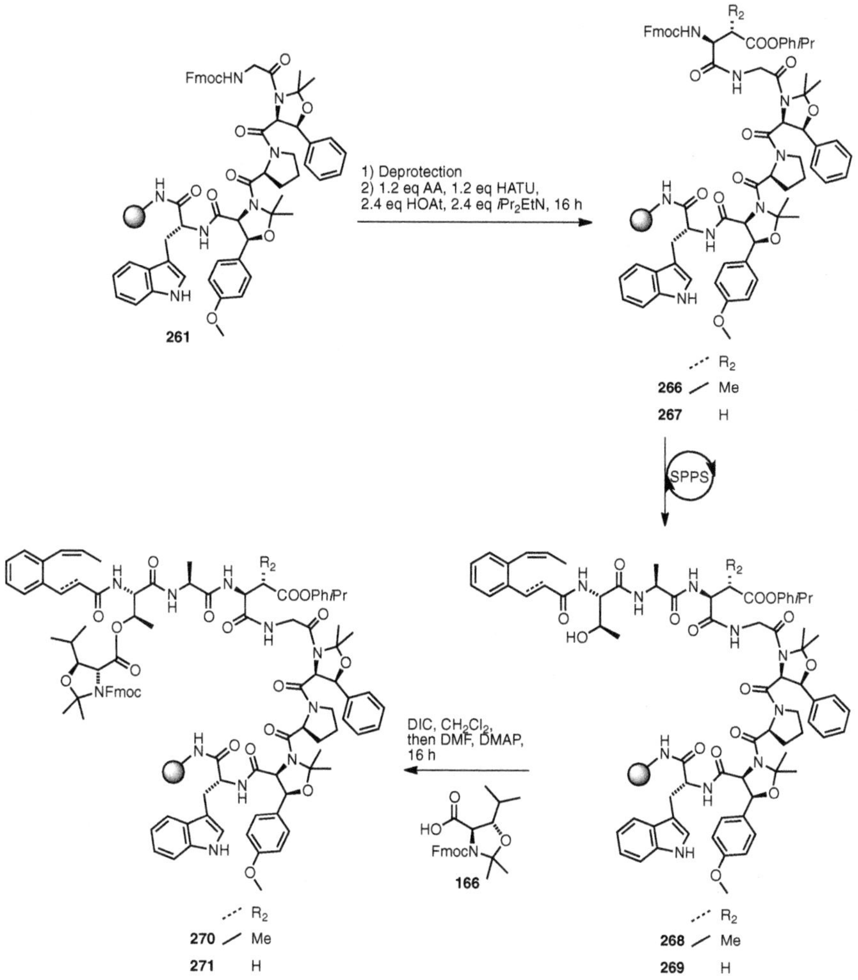

Scheme 4.11 Synthesis of resin-bound branched peptides **270** and **271**

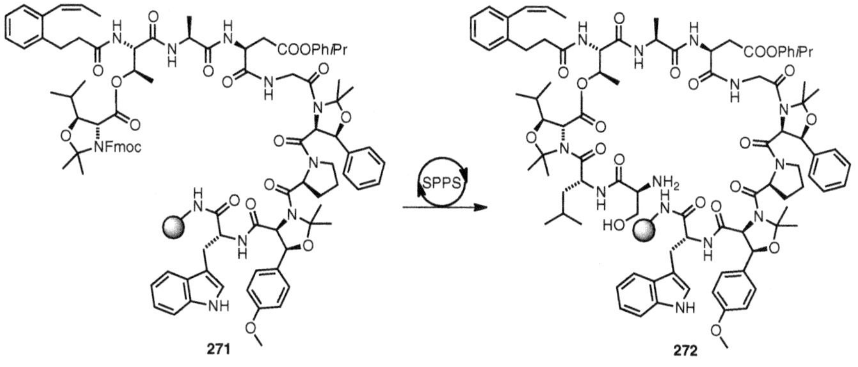

Scheme 4.12 Completion of the synthesis of resin-bound peptide **272**

Scheme 4.13 Attempted microwave assisted coupling of Fmoc-D-Leu-OH to form resin-bound peptide **373**

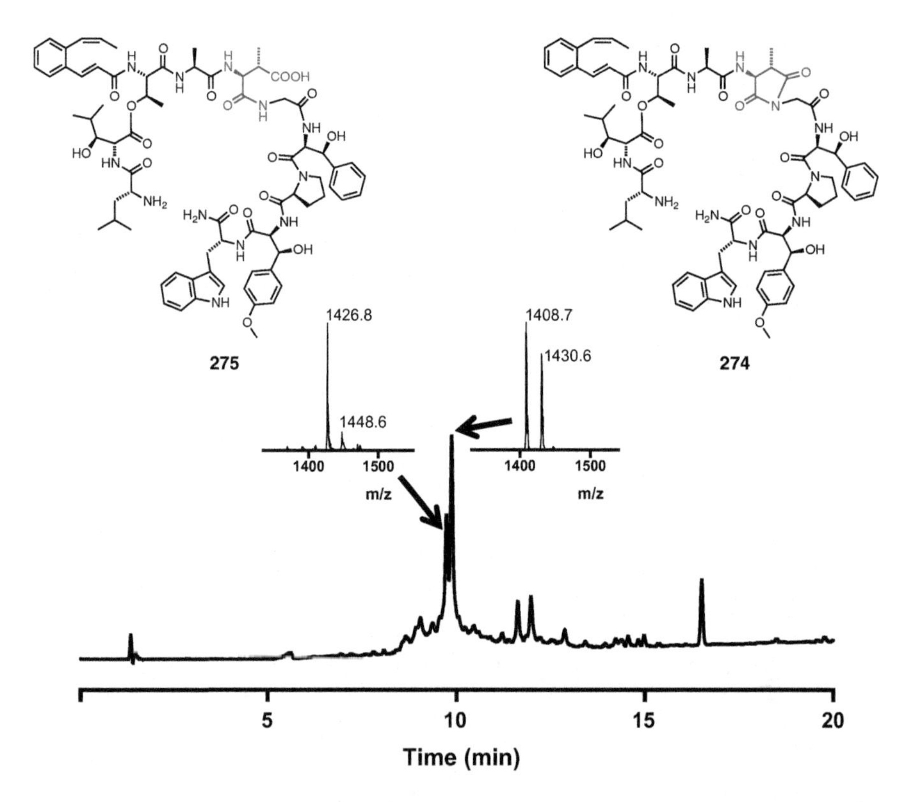

Fig. 4.7 HPLC-MS analysis of an analytical scale resin cleavage of resin-bound peptide **273**. The peak on the left corresponds to the desired product **275** and the peak on the right corresponds to the by-product aspartimide **274**. 0–100% MeCN (0.1% formic acid) in H_2O (0.1% formic acid) over 15 min, $\lambda = 280$ nm. Desired product **275**; *m/z* 1427.7 $(M+H)^+$, 1449.7 $(M+Na)^+$; by-product **274**; *m/z* 1408.7 $(M+H)^+$, 1431.7 $(M+Na)^+$

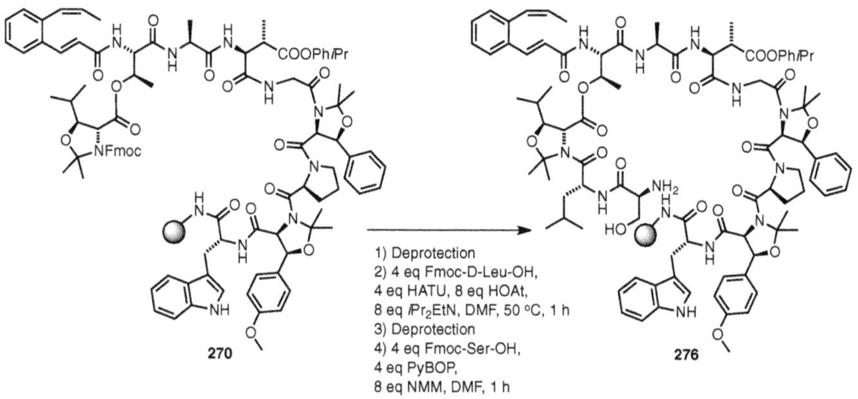

1) Deprotection
2) 4 eq Fmoc-D-Leu-OH, 4 eq HATU, 8 eq HOAt, 8 eq iPr_2EtN, DMF, 50 °C, 1 h
3) Deprotection
4) 4 eq Fmoc-Ser-OH, 4 eq PyBOP, 8 eq NMM, DMF, 1 h

Scheme 4.14 Completion of the synthesis of resin-bound **276** using optimised microwave coupling conditions

in quantitative coupling with no observable aspartimide formation. Final coupling of Fmoc-Ser-OH yielded the completed resin-bound peptide **276** (Scheme 4.14).

4.4.2 Synthesis of Skyllamycin A

Resin-bound peptide **276** was next treated with 1% TFA in CH_2Cl_2 to yield the unprotected linear peptide **250** (Scheme 4.15). The crude peptide was purified by HPLC affording pure linear peptide **250** in 11% yield from the initial resin loading, which corresponds to 92% yield per step. Linear peptide **250** was then treated with $NaIO_4$, in an analogous fashion to the synthesis of aldehyde **248**, to yield aldehyde **247** in 62% yield after HPLC purification. In order to carry out the final cyclisation reaction, aldehyde **247** was incubated at 60 °C in MeCN for 25 h. HPLC-MS analysis of the reaction showed the consumption of starting material and formation of three new peaks with the desired mass of skyllamycin A (**101**). HPLC purification resulted in isolation of the three new peaks, which were sent to the Linington lab for analysis. Peak 3 (pink, Fig. 4.8), isolated in 32% yield, was shown to be the desired synthetic skyllamycin A (**101**), by UHPLC-MS co-injection studies with the authentic isolated natural product (discussed further below).

4.4.3 Synthesis of Skyllamycin C

Skyllamycin C (**103**) was synthesised in the same manner described above for the synthesis of skyllamycins A (**101**) and B (**102**). Resin-bound peptide **272** was cleaved from resin with concomitant deprotection of the Ph*i*Pr group to yield

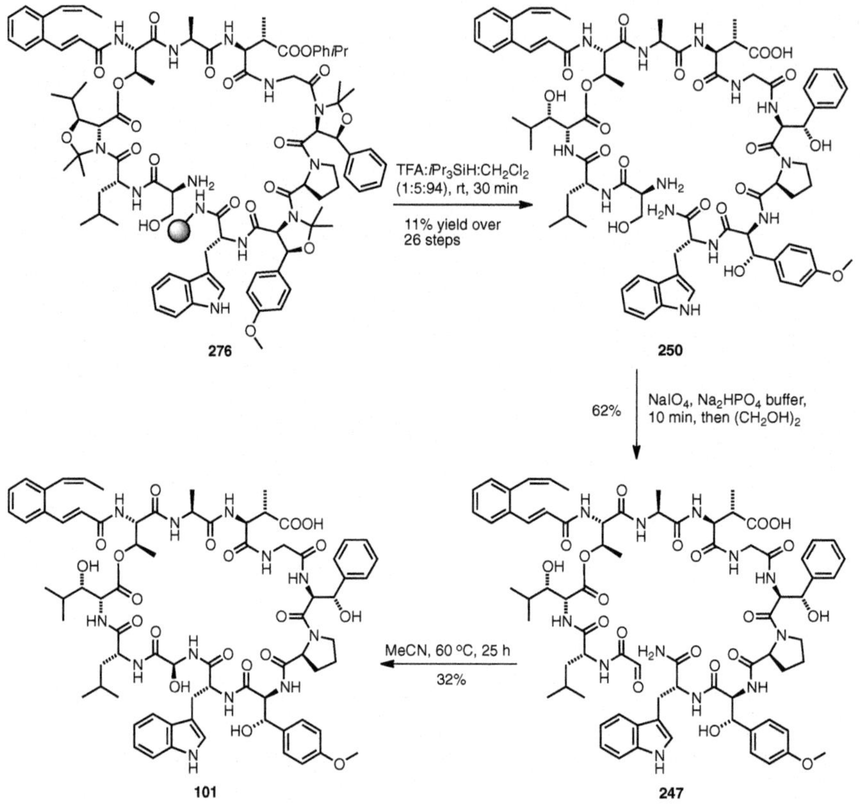

Scheme 4.15 Synthesis of skyllamycin A (**101**)

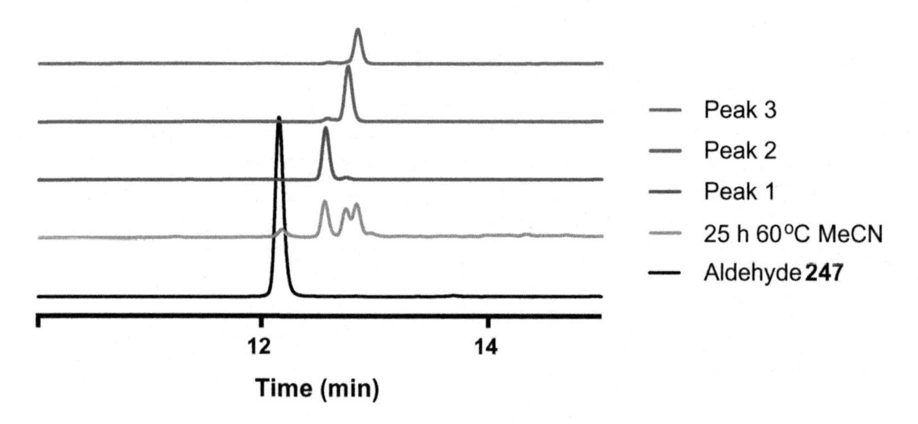

Fig. 4.8 HPLC analysis of the cyclisation reaction of aldehyde **247**. All chromatograms: 0–100% MeCN (0.1% formic acid) in H_2O (0.1% formic acid) over 15 min, λ = 280 nm

unprotected linear peptide **252** (Scheme 4.16). HPLC purification afforded the pure linear peptide **252** in 12% yield from the initial resin loading, corresponding to 92% per step. Linear peptide **252** was subject to oxidative cleavage to yield aldehyde **249** which was purified by HPLC to afford aldehyde **249** in 68% yield. The cyclisation reaction was carried out by incubating aldehyde **249** was in MeCN at 60 °C for 25 h. Interestingly, after the consumption of the starting material, HPLC-MS analysis revealed two peaks with the desired mass of skyllamycin C (**103**), with the latter peak shown to be a mixture of two compounds by ^{1}H-NMR spectroscopy (Peak 2, Fig. 4.9). HPLC purification afforded two separated peaks, with the second peak not separated, but sent to the Linington lab for further analysis. Peak 1 (red, Fig. 4.9), isolated in 33% yield, was shown to be the desired synthetic skyllamycin C (**103**), by UHPLC-MS co-injection studies with the authentic isolated natural product (discussed further below).

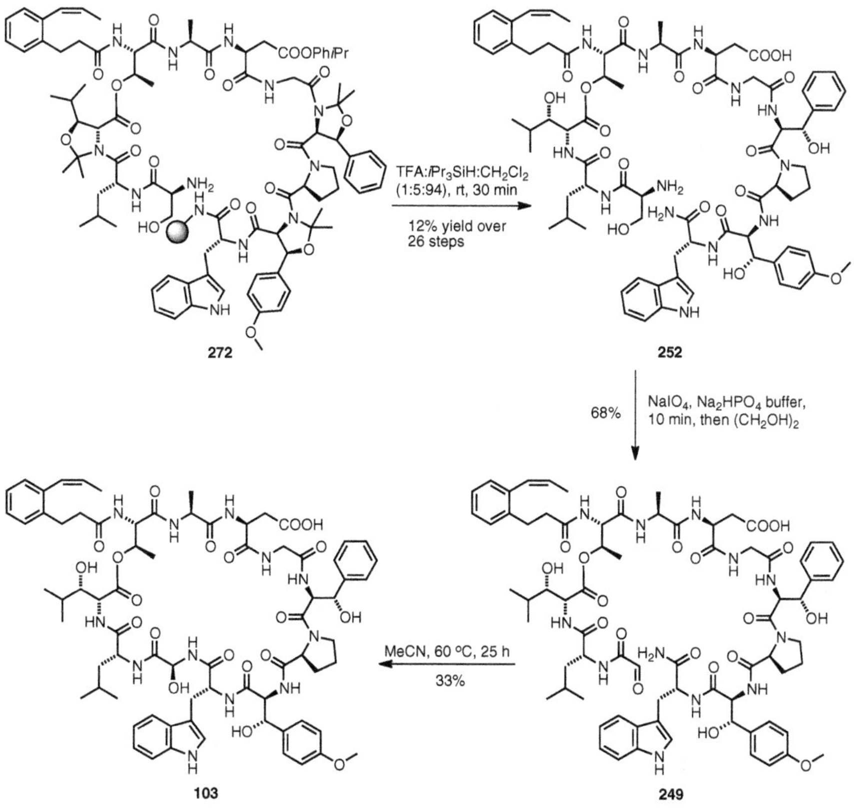

TFA:iPr$_3$SiH:CH$_2$Cl$_2$ (1:5:94), rt, 30 min; 12% yield over 26 steps

NaIO$_4$, Na$_2$HPO$_4$ buffer, 10 min, then (CH$_2$OH)$_2$; 68%

MeCN, 60 °C, 25 h; 33%

272, 252, 249, 103

Scheme 4.16 Synthesis of skyllamycin C (**103**)

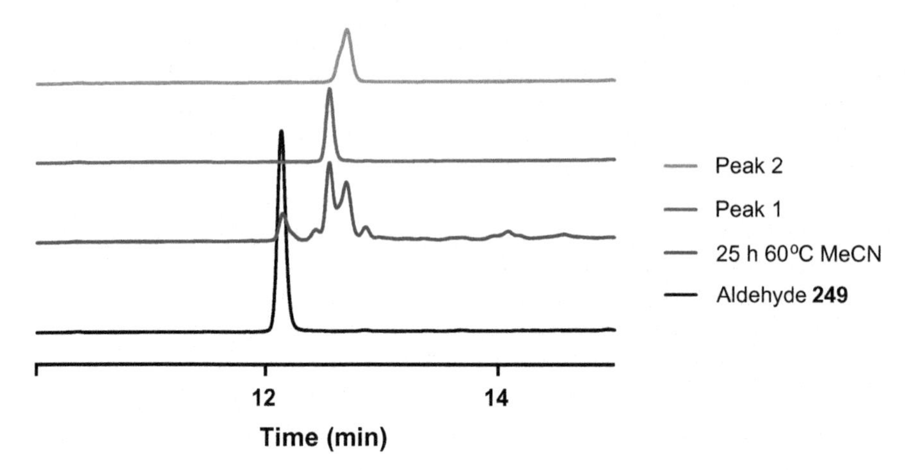

Fig. 4.9 HPLC analysis of the cyclisation reaction of aldehyde **249**. All chromatograms: 0–100% MeCN (0.1% formic acid) in H_2O (0.1% formic acid) over 15 min, λ = 280 nm

4.5 Structural Analysis of Synthetic Skyllamycins A–C

The products formed in the cyclisation reactions described above were sent to the Linington lab for comprehensive structure analysis. The Linington lab previously isolated skyllamycins A–C (**101**–**103**) [17] and, as such, they were able to carry out extensive comparisons between the synthetic samples and the authentic natural products. A number of structural confirmations were carried out, the first being the co-injection studies between the synthetic and authentic isolated natural products using UHPLC-MS. Comprehensive NMR analysis was then carried out on the synthetic material as well as isolated natural skyllamycins. This data, obtained by the Linington lab is presented below.

4.5.1 Skyllamycin A

After the synthesis of skyllamycin A (**101**), three individual compounds were purified and sent to the Linington lab for analysis (Fig. 4.8). The co-injection study carried out with an authentic sample of skyllamycin A (**101**) revealed that peak 3 (pink, Fig. 4.8) was the desired natural product (Fig. 4.10). Both the synthetic and isolated skyllamycin A (**101**) also exhibited comparable circular dichroism (CD) spectra, which confirmed that the synthetic and authentic materials are the same enantiomer (Fig. 4.11).

Synthetic skyllamycin A (**101**) was then analysed extensively by NMR spectroscopy. To begin, the ^{1}H-NMR spectrum of synthetic skyllamycin A (**101**) was obtained and shown to be an extremely good match for the authentic sample

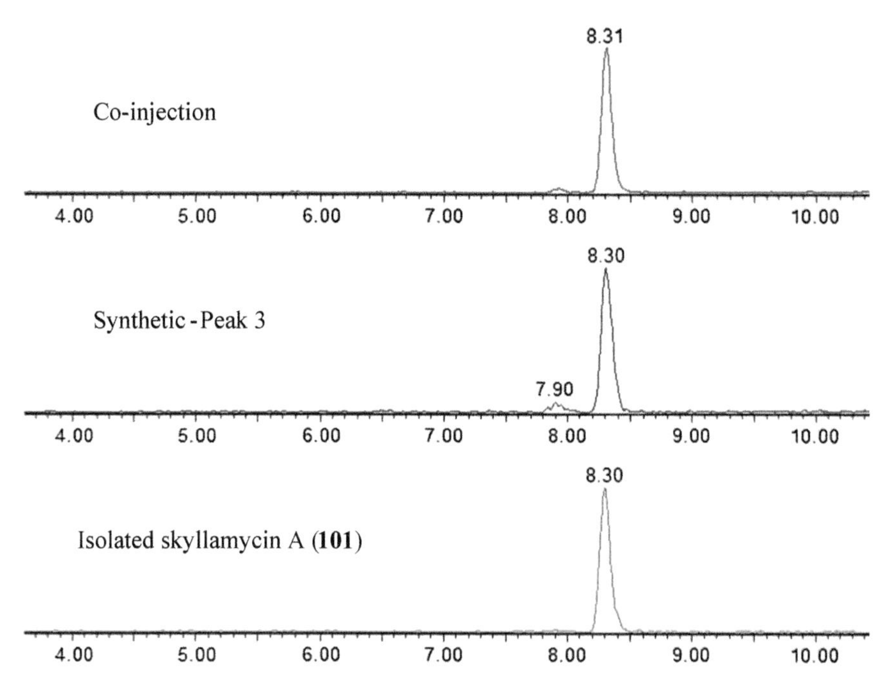

Fig. 4.10 UHPLC-MS co-injection study of synthetic and authentic skyllamycin A (**101**). EIC for 1483.65 $[M+H]^+$ with 50 ppm window

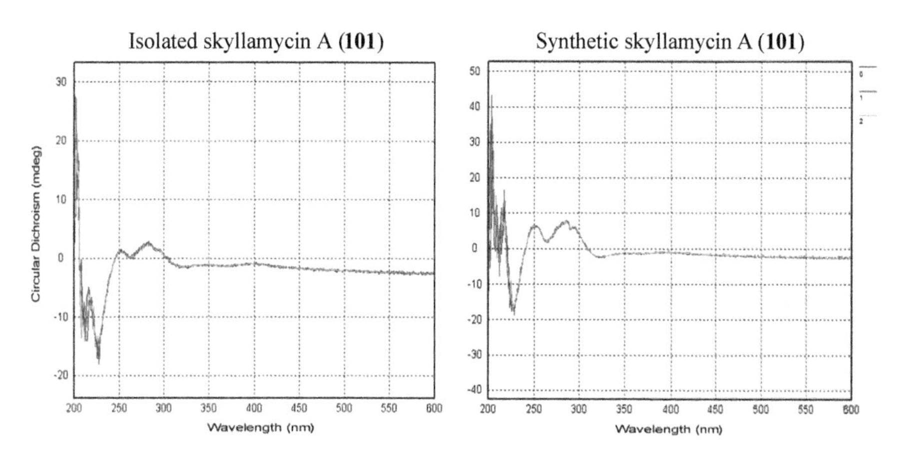

Fig. 4.11 CD spectra of isolated and synthetic skyllamycin A (**101**). Both spectra are an average of three runs

(Fig. 4.12). In particular, the key resonances between 3.0 and 6.0 ppm, which account for the α-protons, as well as many of the β-protons in the modified amino acids, showed very strong overlap. Of particular significance is the presence of the singlet at 5.4 ppm, corresponding to the key α-OH-Gly residue, in both the

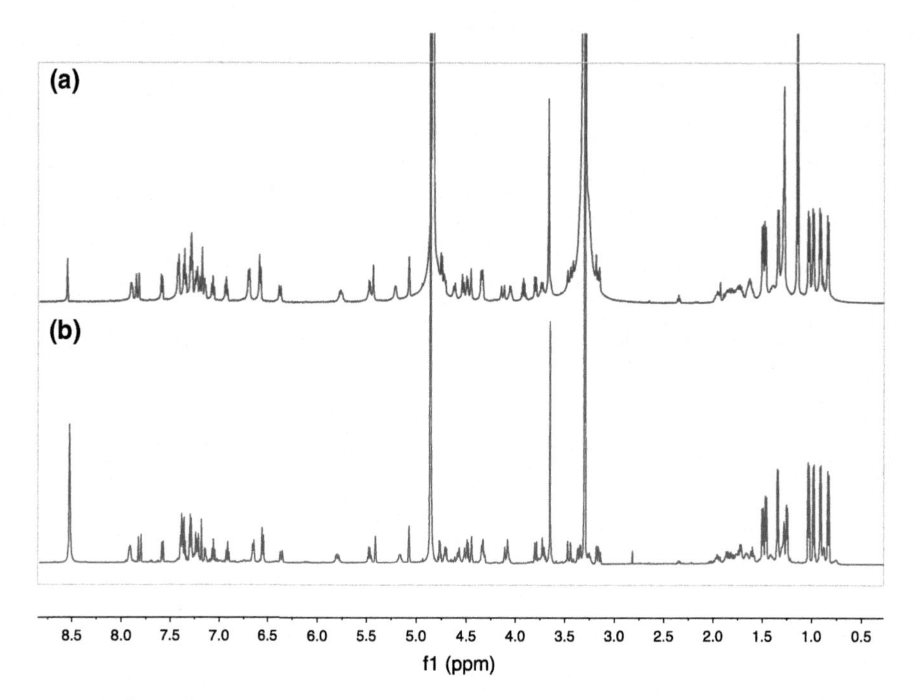

Fig. 4.12 ^{1}H-NMR comparison (600 MHz, CD_3OD) of synthetic skyllamycin A (**101**) (**a**, blue) and authentic skyllamycin A (**101**) (**b**, red)

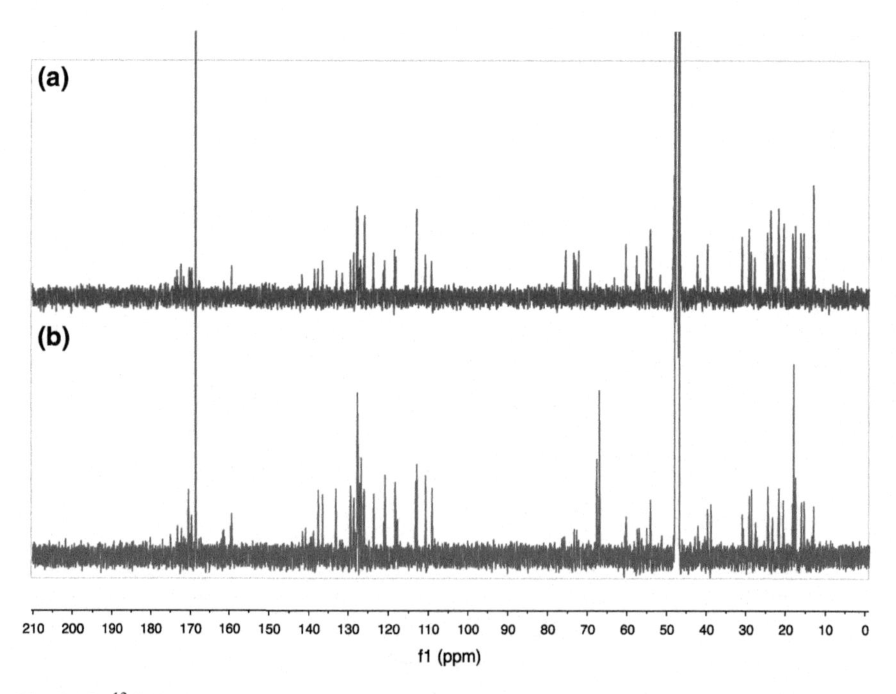

Fig. 4.13 ^{13}C-NMR comparison (150 MHz, CD_3OD) of synthetic skyllamycin A (**101**) (**b**, blue) and authentic skyllamycin A (**101**) (**b**, red)

synthetic and isolated samples. The large doublet at 1.2 ppm and multiplet at 3.8 ppm in the synthetic ^{1}H-NMR spectrum (Fig. 4.12a) are due to an impurity. The ^{13}C-NMR spectra of the authentic and synthetic material also show very strong similarities (Fig. 4.13). It is worth noting that the ^{13}C-spectra obtained for both synthetic and authentic skyllamycin A (**101**) suffer from low signal to noise ratio. Furthermore, the authentic material (Fig. 4.13b) contains a number of impurities, most notably the two larger resonances at 65 ppm. However, these impurities are also present in the ^{13}C-NMR spectra of isolated skyllamycins B (**102**) and C (**103**). Nonetheless, the α-carbon resonances (between 50 and 80 ppm) as well as the carbonyl resonances (between 150 and 180 ppm) show good overlap. Taken together the NMR data, UHPLC-MS co-injections as well as the ORD data provide extremely strong evidence that synthetic skyllamycin A (**101**) is identical to the authentic isolated material.

4.5.2 Skyllamycin B

In the case of skyllamycin B (**102**), the Linington lab carried out separation of the presumed mixture of epimers. After this separation was carried out the UHPLC-MS co-injection study (Fig. 4.14) showed that later eluting compound (Peak 2) was

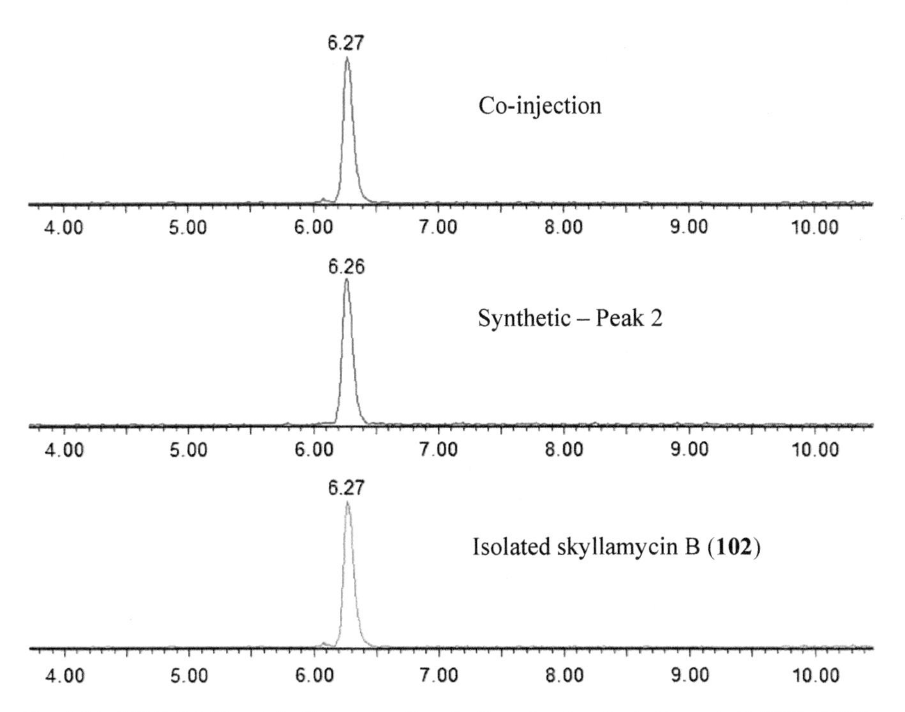

Fig. 4.14 UHPLC-MS co-injection study of synthetic and authentic skyllamycin B (**102**). EIC for m/z 1469.66 $[M+H]^{+}$ with 50 ppm window

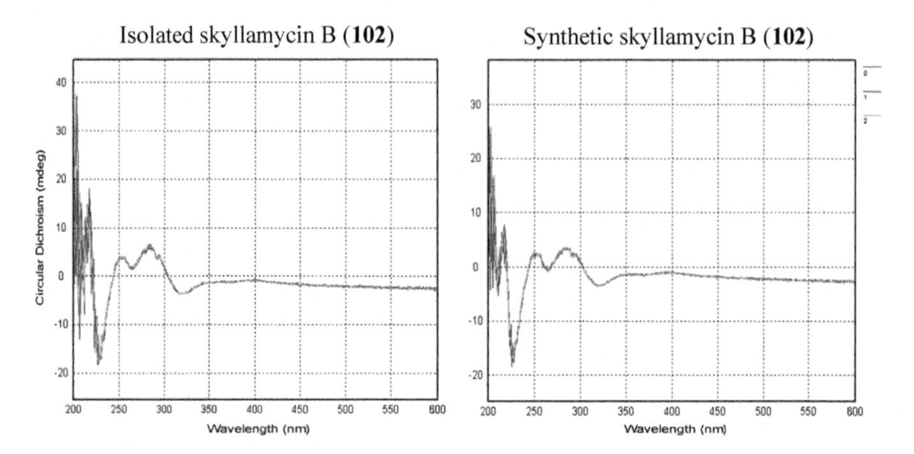

Fig. 4.15 CD spectra of isolated and synthetic skyllamycin B. Both spectra are an average of three runs

indeed identical to the natural product. Both the synthetic and isolated skyllamycin B (**101**) also exhibited comparable CD spectra, which confirmed that the synthetic and authentic materials are the same enantiomer (Fig. 4.15).

Synthetic skyllamycin B (**102**) was then analysed by NMR spectroscopy. The ^{1}H-NMR spectrum of synthetic skyllamycin B (**102**) showed strong overlap with the authentic sample (Fig. 4.16). As was the case with skyllamycin A (**101**), the key resonances between 3.0 and 6.0 ppm, which account for the α-protons, as well as many of the β-protons present in the modified amino acids, showed very strong overlap. In particular the singlet at 5.4 ppm, which corresponds to the α-OH-Gly, is identical to that in the isolated material, verifying the presence of the key α-OH-Gly residue. The ^{13}C-NMR spectra of the authentic and synthetic material were also nearly identical (Fig. 4.17). The authentic material (Fig. 4.17b) contains a number of impurities, most notably the two resonances at 65 ppm. Nonetheless, the α-carbon resonances (between 50 and 80 ppm) as well as the carbonyl resonances (between 150 and 180 ppm) show good overlap. Furthermore, there was strong overlap between the signals between 10 and 30 ppm, which correspond to the aliphatic carbons present in skyllamycin B (**102**). When considered together, all the analytical data, namely the NMR data, UHPLC-MS co-injections as well as the ORD data provide extremely strong evidence that synthetic skyllamycin B (**102**) is identical to the authentic isolated material and that the natural product had been successfully synthesised.

4.5.3 *Skyllamycin C*

Finally, in the case of skyllamycin C (**103**) two separated peaks were sent to the Linington group for analysis. Pleasingly, peak 1 proved to be a positive match in

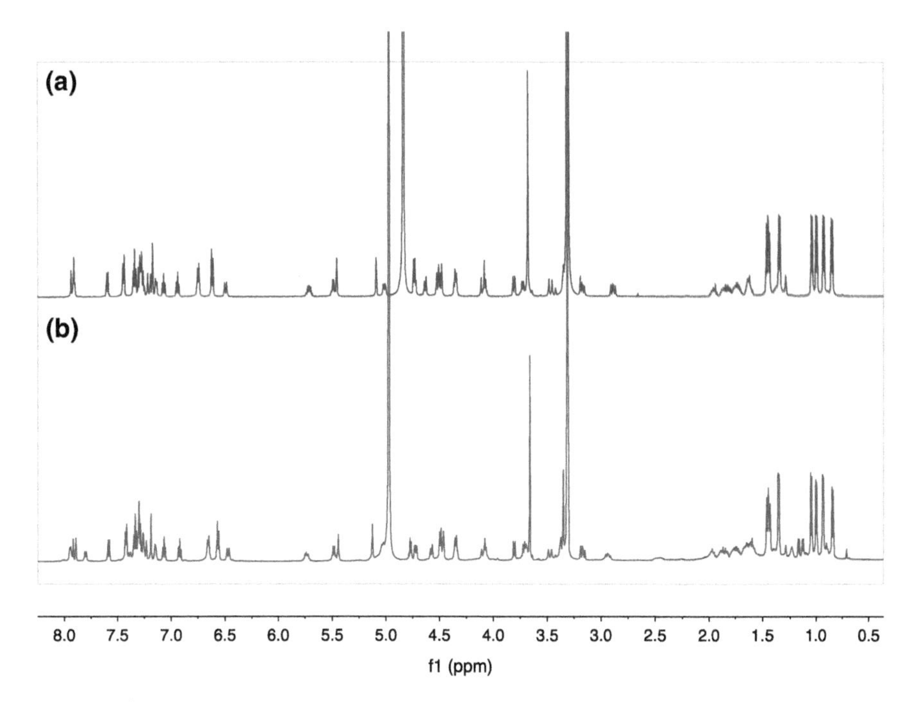

Fig. 4.16 ^{1}H-NMR comparison (600 MHz, CD_3OD) of synthetic skyllamycin B (**102**) (**a**, blue) and authentic skyllamycin B (**102**) (**b**, red)

the UHPLC-MS co-injection study (Fig. 4.18). Furthermore, both the synthetic and isolated skyllamycin C (**103**) exhibited comparable CD spectra, which confirmed that the synthetic and authentic materials are likely the same enantiomer (Fig. 4.19).

Synthetic skyllamycin C (**103**) was then analysed extensively by NMR spectroscopy. The ^{1}H-NMR spectrum of synthetic skyllamycin C (**103**) was very similar to that for the authentic isolated natural product sample (Fig. 4.20). The authentic isolated skyllamycin C spectrum (Fig. 4.20b) contains an impurity at 1.1 ppm, however the rest of this region (0.8–1.5 ppm), containing the aliphatic methyl resonances, shows very strong overlap. Furthermore, the resonances between 3.0 and 6.0 ppm, which account for the α-protons, as well as many of the β-protons present in the modified amino acids, show very strong overlap. Both the synthetic and isolated material possess a singlet at 5.6 ppm, which is the signal from the α-OH-Gly residue, confirming the presence of this key residue. The ^{13}C-NMR spectra of the authentic and synthetic material also show very strong similarities (Fig. 4.21). The synthetic skyllamycin C (**103**) ^{13}C-spectrum suffers from a low signal to noise ratio, and authentic material (Fig. 4.21b) contains a number of impurities, most notably the two resonances at 67 ppm. Nonetheless, the α-carbon

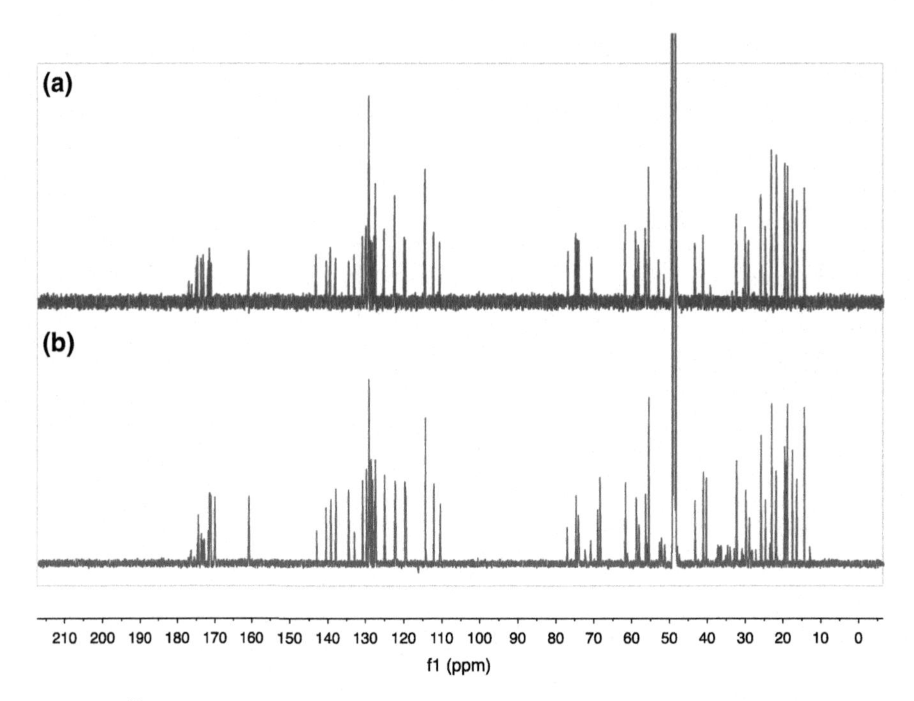

Fig. 4.17 ^{13}C-NMR comparison (150 MHz, CD_3OD) of synthetic skyllamycin B (**102**) (**a**, blue) and authentic skyllamycin B (**102**) (**b**, red)

resonances (between 50 and 80 ppm) as well as the carbonyl resonances (between 150 and 180 ppm) are in very good agreement. Overall, the NMR data, UHPLC-MS co-injections as well as the ORD data provide extremely strong evidence that synthetic skyllamycin C (**103**) is identical to the authentic isolated material and that skyllamycin C (**103**) was successfully synthesised.

In summary, the synthetic skyllamycin samples were thoroughly characterised by NMR, as well as being subject to comparison with the authentic natural products via a number of other analytical techniques including UHPLC-MS co-injections and ORD spectroscopy. Importantly, when compared, all analytical data obtained for both the synthetic and isolated natural products were consistent. Combined with the fact that the isolated skyllamycins are biosynthesised via a single pathway, and that the synthetic skyllamycins were accessed through a divergent SPPS strategy, the analytical data presented above provides extremely strong evidence suggesting that synthetic skyllamycins A–C (**101**–**103**) were identical to the isolated natural products.

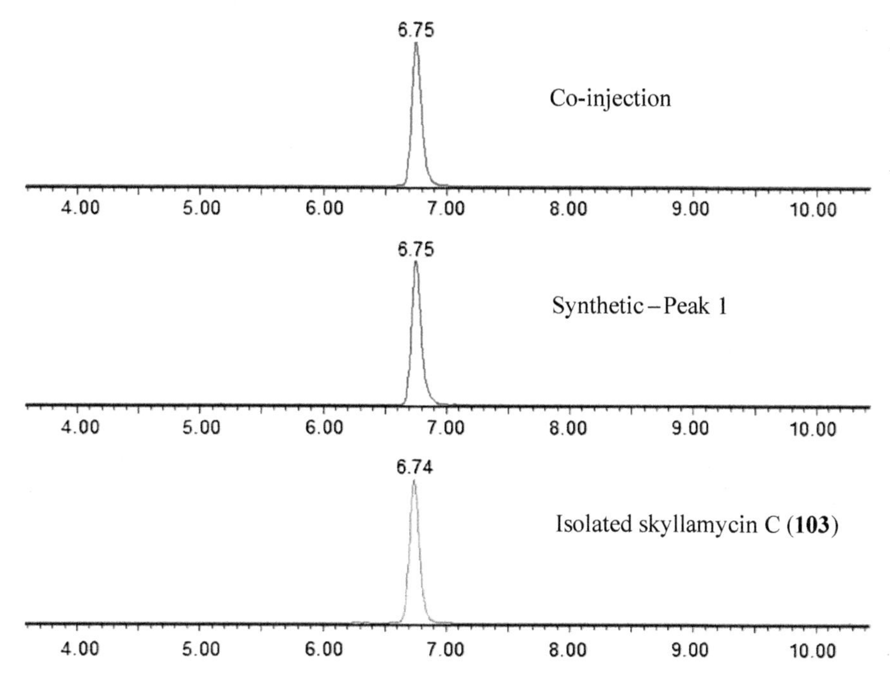

Fig. 4.18 UHPLC-MS co-injection study of synthetic and authentic skyllamycin C (**103**). EIC for m/z 1471.66 [M+H] with 50 ppm window

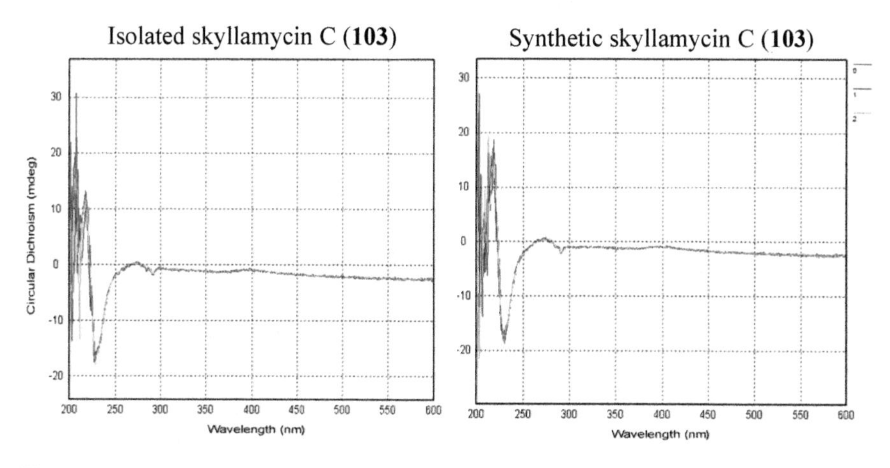

Fig. 4.19 CD spectra of isolated and synthetic skyllamycin C. Both spectra are an average of three runs

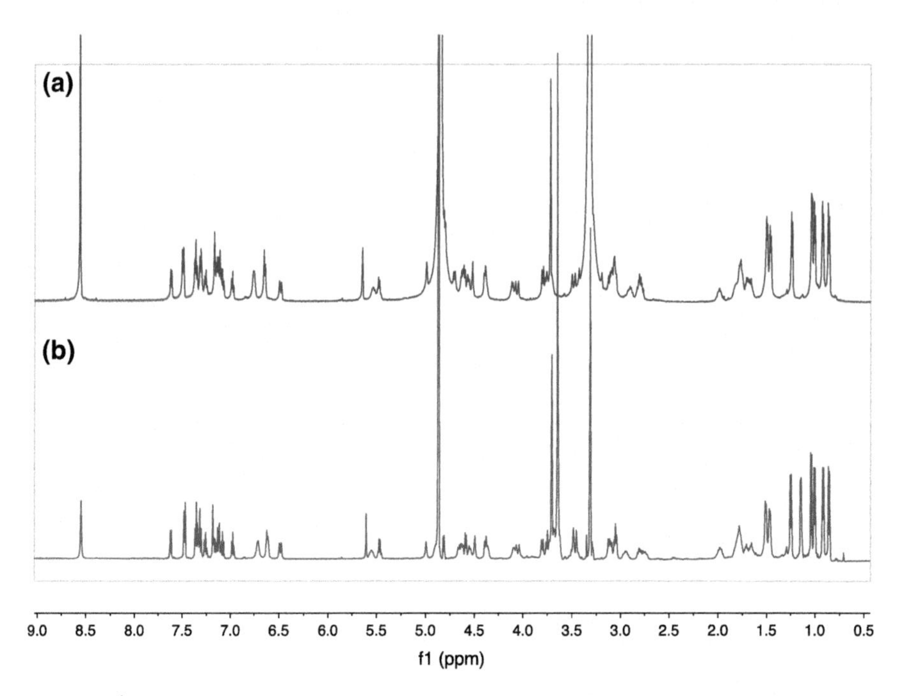

Fig. 4.20 ^{1}H-NMR comparison (600 MHz, CD_3OD) of synthetic skyllamycin C (**103**) (**a**, blue) and authentic skyllamycin C (**103**) (**b**, red)

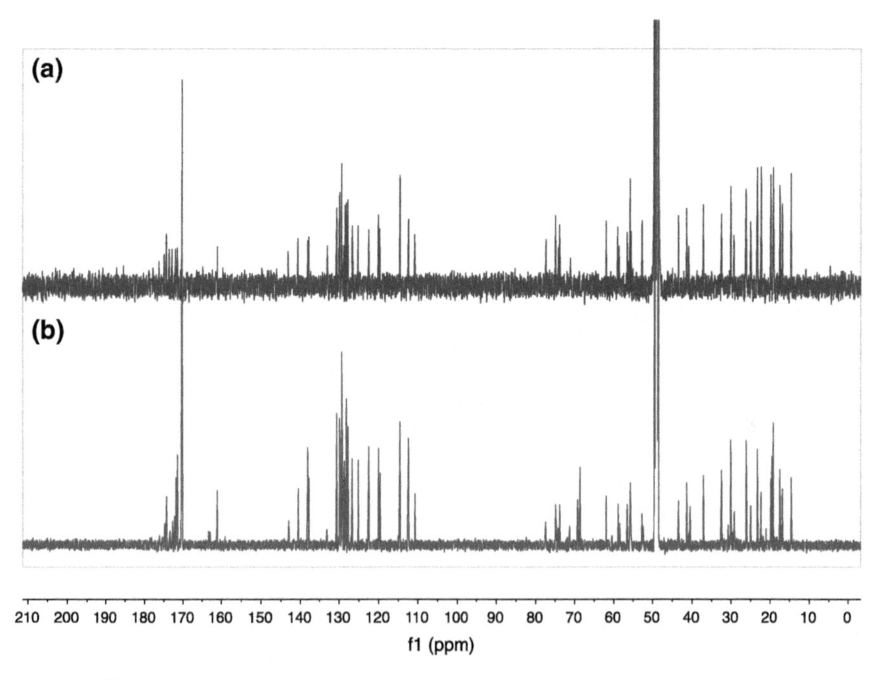

Fig. 4.21 ^{13}C-NMR comparison (150 MHz, CD_3OD) of synthetic skyllamycin C (**103**) (**a**, blue) and authentic skyllamycin C (**103**) (**b**, red)

4.6 Biological Activity of Skyllamycin A–C

With the identity of synthetic skyllamycins A–C (**101–103**) comprehensively confirmed, they were next analysed in the *P. aeruginosa* biofilm inhibition assay alongside the isolated skyllamycin natural product samples. Pleasingly the synthetic natural products showed comparable activity to the isolated material (activity at 250–400 μM). Unfortunately, in this instance both the synthetic and isolated skyllamycins A–C (**101–103**) were not as potent as during the initial isolation [17]. It is possible that this was due to differences in compound formulation prior to the analysis, as new equipment was utilised during this screen. However, these results further substantiate the conclusions drawn from the analytical data, namely that synthetic skyllamycins A–C (**101–103**) are identical to the authentic isolated material.

4.7 Conclusions and Future Directions

This chapter describes the successful total synthesis of skyllamycins A–C (**101–103**). To date there has been no reported total synthesis of these interesting natural products. The synthesis was enabled by a number of critical studies described in the previous chapter, namely the efficient synthesis of the modified amino acid building blocks present in the skyllamycins. An efficient SPPS protocol was developed utilising Sieber amide resin to afford a *C*-terminal carboxamide upon resin cleavage. With the three linear peptides **250–252** in hand a chemoselective oxidative cleavage reaction was employed to yield the three linear peptide aldehydes **247–249**.

The synthesis of the three natural products was enabled by a novel cyclisation with concomitant formation of the rare α-OH-Gly moiety in the final step. This key transformation was optimised after the initial conditions trialled did not afford the natural product. It was determined that incubation of peptide aldehydes **247–249** in MeCN resulted in the formation of the skyllamycin natural products. With the natural products in hand, we worked with our collaborators, the Linington group, (who also reported the isolation of the natural products) who performed a comprehensive structural characterisation of the three synthetic natural products. The combined analytical data including comprehensive NMR analysis, suggested that the synthetic skyllamycins were identical to the isolated natural products. Furthermore the synthetic skyllamycins (**101–103**) showed comparable activity with the authentic material in the *P. aeruginosa* biofilm inhibition screen.

This work lays the foundation for comprehensive studies into the mechanism of action of this unique natural product, as well as the generation of natural analogues with improved biological activity. Future synthetic studies will involve the optimisation of the cyclisation reaction with different solvents and temperatures trialled.

References

1. T. Takeuchi, H. Iinuma, S. Kunimoto, T. Masuda, M. Ishizuka, M. Takeuchi, M. Hamada, H. Naganawa, S. Kondo, H. Umezawa, J. Antibiot. **34**, 1619–1621 (1981)
2. H. Umezawa, S. Kondo, H. Iinuma, S. Kunimoto, Y. Ikeda, H. Iwasawa, D. Ikeda, T. Takeuchi, J. Antibiot. **34**, 1622–1624 (1981)
3. G. Apitz, M. Jäger, S. Jaroch, M. Kratzel, L. Schäffeler, W. Steglich, Tetrahedron **49**, 8223–8232 (1993)
4. G. Apitz, W. Steglich, Tetrahedron Lett. **32**, 3163–3166 (1991)
5. T. Bretschneider, W. Miltz, P. Münster, W. Steglich, Tetrahedron **44**, 5403–5414 (1988)
6. S. Kondo, H. Iwasawa, D. Ikeda, Y. Umeda, Y. Ikeda, H. Iinuma, H. Umezawa, J. Antibiot. **34**, 1625–1627 (1981)
7. Y. Umeda, M. Moriguchi, K. Ikai, H. Kuroda, T. Nakamura, A. Fujii, T. Takeuchi, H. Umezawa, J. Antibiot. **40**, 1316–1324 (1987)
8. P. Durand, P. Richard, P. Renaut, J. Org. Chem. **63**, 9723–9727 (1998)
9. L. Lebreton, J. Annat, P. Derrepas, P. Dutartre, P. Renaut, J. Med. Chem. **42**, 277–290 (1999)
10. V. Schubert, F. Di Meo, P.L. Saaidi, S. Bartoschek, H.P. Fiedler, P. Trouillas, R.D. Süssmuth, Chem. Eur. J. **20**, 4948–4955 (2014)
11. P. Sieber, Tetrahedron Lett. **28**, 2107–2110 (1987)
12. B.H. Nicolet, L.A. Shinn, J. Am. Chem. Soc. **61**, 1615 (1939). 1615
13. H.B.F. Dixon, R. Fields, Methods Enzymol. **25**, 409–419 (1972)
14. H.B.F. Dixon, L.R. Weitkamp, Biochem. J. **84**, 462 (1962)
15. K.F. Geoghegan, J.G. Stroh, Bioconjugate Chem. **3**, 138–146 (1992)
16. O. El-Mahdi, O. Melnyk, Bioconjugate Chem. **24**, 735–765 (2013)
17. G. Navarro, A.T. Cheng, K.C. Peach, W.M. Bray, V.S. Bernan, F.H. Yildiz, R.G. Linington, Antimicrob. Agents Chemother. **58**, 1092–1099 (2014)
18. R. Subirós-Funosas, A. El-Faham, F. Albericio, Tetrahedron **67**, 8595–8606 (2011)

Chapter 5
Experimental

5.1 General Methods and Materials

All reactions were carried out in dried glassware under an argon atmosphere and at room temperature (22 °C) unless aqueous conditions were used or unless otherwise specified. Reactions undertaken at −78 °C utilized a bath of dry ice and acetone. Reactions carried out at 0 °C employed a bath of water and ice. Anhydrous THF, CH_2Cl_2, MeCN, DMF, toluene and MeOH were obtained using a PureSolv® solvent purification system with water detectable only in low ppm levels. Reactions were monitored by thin layer chromatography (TLC) on aluminium backed silica plates (Merck Silica Gel 60 F254). Visualisation of TLC plates was undertaken with an ultraviolet (UV) light at λ = 254 nm and staining with solutions of vanillin, ninhydrin, phosphomolybdic acid (PMA), potassium permanganate or sulfuric acid, followed by exposure of the stained plates to heat. Silica flash column chromatography (Merck Silica Gel 60 40–63 μm) was undertaken to purify crude reaction mixtures using solvents as specified.

All commercially available reagents were used as obtained from Sigma-Aldrich, Merck or Acros Organics. Amino acids, coupling reagents and resins were obtained from NovaBiochem or GL Biochem and peptide synthesis grade DMF was obtained from Merck or Labscan. All non-commercially available reagents were synthesized according to literature procedures as referenced. Microwave assisted peptide couplings were carried out using a Biotage Initiator$^+$ Alstra microwave peptide synthesiser equipped with an inert gas manifold. Fmoc-strategy solid-phase peptide synthesis (Fmoc-SPPS) procedures were employed using HMPB functionalised polyethylene glycol resin (HMPB-NovaPEG), 2-CTC functionalised polystyrene resin or Sieber amide functionalised resin within fritted syringes (purchased from Torviq). All reagent equivalents are in regard to the amount of amino acid loaded to resin.

^{1}H NMR spectra were obtained using a Bruker DRX 400 or DRX 500 at frequencies of 400 MHz or 500 MHz respectively in $CDCl_3$, acetone-d_6, CD_3OD

A. Giltrap, *Total Synthesis of Natural Products with Antimicrobial Activity*,
Springer Theses, https://doi.org/10.1007/978-981-10-8806-3_5

or DMSO-d_6. Chemical shifts are reported in parts per million (ppm) and coupling constants in Hertz (Hz). The residual solvent peaks were used as internal standards without the use of tetramethylsilane (TMS). ^{1}H NMR data is reported as follows: chemical shift values (ppm), multiplicity (s = singlet, d = doublet, t = triplet, q = quartet, m = multiplet, br. = broad, ap. = apparent), coupling constant(s) and relative integral. ^{13}C NMR spectra were obtained using a Bruker DRX 400 or DRX 500 at 100 MHz or 125 MHz in $CDCl_3$, MeOD, acetone-d_6 or DMSO-d_6 unless otherwise specified. ^{13}C NMR data is reported as chemical shift values (ppm). Any rotamers were confirmed by saturation transfer experiments, or the presence of in-phase cross-peaks in a NOESY spectrum. Low-resolution mass spectra for novel compounds were recorded on a Bruker amaZon SL mass spectrometer (ESI) operating in positive mode or on a Shimadzu 2020 (ESI) mass spectrometer operating in positive mode. High resolution mass spectra were recorded on a Bruker-Daltronics Apex Ultra 7.0T Fourier transform (FTICR) mass spectrometer. Circular dichroism spectra were recorded on a Chirascan qCD from 600 nm to 200 nm at a 1 nm resolution at a scan rate of 0.5 scans/s.

LC-MS was performed either on a Shimadzu 2020 LC-MS instrument with an LC-M20A pump, SPD-20A UV/Vis detector and a Shimadzu 2020 (ESI) mass spectrometer operating in positive mode or on a Shimadzu UHPLC-MS equipped with the same modules as above but with an SPD-M30A diode array detector. Separations on the LC-MS system were performed on a Waters Sunfire 5 µm, 2.1 × 150 mm (C18) column. On the UHPLC-MS system, separations were performed on a Waters Acquity 1.7 µm, 2.1 × 50 mm (C18) column. These separations were performed using a mobile phase of 0.1 vol.% formic acid in water (Solvent A) and 0.1 vol.% formic acid in MeCN (Solvent B) using linear gradients. Preparative reverse-phase HPLC was performed using a Waters 500 pump with a 2996 photodiode array detector and a Waters 600 Multisolvent Delivery System.

5.2 Fmoc-SPPS General Protocols

Fmoc Deprotection

A given resin-bound peptide was washed with CH_2Cl_2 (×5) and DMF (×5) before being treated with a solution of 10 vol.% piperidine in DMF (2 × 3 min). The resin was again washed with DMF (×5), CH_2Cl_2 (×5) and DMF (×5).

PyBOP Coupling Conditions

A given resin-bound peptide was washed with CH_2Cl_2 (×5) and DMF (×5) before being treated with a solution of 10 vol.% piperidine in DMF (2 × 3 min). The resin was again washed with DMF (×5), CH_2Cl_2 (×5) and DMF (×5). The resin was shaken for 1 h at room temperature with a solution of the desired Fmoc-protected amino acid (4 equiv.), PyBOP (4 equiv.) and 4-methylmorpholine (NMM)

(8 equiv.) in DMF (0.1 M in regard to loaded peptide). The coupling solution was discharged and the resin washed with DMF (×5), CH_2Cl_2 (×5) and DMF (×5).

HATU Coupling Conditions
A given resin-bound peptide was washed with CH_2Cl_2 (×5) and DMF (×5) before being treated with a solution of 10 vol.% piperidine in DMF (2 × 3 min). The resin was again washed with DMF (×5), CH_2Cl_2 (×5) and DMF (×5). The resin was shaken for 16 h at room temperature with a solution of the desired Fmoc-protected amino acid (1.1–1.5 equiv.), HATU (1.1–1.5 equiv.), HOAt (2.2–3 equiv.) and iPr$_2$NEt (2.2–3 equiv.) in DMF (0.1 M in regard to loaded peptide). The coupling solution was discharged and the resin washed with DMF (×5), CH_2Cl_2 (×5) and DMF (×5).

On-resin Esterification Conditions
To a solution of oxazolidine protected Fmoc-β-OH-D-Leu-OH **166** (4 equiv. compared to the resin-bound peptide) in CH_2Cl_2 (0.06 M) at 0 °C was added DIC (1 equiv. compared to amino-acid **166**). The solution was warmed to room temperature and stirred for 30 min before being concentrated under a stream of N_2. The resultant slurry was dissolved in DMF (0.1 M in regard to loaded peptide) and sucked into the fritted syringe containing resin-bound peptide. Subsequently, a solution of DMAP (catalytic, ~6 small crystals) in DMF (0.15 mL) was sucked up and the resin was shaken at room temperature for 16 h. The coupling solution was discharged and the resin washed with DMF (×5), CH_2Cl_2 (×5) and DMF (×5).

Microwave Coupling Conditions to Oxazolidine Protected Amino Acid
The resin was transferred to a Biotage microwave peptide synthesiser synthesis vessel and treated with a solution of the desired Fmoc-protected amino acid (4 equiv.), HATU (4 equiv.), HOAt (8 equiv.) and iPr$_2$NEt (8 equiv.) in DMF (0.1 M in regard to loaded peptide) under microwave irradiation at 50 °C for 1 h. The resin was then transferred to a fritted syringe, the coupling solution was discharged and the resin washed with DMF (×5), CH_2Cl_2 (×5) and DMF (×5).

5.3 Procedures and Analytical Data for Chapter 2

5.3.1 Synthesis of Goodman's Reagent 54

N,N'-di-Cbz-guanidine (56)

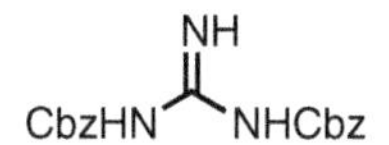

Guanidine hydrochloride **55** (3.82 g, 40.0 mmol) and sodium hydroxide (8.00 g, 200 mmol) were dissolved in a 1:2 v/v mixture of water: CH_2Cl_2 (120 mL) and cooled to 0 °C. Benzyl chloroformate (17.1 mL, 120 mmol) was added dropwise to the mixture and the resulting reaction was stirred at 0 °C for 16 h. Upon completion, the reaction mixture was poured over CH_2Cl_2 (100 mL) and separated with the aqueous phase being collected and re-extracted with CH_2Cl_2 (3 × 50 mL). All organic phases were combined and concentrated in vacuo to afford the crude product as a beige solid which was recrystallized from methanol to afford the title compound **56** (10.2 g, 78%) as a white solid.

^{1}H NMR ($CDCl_3$, 500 MHz) δ (ppm); 8.79 (br. s, 2H), 7.37–7.26 (m, 10H), 5.04 (br. s, 4H); **LRMS**: (+ESI) *m/z* 328 $[M+H]^+$; **mp**. 139–148 °C. These data are in agreement with those previously reported by Feichtinger et al. [1].

***N,N′*-di-Cbz-*N″*-trifluoromethanesulfonate-guanidine (54)**

NTf

CbzHN NHCbz

N,N′-di-Cbz-guanidine **56** (1.5 g, 4.6 mmol) was dissolved in chlorobenzene (45 mL) and cooled to 0 °C prior to addition of NaH (370 mg, 9.2 mmol, 60% by weight in oil). The reaction mixture was stirred for 1 h at 0 °C, then cooled to −45 ° C. Trifluoromethanesulfonic anhydride (770 μL, 4.6 mmol) was added dropwise to the solution which was then warmed to room temperature and stirred for 16 h. The reaction mixture was quenched with water and concentrated in vacuo. The resulting crude solid was redissolved in EtOAc (100 mL) and washed with 2 M $NaHSO_4$ (2 × 100 mL), water (100 mL) and saturated aqueous NaCl (100 mL). The washed organic phase was then dried ($MgSO_4$), filtered and concentrated in vacuo to give a crude oil which was purified by flash chromatography (eluent: 95:5 v/v CH_2Cl_2: Et_2O), affording the title compound **54** (2.0 g, 96%) as a colourless oil.

^{1}H NMR ($CDCl_3$, 500 MHz) δ (ppm); 10.31 (br. s, 2H), 7.40 (br. s, 10H), 5.27 (br. s, 4H); **^{19}F-NMR** ($CDCl_3$, 470 MHz) δ (ppm); −78.64; **LRMS**: (+ESI) *m/z* 460 $[M+H]^+$. These data are in agreement with those previously reported by Feichtinger et al. [1].

5.3.2 Synthesis of Fmoc-L-allo-Enduracididine (Cbz)$_2$-OH (58)

(2*S*,4*R*)-*tert*-butyl 2-((*tert*-butoxycarbonyl)amino)-4-hydroxy-5-nitropentanoate (49)

OH

NO_2

O*t*Bu

BocHN

O

Boc-L-Asp-O*t*Bu **51** (3.00 g, 10.4 mmol) and 1,1′-carbonyldiimidazole (1.68 g, 10.4 mmol) were dried in vacuo for 1 h then dissolved in nitromethane (53 mL). The reaction mixture was stirred at room temperature for 45 min, at which point potassium *tert*-butoxide (2.24 g, 20.8 mmol) was added to the reaction mixture. The reaction mixture was stirred at room temperature for an additional 2.5 h, then quenched with 50 vol.% glacial acetic acid in water (50 mL) and extracted with EtOAc (3 × 50 mL). The combined organic phases were washed with water (100 mL), saturated aqueous $NaHCO_3$ solution (100 mL), water (100 mL) and saturated aqueous NaCl (100 mL). The washed organic phase was dried over $MgSO_4$, filtered and concentrated in vacuo. The crude residue was azeotroped with toluene (×3) and concentrated in vacuo to afford nitroketone **50** which was used without purification.

Crude nitroketone **50** was dissolved in anhydrous THF (150 mL) and cooled to −78 °C. To this solution was slowly added a 1 M solution of L-Selectride® in THF (6 mL, 6.00 mmol), the resulting reaction mixture was stirred at −78 °C for 3 h. The reaction mixture was then poured onto saturated aqueous NH_4Cl solution (150 mL) and diluted with water (50 mL). The resulting mixture was extracted with EtOAc (3 × 150 mL) and the combined organic phases were washed with saturated aqueous NaCl (300 mL), dried ($MgSO_4$), filtered and concentrated in vacuo to give a crude yellow oil which was purified by flash chromatography (eluent: 95:5 v/v CH_2Cl_2:Et_2O), affording nitro-alcohol **49** (1.80 g, 52% over 2 steps) as a white solid.

^{1}H NMR: ($CDCl_3$, 500 MHz) δ (ppm); 5.46 (br. d, J = 6.4 Hz, 1H), 4.57–4.37 (m, 3H), 4.30–4.19 (m, 1H), 3.55 (br. s, 1H), 2.11–1.81 (m, 2H), 1.46 (s, 9H), 1.43 (s, 9H); **LRMS**: (+ESI) *m/z* 335 $[M+H]^+$; **$[\alpha]_D$**: +21.6° (*c* 0.3, CH_2Cl_2); **mp.** 106–120 °C. These data are in agreement with those previously reported by Rudolph et al. [2].

(2*S*,4*R*)-*tert*-butyl 5-(2,3-bis((benzyloxy)carbonyl)guanidino)-2-((*tert*-butoxycarbonyl)amino)-4-hydroxypentanoate (48)

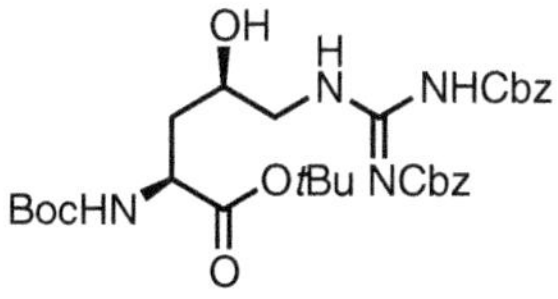

Nitro-alcohol **49** (1.80 g, 5.38 mmol) was dissolved in anhydrous MeOH (54 mL) and to this solution was added 10% w/w palladium on activated carbon (575 mg, 540 μmol palladium), and glacial acetic acid (308 μL, 5.38 mmol). The reaction vessel was evacuated and flushed with nitrogen (×3) then filled with an atmosphere of H_2 (1 atm). The reaction was stirred at room temperature for 18 h, then evacuated and flushed with nitrogen (×3), and filtered through Celite®. The filtrate was concentrated in vacuo to afford a crude beige foam which was azeotroped with toluene (×3) and redissolved in CH_2Cl_2 (15 mL). To this solution was added a solution of guanidinylating reagent **5** (2.48 g, 5.40 mmol) in MeCN (15 mL) via canula. Et_3N (313 μL, 1.85 mmol) was added and the reaction mixture was stirred at 40 °C for 18 h, then poured onto a saturated aqueous NH_4Cl solution (30 mL). The mixture was extracted with CH_2Cl_2 (3 × 50 mL) and the combined organic phases were dried ($MgSO_4$), filtered, and concentrated in vacuo to give a crude oil which was purified by flash chromatography (eluent: 35:65 → 40:60 v/v EtOAc/ petroleum benzines), affording compound **48** (2.39 g, 72%) as a white solid.

^{1}H NMR: ($CDCl_3$, 400 MHz) δ (ppm); 11.69 (br. s, 1H), 8.69 (t, *J* = 5.41 Hz, 1H), 7.42–7.24 (m, 10H), 5.49–5.37 (m, 1H), 5.18 (s, 2H), 5.10 (s, 2H), 4.27–4.16 (m, 1H), 3.99–3.90 (m, 1H), 3.63 (ddd, *J* = 14.0, 5.7, 2.5 Hz, 1H), 3.40 (ddd, *J* = 14.0, 7.5, 5.0 Hz, 1H), 2.01–1.90 (m, 1H), 1.82 (ddd, *J* = 14.3, 9.0, 6.6 Hz, 1H), 1.43 (s, 18H); **^{13}C NMR** ($CDCl_3$, 100 MHz) δ (ppm); 171.5, 163.2, 157.0, 155.7, 153.6, 136.5, 134.5, 128.8, 128.7, 128.5, 128.4, 128.1, 128.0, 82.2, 80.0, 68.6, 68.3, 67.1, 52.1, 47.2, 37.8, 28.3, 27.9; **LRMS**: (+ESI) *m/z* 637 $[M+Na]^+$; **HRMS**: (+ESI) Calc. for $C_{31}H_{42}N_4O_9$: 615.3025 $[M+H]^+$, Found: 615.3032 $[M+H]^+$; **IR (ATR)**: ν_{max} = 3335, 3306, 2958, 2926, 2854, 1732, 1643, 1625, 1571, 1499, 1455, 1382, 1368, 1352, 1326, 1257, 1214 cm^{-1}; **$[\alpha]_D$**: +6.0° (*c* 0.3, CH_2Cl_2); **mp.** 118–130 °C.

(*S*)-benzyl 2-(((benzyloxy)carbonyl)imino)-5-((*S*)-3-(*tert*-butoxy)-2-((*tert*-butoxycarbonyl)amino)-3-oxopropyl)imidazolidine-1-carboxylate (47)

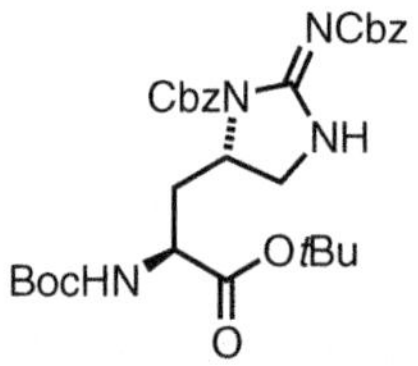

Guanidinylated compound **48** (2.39 g, 3.89 mmol) was dissolved in anhydrous CH_2Cl_2 (140 mL) and cooled to −78 °C. To this solution was added *i*Pr_2NEt (3.24 mL, 18.6 mmol) followed by dropwise addition of trifluoromethanesulfonic anhydride (669 μL, 4.66 mmol). The resulting reaction mixture was stirred at −78 °C for 3 h, then warmed to room temperature for 15 min and quenched with a saturated aqueous NH_4Cl solution (140 mL). This mixture was extracted with CH_2Cl_2 (2 × 100 mL) and the combined organic phases were washed with a saturated aqueous $NaHCO_3$ solution (100 mL). The washed organic phase was dried ($MgSO_4$), filtered and concentrated in vacuo to give a brown oil which was purified by flash chromatography (eluent: 30:70 → 50:50 v/v EtOAc/petroleum benzines), affording Boc-L-*allo*-End$(Cbz)_2$-O*t*Bu **47** (1.83 g, 79%) as a white foam.

^{1}H NMR: (acetone-d_6, 400 MHz) δ (ppm); 7.63–7.57 (m, 2H) 7.44–7.25 (m, 8H), 6.24 (d, *J* = 4 Hz, 1H), 5.33–5.24 (m, 2H), 5.16–5.06 (m, 2H), 4.62–4.53 (m, 1H), 4.15 (ap. q, *J* = 7.2 Hz, 1H), 3.90 (dd, *J* = 10.6, 9.1 Hz, 1H), 3.65 (dd, *J* = 7.6, 3.0 Hz, 1H), 2.37 (ddd, *J* = 13.5, 6.9, 3.3 Hz, 1H), 2.08–1.99 (m, 1H, obscured by residual solvent), 1.46 (s, 9H), 1.41 (s, 9H); **^{13}C NMR** (acetone-d_6, 100 MHz) δ (ppm); 171.7, 164.1, 159.5, 156.2, 152.1, 138.6, 137.0, 129.2, 129.1, 128.7, 128.6, 128.4, 128.4, 82.1, 79.5, 68.4, 67.4, 54.9, 52.9, 47.0, 36.5, 28.6, 28.1; **LRMS**: (+ESI) *m/z* 619 $[M+Na]^+$; **HRMS**: (+ESI) Calc. for $C_{31}H_{40}N_4O_8$: 597.2919 $[M+H]^+$, Found: 597.2923 $[M+H]^+$; **IR (ATR)**: ν_{max} = 3349, 2976, 2923, 2854, 2162, 1713, 1654, 1617, 1498, 1440, 1393, 1368, 1306, 1258 cm^{-1}; **$[\alpha]_D$**: +17.3° (*c* 0.3, CH_2Cl_2).

(*S*)-2-((((9*H*-fluoren-9-yl)methoxy)carbonyl)amino)-3-((*S*)-3-((benzyloxy)carbonyl)-2-(((benzyloxy)carbonyl)imino)imidazolidin-4-yl)propanoic acid (58)

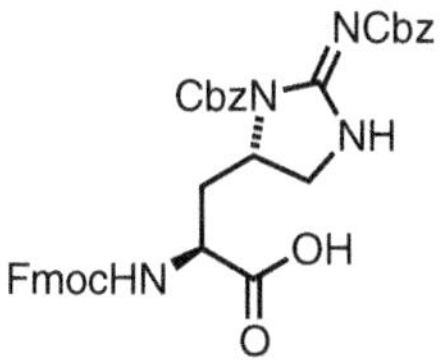

Boc-L-*allo*-End(Cbz)$_2$-O*t*Bu **47** (263 mg, 441 μmol) was dissolved in a mixture of TFA (4.5 mL) and water (0.45 mL). The mixture was stirred at room temperature for 3 h, then concentrated under a stream of nitrogen. The resulting crude oil was azeotroped with toluene (3 × 10 mL) and concentrated in vacuo to remove residual TFA. The concentrated crude material was then dissolved in a mixture of THF (4 mL) and saturated aqueous $NaHCO_3$ solution (2.5 mL). Fmoc-OSu (156 mg, 459 μmol) was added to this mixture and the reaction was stirred at room temperature for 20 h. The reaction mixture was acidified to pH = 2 with aqueous HCl (2 M), then extracted with EtOAc (3 × 20 mL). The combined EtOAc phases were dried (Na_2SO_4), filtered and concentrated in vacuo to give a crude white foam that was purified by flash chromatography (eluent: 0:100 → 10:90 v/v MeOH/EtOAc/) to afford Fmoc-L-*allo*-End(Cbz)$_2$-OH (**58**) (168 mg, 57%) as a white foam.

^{1}H NMR (DMSO-d_6, 400 MHz) δ (ppm); 7.89 (d, J = 7.9 Hz, 2H), 7.78 (d, J = 8.2 Hz, 1H), 7.71 (dd, J = 2.4, 7.7 Hz, 2H), 7.52–7.24 (m, 14H), 5.26–5.14 (m, 2H), 5.09–4.99 (m, 2H), 4.46–4.16 (m, 4H), 4.14–4.04 (m, 1H), 3.68 (dd, J = 10.4, 9.4 Hz, 1H), 3.39 (dd, J = 10.7, 2.4 Hz, 1H), 2.29–2.19 (m, 1H), 1.97–1.85 (m, 1H) **^{13}C NMR** (DMSO-d_6, 100 MHz) δ (ppm); 173.1, 162.4, 157.7, 156.0, 150.6, 143.7, 140.7, 137.2, 135.8, 128.2, 128.0, 127.7, 127.6, 127.6, 127.4, 127.1, 125.3, 125.2, 120.1, 67.2, 66.1, 65.7, 53.9, 51.5, 46.6, 45.8, 34.9; **LRMS**: (+ESI) *m/z* 685 [M+Na]$^+$; **HRMS**: (+ESI) Calc. for $C_{37}H_{34}N_4O_8$: 663.2449 [M +H]$^+$, Found: 663.2460 [M+H]$^+$; **IR (ATR)**: ν_{max} = 2950, 2925, 2855, 2163, 1706, 1218 cm^{-1}; **[α]$_D$**: +4.0° (*c* 0.1, CH_2Cl_2).

5.3.3 Synthesis of Protected Fmoc-D-Thr-OH

(2*R*,3*S*)-2-((((9*H*-fluoren-9-yl)methoxy)carbonyl)amino)-3-(allyloxy)butanoic acid (73)

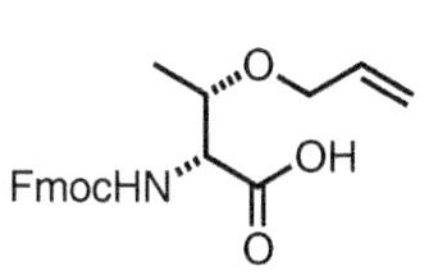

To a suspension of NaH (0.72 g, 18.2 mmol, 60% by weight in oil) in DMF (12.5 mL) at 0 °C was added a solution of Boc-D-Thr-OH (1.60 g, 7.30 mmol) in DMF (25 mL) dropwise, followed by allyl bromide (0.63 mL, 7.30 mmol). The reaction mixture was warmed to room temperature and stirred for 3 h before being quenched with saturated aqueous NH_4Cl (10 mL) and diluted with EtOAc (100 mL). The organic phase was washed with aqueous HCl (5 × 50 mL, 0.5 M), dried ($MgSO_4$), filtered and concentrated in vacuo to yield crude allyl ether **72** which was used without further purification.

To a solution of crude Boc amino acid **72** (1.42 g, 5.47 mmol) in CH_2Cl_2 (35 mL) was added HCl in dioxane (13.7 ml, 54.8 mmol). The reaction mixture was stirred at room temperature for 2 h at which point a white precipitate formed before being concentrated in vacuo. To a solution of the crude solid in THF (35 mL) and saturated aqueous $NaHCO_3$ (17 mL) was added Fmoc-OSu (1.94 g, 5.75 mmol). The reaction mixture was stirred at room temperature for 16 h before being diluted with water (20 mL) and washed with diethyl ether (2 × 50 mL). The aqueous layer was acidified to pH = 1 with aqueous HCl (1 M) and then extracted with EtOAc (3 × 70 mL) and the combined organic phase was washed with saturated aqueous NaCl (100 mL), dried over $MgSO_4$, filtered and concentrated in vacuo to afford the desired Fmoc-D-Thr(Allyl)-OH (**73**) (1.69 g, 82%) as a white foam that was deemed suitable for further use without purification.

^{1}H NMR ($CDCl_3$, 300 MHz) δ (ppm); 8.34 (br. s, 1H), 7.77 (d, J = 7.6 Hz, 2H), 7.66–7.58 (m, 2H), 7.45–7.29 (m, 4H), 5.94–5.78 (m, 1H), 5.63 (d, J = 9.3 Hz, 1H), 5.26 (dd, J = 17.3, 1.5 Hz, 1H), 5.18 (dd, J = 10.4, 1.2 Hz, 1H) 4.50–4.34 (m, 3H), 4.33–3.89 (m, 4H), 1.24 (d, J = 6.3 Hz, 3H); **^{13}C NMR** ($CDCl_3$, 75 MHz) δ (ppm) 175.3, 156.8, 143.9, 141.3, 134.1, 127.7, 127.1, 125.2, 120.0, 117.6, 74.1, 70.2, 67.4, 58.4, 47.1, 16.2; **LRMS**: (+ESI) m/z 382 $[M+H]^+$.

(2*R*,3*S*)-2-((((9*H*-fluoren-9-yl)methoxy)carbonyl)amino)-3-((*tert*-butyldimethylsilyl)oxy)butanoic acid (79)

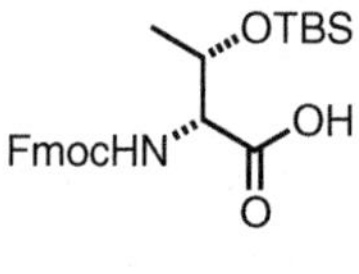

To a solution of Fmoc-D-Thr-OH (**78**) (500 mg, 1.45 mmol) in DMF (3 mL) was added iPr_2NEt (0.81 mL, 4.64 mmol) followed by *tert*-butyldimethylsilyl chloride (441 mg, 2.93 mmol) in two equal portions and the solution was stirred at room temperature for 16 h. The reaction mixture was quenched with water (10 mL) and then acidified to pH = 3 with aqueous HCl (1 M) before being extracted with EtOAc (3 × 15 mL) and the combined organic extracts were was with saturated aqueous NaCl (2 × 15 mL), dried (Mg_2SO_4), filtered and concentrated in vacuo. The crude product was purified by flash chromatography (eluent: 14:86 → 66:34:1 v/v/v EtOAc/petroluem benzines/AcOH) to yield TBS protected amino acid **79** as a white foam (479 mg, 73%).

^{1}H NMR ($CDCl_3$, 500 MHz) δ (ppm); 8.76 (brs, 1H), 7.78 (d, J = 7.3 Hz, 2H), 7.63 (ap. t, J = 7.3 Hz, 2H), 7.41 (d, J = 7.5 Hz, 2H), 7.32 (d, J = 7.5 Hz, 2H), 5.56 (d, J = 9.0 Hz, 1H), 4.54–4.37 (m, 3H), 4.35 (dd, J = 8.8, 2.5 Hz, 1H), 4.28 (t, J = 7.3 Hz, 1H), 1.23 (d, J = 6.5 Hz, 3H), 0.91 (s, 9H), 0.12 (s, 3H), 0.09

(s, 3H); **LRMS**: (+ESI) *m/z* 456 [M+H]. These data are in agreement with those previously reported by Huang et al. [3].

(2*R*,3*S*)-2-((((9*H*-fluoren-9-yl)methoxy)carbonyl)amino)-3-((triethylsilyl)oxy) butanoic acid (89)

D-Threonine (5.00 g, 42.0 mmol) **71** and Fmoc-OSu (14.9 g, 44.1 mmol) were dissolved in a 2:1 v/v mixture of THF: saturated aqueous $NaHCO_3$ (100 mL). The reaction mixture was stirred at room temperature for 16 h. The reaction was then diluted with water (50 mL) and the pH of the mixture was adjusted to pH = 9 with saturated aqueous $NaHCO_3$. The mixture was extracted with diethyl ether (3 × 50 mL) and the aqueous layer was acidified to pH = 1 with aqueous HCl (1 M). The acidic aqueous mixture was extracted with EtOAc (3 × 100 mL) and the combined organic extracts were washed with saturated aqueous NaCl (100 mL), dried over Na_2SO_4, filtered and concentrated in vacuo to afford crude Fmoc-D-Thr-OH **78** (14.3 g) as a white foam which was deemed to be sufficiently pure and used without further purification. A portion of the crude Fmoc-D-Thr-OH (3.00 g, 8.79 mmol) was dissolved in DMF (20 mL) and cooled to 0 °C. To this cooled solution was added *i*Pr_2NEt (4.90 mL, 28.1 mmol) followed by chlorotriethylsilane (1.48 mL, 17.6 mmol) dropwise. The reaction mixture was stirred at 0 °C for 20 min then warmed to room temperature and stirred for an additional 16 h. The reaction mixture was then cooled to 0 °C, diluted with water (20 mL) and poured onto saturated aqueous NH_4Cl (20 mL). The mixture was washed with EtOAc (2 × 50 mL), acidified to pH = 2 with aqueous HCl (1 M) and extracted with EtOAc (3 × 50 mL). The combined organic extracts were washed with saturated aqueous NH_4Cl (100 mL), water (100 mL) and saturated aqueous NaCl (100 mL), dried (Na_2SO_4), filtered and concentrated in vacuo to afford a crude colourless oil which was purified by flash chromatography (eluent: 20:80 → 50:50 v/v EtOAc/ petroleum benzines) to afford Fmoc-D-Thr(TES)-OH (**89**) (1.47 g, 37% over two steps) as a colourless oil.

^{1}H NMR ($CDCl_3$, 400 MHz) δ (ppm); 7.77 (d, *J* = 7.5 Hz, 2H), 7.62 (ap. t, *J* = 6.8 Hz, 2H), 7.40 (ap. t, *J* = 7.5 Hz, 2H), 7.32 (ap. t, *J* = 7.5 Hz, 2H), 5.62 (d, *J* = 8.5 Hz, 1H), 4.55–4.37 (m, 3H), 4.35 (dd, *J* = 8.2, 2.2 Hz, 1H), 4.27 (t, *J* = 7.2 Hz, 1H), 1.23 (d, *J* = 6.3 Hz, 3H), 0.98 (t, *J* = 8.1 Hz, 9H), 0.64 (q, *J* = 8.2 Hz, 6H); **^{13}C NMR** ($CDCl_3$, 100 MHz) δ (ppm); 174.5, 156.6, 144.0, 143.8, 141.5, 127.9, 127.2, 125.3, 120.1, 68.4, 67.4, 59.4, 47.3, 20.0, 6.8, 4.8 (1 extra signal due to restricted rotation about the Fmoc group); **LRMS**: (+ESI) *m/z* 478 $[M+Na]^+$; **HRMS**: (+ESI) Calc. for $C_{25}H_{33}NO_5Si$: 478.2020 $[M+Na]^+$, Found: 478.2024 $[M+Na]^+$; **IR (ATR)**: ν_{max} = 3437, 3067, 2955, 2911, 2876, 1719, 1510, 1478, 1450, 1413, 1378, 1341, 1310, 1209 cm^{-1}; **$[\alpha]_D$**: −1.5° (*c* 10.0, CH_2Cl_2).

5.3.4 *Synthesis of Alloc-L-Ile-OH (88)*

(2*S*,3*S*)-2-(((allyloxy)carbonyl)amino)-3-methylpentanoic acid (88)

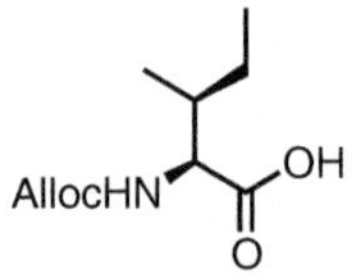

L-Isoleucine **87** (1.00 g, 7.62 mmol) was suspended in saturated aqueous Na_2CO_3 (13 mL) and cooled to 0 °C. A mixture of allyl chloroformate (891 μL, 8.38 mmol) and 1,4-dioxane (28 mL) was added dropwise. The resulting reaction mixture was stirred at 0 °C for 2 h, then diluted with water (150 mL), washed with Et_2O (3 × 150 mL) and acidified to pH = 2 with aqueous HCl (1 M). The acidified mixture was extracted with EtOAc (3 × 200 mL) and the combined organic extracts were washed with saturated aqueous NaCl (200 mL), dried over Na_2SO_4, filtered and concentrated in vacuo to afford Alloc-Ile-OH (**88**) as a colourless oil (1.57 g, 95%) which was used without purification.

^{1}H NMR ($CDCl_3$, 400 MHz) δ (ppm): 7.70 (br. s, 1H), 5.96–5.83 (m, 1H), 5.40–5.25 (m, 2H), 5.21 (dq, *J* = 10.5, 1.2 Hz, 1H), 4.63–4.52 (m, 2H), 4.35 (dd, *J* = 9.0, 4.5 Hz, 1H), 2.00–1.83 (m, 1H), 1.54–1.40 (m, 1H), 1.28–1.13 (m, 1H), 0.96 (d, *J* = 7.0 Hz, 3H), 0.92 (t, *J* = 7.4 Hz, 3H); **LRMS**: (+ESI) *m/z* 216 $[M+H]^+$; **$[\alpha]_D$**: +9.6° (c 0.3, CH_2Cl_2). These data are in agreement with those previously reported by Jad et al. [4].

5.3.5 *Solid-Phase Synthesis of Teixobactin (28)*

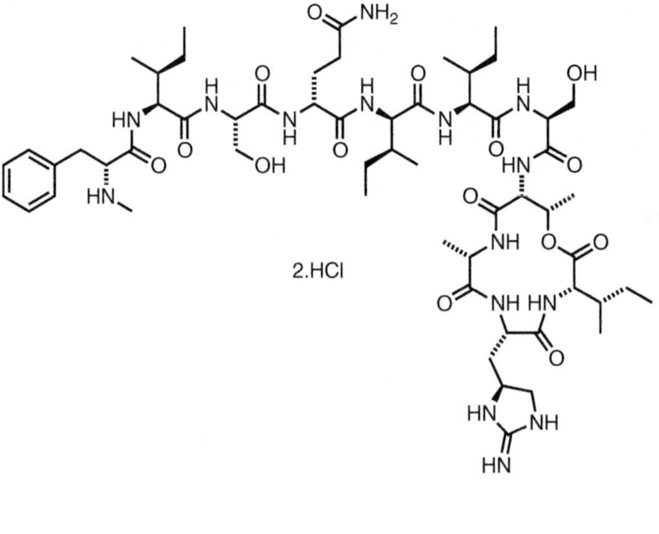

Fmoc-D-Thr(TES)-OH (**89**) (549 mg, 1.2 mmol, 10 equiv.) was dissolved in anhydrous CH_2Cl_2 (6 mL) and cooled to 0 °C. DIC (94 μL, 0.60 mmol, 5 equiv.) was then added to the cooled solution, which was then warmed to room temperature and stirred under an atmosphere of argon for 30 min. The reaction mixture was concentrated under a stream of nitrogen and the resulting crude solid was redissolved in a 1:1 v/v mixture of CH_2Cl_2:DMF (1.2 mL). This mixture, along with a solution of DMAP (catalytic, ~6 crystals) in DMF (0.1 mL), was shaken for 16 h at room temperature with HMPB-NovaPEG resin (234 mg, 0.64 mmol g^{-1}) in a fritted syringe, which had been swollen in CH_2Cl_2 for 30 min and washed with CH_2Cl_2 (×5).

The loading mixture was discharged from the fritted syringe and the resin was washed with CH_2Cl_2 (×5) and DMF (×5). The loaded resin was then capped via treatment with 10 vol.% acetic anhydride in pyridine (3 mL) with dissolved DMAP (catalytic ~6 crystals) for 45 min at room temperature. The resin was again washed with DMF (×5), CH_2Cl_2 (×5) and DMF (×5). Resin loading was determined after Fmoc-deprotection of the loaded amino acid, in which the resin was treated with a solution of 10 vol.% piperidine in DMF (2 × 3 min) then washed with DMF (×5), CH_2Cl_2 (×5) and DMF (×5). Combined deprotection solutions were made up to 10 mL with 10 vol.% piperidine in DMF and diluted 1:100 with 10 vol.% piperidine in DMF. Resin loading was determined to be 115 μmol by measurement of the UV absorbance at λ = 301 nm of the diluted deprotection solution. The resin was then subjected to a mixture of 1 M tetrabutylammonium fluoride (TBAF) in THF (2.30 mL, 2.30 mmol, 20 equiv.), glacial acetic acid (131 μL, 2.30 mmol, 20 equiv.) and CH_2Cl_2 (2.3 mL) at room temperature for 2 h (×2). The resin was washed with CH_2Cl_2 (×5), DMF (×5) and CH_2Cl_2 (×5).

Fmoc-Ser(*t*Bu)-OH was coupled according to standard PyBOP coupling conditions (Sect. 5.2). Alloc-Ile-OH (247 mg, 1.15 mmol, 10 equiv.) was dissolved in anhydrous CH_2Cl_2 (5.75 mL) and cooled to 0 °C. DIC (90 μL, 575 μmol, 5 equiv.) was added to this solution which was then warmed to room temperature and stirred for 30 min. The reaction mixture was concentrated under a stream of nitrogen and subsequently redissolved in a 1:1 v/v mixture of CH_2Cl_2:DMF (1.2 mL). This solution, along with a solution of DMAP (catalytic ~6 crystals) in DMF (0.1 mL), was shaken with the resin and then washed with CH_2Cl_2 (×5), DMF (×5) and CH_2Cl_2 (×5). The linear peptide was elongated using standard PyBOP coupling conditions (Sect. 5.2), incorporating the commercially available amino acids Fmoc-Ile-OH, Fmoc-D-*allo*-Ile-OH, Fmoc-D-Gln(Trt)-OH, Fmoc-Ser(*t*Bu)-OH, Fmoc-Ile-OH and *N*-methyl-Boc-D-Phe-OH, to afford resin-bound depsipeptide **93**.

Alloc deprotection was carried out via treatment with a solution of $Pd(PPh_3)_4$ (27 mg, 23 μmol, 0.2 equiv.) and $PhSiH_3$ (283 μL, 2.3 mmol, 20 equiv.) in CH_2Cl_2 (1.2 mL) for 20 min (×2) at room temperature. The deprotection solution was discharged and the resin was washed with CH_2Cl_2 (×5) and DMF (×5) before coupling of Fmoc-L-*allo*-End(Cbz)$_2$-OH (**58**) under standard HATU conditions. After resin washing, optimised Fmoc-deprotection conditions of 30 s treatment with 10 vol.% piperidine in DMF was carried out followed immediate coupling of Fmoc-Ala-OH under standard PyBOP coupling conditions. The resin was then Fmoc-deprotected via treatment with 10 vol.% piperidine in DMF (2 × 3 min) and thoroughly washed with DMF (×5) and CH_2Cl_2 (×20). The resin was treated with 1 vol.% TFA in CH_2Cl_2 (4 × 20 min) and the solutions were combined and concentrated under a stream of nitrogen, azeotroped with toluene (×3) and concentrated in vacuo to afford the protected linear peptide **85** as a crude solid.

Crude **85** was used without purification and dissolved in DMF (11.5 mL) to a concentration of 10 mM. To this solution was added DMTMM.BF_4 (56.0 mg, 173 μmol) and *i*Pr_2NEt (60.0 μL, 345 μmol). The reaction was stirred at room temperature for 16 h and monitored by HPLC-MS. Upon completion, the reaction mixture was concentrated under a stream of nitrogen and re-dissolved in a mixture of 70:10:12:8 v/v/v/v TFA:thioanisole:TfOH:*m*-cresol (1 mL) at 0 °C. The reaction was stirred at 0 °C for 1 h before being poured onto cold Et_2O (50 mL) and centrifuged. The Et_2O was then removed and the crude solid was dissolved in a mixture of MeCN and H_2O and lyophilised to yield crude teixobactin (**28**). The crude product was purified by RP-HPLC using a Waters XBridge Prep OBD 5 μm 19 × 150 mm (C18) column using a 0–50 vol.% MeCN in H_2O (0.1% TFA) focussed gradient (0–30 vol.% MeCN over 2 min, 30–50 vol.% over 20 min) at a flow rate flow rate of 15 mL min^{-1} and lyophilized to give pure teixobactin (**28**) as a TFA salt. Re-lyophilisation in the presence of 5 mM HCl (×3) afforded teixobactin (**28**) (4.99 mg, 3.80 μmol) as its *bis*-HCl salt in 3.3% yield over 24 linear steps.

LRMS: (+ESI) m/z calculated mass 621.9 $[M+2H]^{2+}$, 1242.7 $[M+H]^+$, 1264.7 $[M+Na]^+$: m/z observed 622.4 $[M+2H]^{2+}$, 1243.6 $[M+H]^+$, 1265.6 $[M+Na]^+$; **HRMS**: (+ESI) Calc. for $C_{58}H_{95}N_{15}O_{15}$: 1242.7205 $[M+H]^+$, Found: 1242.7201 $[M+H]^+$; **IR (ATR)**: ν_{max} = 3281, 2964, 2932, 2877, 1742, 1662, 1631, 1526, 1458, 1384, 1301, 1260 cm^{-1}; **Analytical UHPLC** R_t 4.6 min (0–100% MeCN (0.1% TFA) in H_2O (0.1% TFA) over 5 min, λ = 214 nm). ^{1}H-NMR and ^{13}C-NMR data is listed below in Table 5.1 and 5.2, respectively.

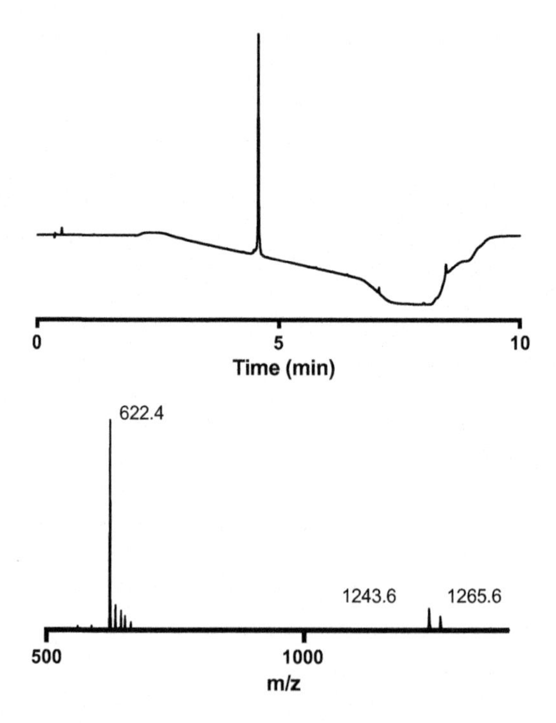

Table 5.1 ^{1}H-NMR comparison of natural and synthetic teixobactin (**28**) in DMSO-d_6 referenced at 2.50 ppm. All assignments were made based on COSY, TOCSY, HSQC and HMBC data in comparison with the isolated material (see Fig. 5.1 for positions of carbon centres)

Position	Natural δ ^{1}H/ppm (no. H, mult, *J* Hz)	Synthetic δ ^{1}H/ppm (no. H, mult, *J* Hz)	Δδ/ ppm	Position	Natural δ ^{1}H/ppm (no. H, mult, *J* Hz)	Synthetic δ ^{1}H/ppm (no. H, mult, *J* Hz)	Δδ/ ppm
1	2.5 (3H, brs)	2.47 (3H, br t, 4.4)	0.03	29	4.29 (1H, m)	4.39 (1H, m)	−0.1
2	4.21 (1H, dd, 9.4, 5.3)	4.27 (1H, m)	−0.06	29-NH	7.78 (1H, d, 8.8)	7.89 (1H, d, 8.1)	−0.11
2-NH	9.3, 9.0 (2H, v br s)	10.09, 9.08 (v br s)	na	30	1.83 (1H, m)	1.82 (1H, m)	0.01
3	3.00 (1H, dd, 13.2, 9.4)	2.95 (1H, dd, 12.7, 10.8)	0.05	31	0.84 (3H, m)	0.88 (3H, d, 6.7)	−0.04
	3.15 (1H, 13.2, 5.3)	3.29 (1H, dd, 12.8, 4.5)	−0.14	32	1.11 (1H, m)	1.08 (1H, m)	0.03
4					1.42 (1H, m)	1.44 (1H, m)	−0.02
5,5′	7.24 (2H, m)	7.21 (m, 2H)	0.03	33	0.85 (3H, m)	0.77 (3H, m)	−0.07
6,6′	7.31 (2H, m)	7.28 (m, 2H)	0.03	34			
7	7.27 (1H, m)	7.22 (m, 1H)	0.05	35	4.47 (1H, dt, 5.0, 5.2)	4.64[b] (1H, m)	−0.19
8				35-NH	8.37 (1H, d, 5.2)	8.99 (1H, d, 8.7)	−0.62
9	4.12 (1H, dd, 7.8, 7.2)	4.07 (1H, ap t, 7.3)	0.05	36	3.64 (1H, m)	3.56 (1H, m)	0.08
9-NH	8.43 (1H, d, 7.2)	8.64 (1H, d, 8.3)	−0.21		3.80 (1H, dd, 10.8, 5.0)	3.87 (1H, m)	−0.07
10	1.56 (1H, m)	1.56 (1H, m)	0	36-OH	Exchanged	Exchanged	
11	0.62 (3H, d, 6.7)	0.53 (3H, d, 6.6)	0.09	37			
12	0.76 (1H, m)	0.72 (1H, m)	0.04	38	4.64 (1H, dd, 9.5, 2.2)	4.69 (1H, ap. d, 11.0)	−0.05
				38-NH	Not reported	8.93 (1H, d, 9.9)	na
	1.07 (1H, m)	1.02 (1H, m)	0.05	39	5.36 (1H, dq, 2.2, 6.4)	5.37 (1H, dq, 2.0, 6.2)	−0.01
13	0.66 (3H, t, 7.1)	0.61 (3H, t, 7.1)	0.05	40	1.13 (3H, d, 6.4)	1.05 (3H, 6.4)	0.08
14				41			
15	4.34 (1H, m)	4.30 (1H, m)	0.04	42	3.97 (1H, dq, 5.1, 7.5)	3.89 (1H, m)	0.08
15-NH	7.88 (1H, d, 7.9)	8.09 (1H, d, 7.6)	−0.21	42-NH	8.05 (1H, d, 5.1)	8.16 (1H, d, 5.2)	−0.11

(continued)

Table 5.1 (continued)

Position	Natural δ ^{1}H/ppm (no. H, mult, *J* Hz)	Synthetic δ ^{1}H/ppm (no. H, mult, *J* Hz)	Δδ/ppm	Position	Natural δ ^{1}H/ppm (no. H, mult, *J* Hz)	Synthetic δ ^{1}H/ppm (no. H, mult, *J* Hz)	Δδ/ppm
16	3.57 (1H, dd, 10.8, 5.6)	3.54 (1H, m)	0.03	43	1.34 (3H, d, 7.5)	1.26 (3H, d, 7.3)	0.08
	3.63 (1H, m)	3.62 (1H, m)	0.01	44			
16-OH	Exchanged			45	4.38 (1H, m)	4.35 (1H, m)	0.03
17				45-NH	8.32 (1H, d, 9.1)	8.85 (1H, d, 10)	−0.53
18	4.33 (1H, m)	4.30 (1H, m)	0.03	46	2.03 (2H, m)	2.13 (2H, m)[c]	−0.1
18-NH	7.85 (1H, d, 7.9)	8.02 (1H, d 8.0)	−0.17	47	3.90 (1H, m)	3.82 (1H, m)	0.08
19[a]	1.74 (1H, m)	1.71 (1H, m)	0.03	47-NH	7.95 (1H, br s)	8.00 (1H, br s)	−0.05
	1.92 (1H, m)	1.87 (1H, m)	0.05	48	3.36 (1H, dd, 9.4, 7.7)	3.44 (1H, ap t, 8.0)	−0.08
20[a]	2.10 (2H, m)	2.08 (2H, m)	0.02		3.66 (1H, t, 9.4)	3.60 (1H, m)	0.06
21				48-NH	8.1 (1H, br s)	8.05 (1H, br s)	0.05
21-NH2	6.63 (1H, br s)	6.76 (1H, br s)	−0.13	49			
	7.11 (1H, br s)	7.26 (1H, br s)	−0.15	49-NH	7.76 (2H, br s)	7.80 (2H, br s)	−0.04
22				50			
23	4.36 (1H, m)	4.37 (1H, m)	−0.01	51	4.03 (1H, t, 9.4)	4.01 (1H, t, 9.8)	0.02
23-NH	7.70 (1H, d, 8.8)	7.75 (1H, d, 9)	−0.05	51-NH	8.01 (1H, d 9.4)	8.75 (1H, d, 9.8)	−0.74
24	1.8 (2H, m)	1.80 (1H, m)	0.0	52	1.77 (1H, m)	1.88 (1H, m)	−0.11
25	0.82 (3H, m)	0.77 (3H, m)	0.05	53	0.81 (3H, m)	0.78 (3H, m)	0.03
26	1.09 (1H, m)	1.06 (1H, m)	0.03	54	0.77 (1H, m)	1.13 (1H, m)	−0.36
	1.32 (1H, m)	1.28 (1H, m)	0.04		1.07 (1H, m)	1.41 (1H, m)	−0.34
27	0.82 (3H, m)	0.82 (3H, m)	0.0	55	0.82 (3H, m)	0.80 (3H, m)	0.02
28				56			

Note Discrepencies in NH chemical shifts are attributed to differences in pH and concentration. [a]Methylene protons at C-19 and C-20 were misassigned in the isolation paper by Ling et al. [5]. These assignments have been corrected in our data. [b]The data in the isolation paper appears to be quoted with the incorrect chemical shift. [c]Methylene protons appear as two separate signals in our data at 2.04 and 2.22 ppm each 1H, with the average (2.13) presented in the table

Table 5.2 ^{13}C-NMR comparison of natural and synthetic teixobactin (**28**) in DMSO-d_6 referenced to 39.52 ppm. ^{13}C were extracted from the HSQC and HMBC spectra. All assignments were made based on COSY, TOCSY, HSQC and HMBC data in comparison with the isolated material (see Fig. 5.1 for positions of carbon centres)

Position	Natural δ ^{13}C/ppm	Synthetic δ ^{13}C/ppm	Δδ/ ppm	Position	Natural δ ^{13}C/ppm	Synthetic δ ^{13}C/ppm	Δδ/ ppm
1	31.9	30.9	1	29	57.3	56.6	0.7
2	61.9	61.0	0.9	29-NH			
2-NH				30	36.9	36.7[b]	0.2
3	36.4	35.5	0.9	31	15.4	15.4[d]	0.0
				32	25.3	24.1	1.2
4	135.0	134.7	0.3				
5,5′	129.7	129.0	0.7	33	11.2	11.4[e]	−0.2
6,6′	128.9	128.3	0.6	34	171.6	170.7[c]	0.9
7	127.5	126.8	0.7	35	56.5	55.0	1.5
8	167.1	166.6	0.5	35-NH			
9	57.9	57.4	0.5	36	62.7	63.5	−0.8
9-NH							
10	36.5	35.8	0.7	36-OH			
11	15.5	14.9	0.6	37	171.7	171.4[c]	0.3
12	24.4	23.8	0.6	38	56.2	55.3	0.9
				38-NH			
				39	71.2	70.0	1.2
13	11.3	10.8	0.5	40	15.9	15.3	0.6
14	170.6	170.1	0.5	41	168.9	167.9	1.0
15	55.6	55.2	0.4	42	52.2	51.6	0.6
15-NH				42-NH			
16	62.4	61.7	0.7	43	17.1	16.5	0.6
				44	173.1	172.5	0.6
16-OH				45	52.2	51.9	0.3
17	170.2	169.7	0.5	45-NH			
18	52.7	52.0	0.7	46	37.2	36.2	1
18-NH				47	53.5	53.2	0.3
19[a]	28.4	27.9	0.5	47-NH			
20[a]	31.9	31.4	0.5	48	48.3	47.7	0.6
21	174.4	173.9	0.5	48-NH			
21-NH2				49	160.0	159.0	1
				49-NH			
22	170.9	170.9[c]	0.0	50	171.8	172.5[c]	−0.7

(continued)

Table 5.2 (continued)

Position	Natural δ ^{13}C/ppm	Synthetic δ ^{13}C/ppm	Δδ/ ppm	Position	Natural δ ^{13}C/ppm	Synthetic δ ^{13}C/ppm	Δδ/ ppm
23	56.8	55.5	1.3	51	57.8	57.0	0.8
23-NH				51-NH			
24	37.4	36.7[b]	0.7	52	36.3	35.2	1.1
25	14.7	14.3[d]	0.4	53	16.0	15.0	1.0
26	26.2	25.6	0.8	54	24.5	24.8	−0.3
27	10.6	11.2[e]	−0.6	55	11.8	10.1[e]	1.7
28	171.4	170.8[c]	0.6	56	169.3	169.3	0.0

Note [a]Methylene carbons at C-19 and C-20 were misassigned in the isolation paper by Ling et al. [5]. These assignments have been corrected in our data. [b]Assignment difficult due to signal overlap. [c,d,e]Correspond to signals in the isolation paper in which the 'assignments may be switched due to overlap' and were similarly difficult to assign in this case

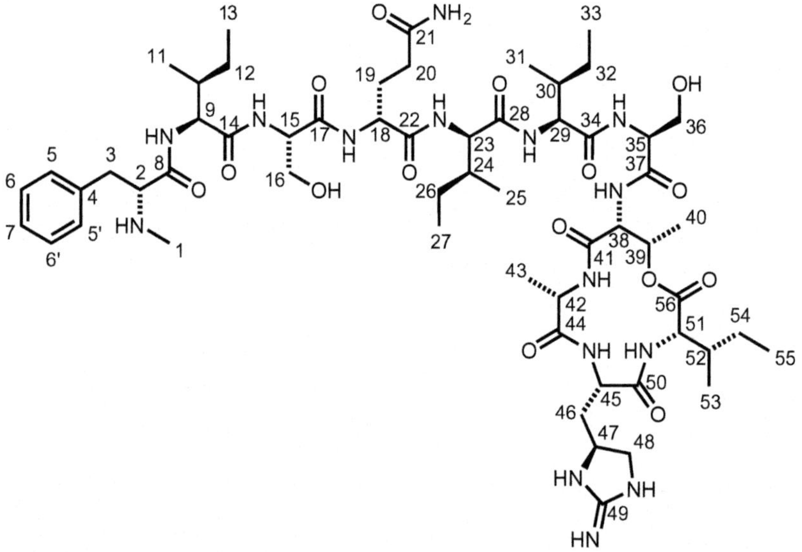

Fig. 5.1 Structure of teixobactin (**28**) with numbered carbon centres for use in NMR analysis

5.3.6 *Antimicrobial Screening of Teixobactin (28)*

Resazurin Assay for Mtb [6]

The compounds were originally stored as 10 mM stock solutions in 100% DMSO. Two fold serial dilutions of the compounds were made in a 96 well plate using

Middlebrook 7H9 medium supplemented with ADC (0.5% v/v glycerol and 0.05% v/v Tween-80). *M. tuberculosis* H37Rv was grown to mid-exponential phase to an OD_{600} of 0.6–0.8 in 7H9 media at 37 °C. On the day of the assay, culture was diluted to an OD_{600} of 0.002 and 100 µl of bacterial suspension was added to the 96 well plate containing 100 µL of the diluted compounds. The plate was incubated for 5 days at 37 °C in a humidified incubator and 30 µL of Resazurin (0.02% w/v) and 12.5 µL of Tween-80 was added to each well and incubated for further 24 h. On day 6, the fluorescence was read using a BMG Labtech Polarstar plate reader (excitation 530 nm and emission 590 nm). The results are presented as *M. tuberculosis* survival as a percentage of negative control (no drug controls).

High-Throughput Antibacterial Inhibition Assay
Bacterial test strains were grown on fresh agar plates and individual colonies used to inoculate 3 mL of sterile media. All staphylococcal strains were grown in tryptic soy broth (17 g tryptone, 3 g soytone, 2.5 g dextrose, 5 g NaCl and 2.5 g dipotassium phosphate in 1 L distilled water; pH = 7.5). *P. alcalifaciens, O. anthropi, E. aerogenes and A. baumanii* were grown in nutrient broth (Difco, USA) while *B. subtilis, E. coli, V. cholerae, S. typhimurium, P. aeruginosa* and *Y. pseudotuberculosis* cultures were grown in Luria Broth (10 g tryptone, 5 g yeast extract and 10 g NaCl in 1 L distilled water; pH = 7.5). All three media were autoclaved at 121 °C for 30 min. Inoculated cultures were grown overnight with shaking (200 rpm; 30°C). Saturated overnight cultures were diluted 1:1000 or 1:100 according to turbidity and dispensed into sterile clear polypropylene 384 well plates (30 µL screening volume). Optical density (OD_{600}) of cultures at a 1:100 dilution were recorded (Shimadzu UV-Visible Spectrophotometer) and further diluted on agar plates to calculate colony forming units (CFU) per milliliter of culture. DMSO solutions of test compounds (200 nL) were pinned into each well at t_0 using a high-throughput pinning robot (Perkin Elmer Janus MDT). In the 384 well plate lanes 1 and 2 were reserved for DMSO vehicle negative controls, while lanes 23 and 24 contained only culture medium and test organisms. After compound addition, screening plates were stacked in an automated plate reader/shaker (Perkin Elmer EnVision) and a OD_{600} reading was collected every 1 h for 16–20 h. The resulting growth curves for each dilution series were used to determine MIC values for all test compounds following standard procedures [7].

Bacterial Strains
Gram-positive: *Bacillus subtilis* 168, Methicillin susceptible *Staphylococcus aureus* (MSSA) (ATCC 29213), Methicillin resistant *S. aureus* (MRSA) (BAA-44).

Gram-negative: *Escherichia coli* K12 (BW 25113), *Providencia alcalifaciens* (ATCC 9886), *Ochrobactrum anthropi* (ATCC 49687), *Enterobacter aerogenes* (ATCC 35029), *Acinetobacter baumanii* (NCIMB 12457, *Vibrio cholerae* O1

Table 5.3 Average MIC values (μM) for teixobactin (**28**) and clinically relevant antibiotics derived from high-throughput antibacterial screening (see above) for select Gram-negative and Gram-positive bacterial strains

	Screening dilution	Average OD	Average CFU	Teixobactin (**28**) MIC (μM)	Vancomycin MIC (μM)	Linezolid MIC (μM)	Ciprofloxacin MIC (μM)
S. aureus (MSSA)	1000	0.55	3.3E+09	1.1	0.69	1.4	0.69
S. aureus (MRSA)	1000	0.47	4.9E+09	1.1	0.87	1.2	>66
E. coli	1000	0.58	3.2E+09	>27	>66	>66	0.013
B. subtilis	1000	0.43	3.0E+08	0.21	0.17	0.22	0.13
P. alcalifaciens	100	0.11	1.9E+10	>27	>66	>66	0.027
O. anthropi	100	0.17	7.0E+07	>27	>66	>66	0.85
E. aerogenes	100	0.2	3.8E+10	>27	>66	>66	0.022
A. baumanii	100	0.19	1.8E+10	>27	>66	>66	2.4
V. cholerae	1000	0.43	1.4E+11	>27	>66	>66	0.016
S. typhimurium	1000	0.4	5.0E+09	>27	>66	>66	0.027
P. aeruginosa	1000	0.48	2.0E+07	>27	>66	>66	1.4
Y. pseudotuberculosis	1000	0.42	2.0E+07	>27	>66	>66	0.0081

(biotype El Tor A1552), *Salmonella typhimurium* LT2, *Pseudomonas aeruginosa* (ATCC 27853), *Yersinia pseudotuberculosis* (IP2666 pIBI).

High-Throughput Antibacterial Inhibition Assay Results

See Table 5.3.

5.4 General Procedures for Chapter 3

5.4.1 Fmoc-SPPS Protocols for Chapter 3

2-CTC Resin Loading
2-CTC resin (maximum loading 0.9 mmol/g) in a fritted syringe was swollen in CH_2Cl_2 (5 mL) for 30 min before being washed with a solution of 20 vol.% *i*Pr_2NEt in CH_2Cl_2 (2 × 5 mL). The resin was then shaken with a solution of Fmoc-Gly-OH (0.6 equiv. relative to resin functionalisation) and *i*Pr_2NEt (1.2 equiv. relative to resin functionalisation) in CH_2Cl_2 at room temperature for 16 h. The resin was then washed CH_2Cl_2 (×5), DMF (×5) and CH_2Cl_2 (×5), and then treated with a solution of CH_2Cl_2:MeOH:*i*Pr_2NEt (v/v/v, 17:2:1, 5 mL) for 30 min. The resin was then washed CH_2Cl_2 (×5), DMF (×5) and CH_2Cl_2 (×10) before being dried under high vacuum and accurately split by weighing before the resin loading was determined. Loaded resin was treated with a solution of 10 vol.% piperidine in DMF (2 × 3 min) which was then diluted (100 μL in 10 mL) and the absorbance analysed at $\lambda = 301$ nm to determine number of μmol of amino acid loaded to resin.

2-CTC Resin Cleavage Conditions
Resin-bound peptide was washed with CH_2Cl_2 (×5) and DMF (×5) before being treated with a solution of 10 vol.% piperidine in DMF (2 × 3 min). The resin was again washed with DMF (×5), CH_2Cl_2 (×5), DMF (×5) and CH_2Cl_2 (×20). The resin was treated with a 20 vol.% solution of HFIP in CH_2Cl_2 (~5 mL, 4 × 4 min) and the solution transferred to a round bottom flask and diluted with CH_2Cl_2 (30 mL) before being concentrated under a stream of N_2 and dried in vacuo.

5.4.2 General Procedures for Modified Amino Acid Synthesis

General Procedure A: Selective Oxidation of the Primary Alcohol
To a solution of amino-diol (1 equiv.) in acetone (0.2 M) was added TEMPO (0.2 equiv.) followed by aqueous $NaHCO_3$ (5% g/100 mL) to bring the solution to

0.1 M. The mixture was cooled to 0 °C and sodium hypochlorite (3.5 equiv., 10–15% by weight) was added portionwise over 25 min. The mixture was warmed to room temperature and stirred for 3–16 h (until judged complete by TLC analysis). The reaction mixture was diluted with water and the aqueous phase was washed with Et_2O (2×) then acidified to pH = 2 with aqueous HCl (1 M). The aqueous phase was extracted with EtOAc (3×) and the combined organic layers were dried ($MgSO_4$), filtered and concentrated in vacuo.

General Procedure B: Boc Deprotection Followed by Fmoc Protection
To a solution of acid (1 equiv.) in CH_2Cl_2 (0.16 M) was added HCl in dioxane (10 equiv., 4 M). The solution was stirred at room temperature for 1.5–3 h (until judged complete by TLC analysis) and the solvent was removed in vacuo. To a solution of the crude residue in a mixture of THF: saturated aqueous $NaHCO_3$ (2:1 v/v, 0.1 M) was added Fmoc-OSu (1.05 equiv.). The reaction was stirred for 16 h at room temperature before being poured onto water and washed with Et_2O (2×). The aqueous phase was then acidified to pH = 1 with aqueous HCl (1 M) and extracted with EtOAc (3×) and the combined organic layers were dried ($MgSO_4$), filtered and concentrated in vacuo.

General Procedure C: Oxazolidine Protection of β-OH Amino Acid.
To a solution of alcohol (1 equiv.) in acetone (0.14 M) was added 2,2-dimethoxypropane (10 equiv.) followed by BF_3•OEt_2 (0.1 equiv.). The reaction mixture was stirred at room temperature for 5–16 h (until judged complete by TLC analysis). The reaction mixture was poured onto saturated aqueous NH_4Cl and extracted with EtOAc (3×) and the combined organic layers were dried ($MgSO_4$), filtered and concentrated in vacuo.

5.4.3 General Procedures for Peptide Cyclisation

General Procedure D: Peptide Cyclisation Conditions
Crude cleaved peptide was suspended in DMF (0.01 M) before a solution of DMTMM.BF_4 (1.5 equiv.) in DMF (0.01 M) with respect to peptide) was added, to give an overall concentration of 0.005 M of crude peptide, followed by $i$$Pr_2$NEt (2 equiv.) and stirred at room temperature for 16 h. The reaction mixture was analysed by HPLC-MS and concentrated under a stream of N_2. The crude solid was diluted with CH_2Cl_2 (~30 mL) and further concentration by a stream of N_2 and dried in vacuo.

General Procedure E: Cyclic Peptide Deprotection Conditions
To a solution of crude cyclised peptide in CH_2Cl_2 (0.001 M wrt. crude linear peptide) was added $i$$Pr_3$SiH (5% v/v of CH_2Cl_2) followed by TFA (1% v/v of CH_2Cl_2). The reaction mixture was stirred for 30 min, then diluted with CH_2Cl_2

(equal to initial volume) and concentrated under a stream of N_2 and dried in vacuo. Cyclic peptides were purified via HPLC using a Waters Sunfire C18 OBD 19 × 150 mm column, using a 0–60 vol.% MeCN in H_2O (0.1% TFA) focussed gradient (0–50 vol.% MeCN over 5 min, 50–60 vol.% over 15 min) at a flow rate flow rate of 16 mL min^{-1}.

5.5 Procedures and Analytical Data for Chapter 3

5.5.1 Synthesis of Cinnamoyl Moiety 130

2-iodobenzaldehyde (138)

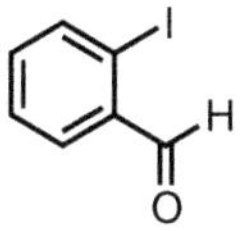

To a solution of *m*-iodo benzyl alcohol **137** (3.00 g, 12.8 mmol) in CH_2Cl_2 (30 mL) was added pyridinium chlorochromate (3.04 g, 15.14 mmol). The mixture was left to stir at room temperature for 3.5 h at which point TLC showed the presence of starting material. A further portion of pyridinium chlorochromate (0.5 g, 2.5 mmol) was added and the mixture was stirred for a further 1 h. The mixture was concentrated in vacuo and crude residue was purified by flash chromatography (eluent: 3:7 v/v EtOAc/petroleum benzines) to yield aldehyde **138** as a pale yellow solid (2.42 g, 76%).

^{1}H-NMR ($CDCl_3$, 500 MHz) δ (ppm) 10.07 (d, J = 0.7 Hz, 1H), 7.95 (dd, J = 7.90, 0.75 Hz 1H), 7.88 (dd, J = 7.7, 1.8 Hz, 1H), 7.46 (t, J = 7.5 Hz, 1H), 7.28 (td, J = 7.6, 1.8 Hz, 1H); **LRMS** (+ESI) *m/z* 233 $[M+Na]^+$; **mp.** 38.2–39.9 °C. These data are in agreement with those previously reported by Tummatorn and Dudley [8].

(*E*)-ethyl 3-(2-iodophenyl)acrylate (140)

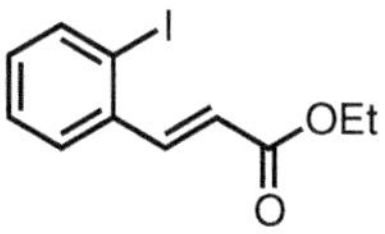

To a solution of NaH (265 mg, 6.65 mmol, 60% dispersion in oil) in CH_2Cl_2 (24 mL) at 0 °C was added triethyl phosphonoacetate **139** (0.96 mL, 4.87 mmol) dropwise. The mixture was stirred at this temperature for 5 min. A solution of aldehyde **138** (1.1 g, 4.43 mmol) in CH_2Cl_2 (19 mL) was added dropwise to the above mixture and the resultant mixture was stirred at 0 °C for 2 h. The reaction mixture was quenched with cold H_2O (10 mL) and poured onto H_2O (20 mL). The aqueous layer was extracted with EtOAc (4 × 20 mL). The combined organic layers were dried ($MgSO_4$), filtered and concentrated in vacuo. The crude product was purified by flash chromatography (eluent: 3:97 → 5:95 v/v EtOAc/petroleum benzines) to yield iodide **140** as a yellow oil which solidified on freezing (1.2 g, 89%).

^{1}H-NMR ($CDCl_3$, 400 MHz) δ (ppm) 7.90 (d, J = 15.8 Hz, 1H), 7.90 (dd, J = 8.0, 1.2 Hz, 1H), 7.56 (dd, J = 7.8, 1.6 Hz, 1H), 7.38–7.34 (m, 1H), 7.05 (td, J = 7.6, 1.6 Hz, 1H), 6.31 (d, J = 15.8 Hz, 1H), 4.29 (q, J = 7.1 Hz, 2H), 1.35 (t, J = 7.1 Hz, 3H); **LRMS** (+ESI) *m/z* 325 $[M+Na]^+$. These data are in agreement with those previously reported by Sun et al. [9].

(*E*)-ethyl 3-(2-(prop-1-yn-1-yl)phenyl)acrylate (142)

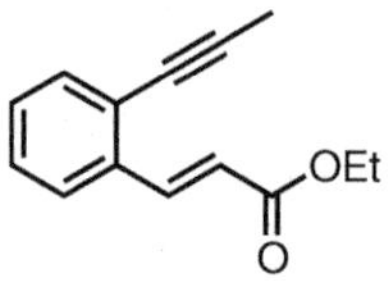

OEt
O

To a solution of 1-bromo-1-propene **143** (0.46 mL, 5.35 mmol) in THF (7.1 mL) at −78 °C was added *n*-BuLi (2.8 mL, 7.08 mmol, 2.5 M in hexanes) dropwise. The solution was stirred at this temperature for 45 min. H_2O (128 μL, 7.08 mmol) was added and the resultant solution was warmed to room temperature. After 45 min, a solution of iodide **140** (535 mg, 1.77 mmol) in THF (2.2 mL) was added dropwise, followed by CuI (34 mg, 0.18 mmol), *i*Pr_2NH (5.2 mL) and $Pd(PPh_3)_2Cl_2$ (62 mg, 0.089 mmol). The resultant mixture was stirred at room temperature for 16 h. The reaction mixture was then poured onto saturated aqueous NH_4Cl (30 mL), and extracted with Et_2O (3 × 30 mL). The combined organic layers were dried ($MgSO_4$), filtered and concentrated in vacuo. The crude product was purified by flash chromatography (eluent: 2:98 → 5:95 v/v EtOAc/petroleum benzines) to yield alkyne **142** as a yellow solid (344 mg, 91%).

^{1}H-NMR ($CDCl_3$, 500 MHz) δ (ppm) 8.21 (d, J = 16.1 Hz, 1H), 7.62–7.61 (m, 1H), 7.47–7.45 (m, 1H), 7.31–7.29 (m, 2H), 6.51 (d, J = 16.1 Hz, 1H), 4.29 (q, J = 7.1 Hz, 2H), 2.15 (s, 3H), 1.37 (t, J = 7.1 Hz, 3H); **^{13}C-NMR** ($CDCl_3$, 125 MHz) δ (ppm) 167.1, 143.0, 135.8, 133.1, 129.7, 127.9, 126.1, 125.1, 119.5, 92.5, 77.2, 60.6, 14.5, 4.7; **LRMS** (+ESI) *m/z* 215 $[M+H]^+$; **HRMS** (+ESI) Calc. for $C_{14}H_{14}O_2$ $[M+Na]^+$: 237.0886, Found: 237.0886; **IR** ν_{max} (ATR) 2980, 2916, 2247, 2212, 1710, 1633, 1478, 1315, 1268, 1176 cm^{-1}; **mp**. 37.6–39.4 °C.

(*E*)-ethyl 3-(2-((*Z*)-prop-1-en-1-yl)phenyl)acrylate (149)

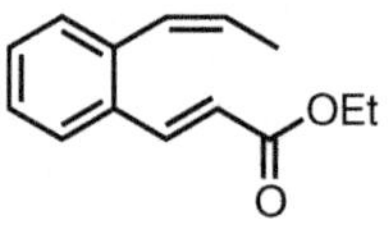

OEt
O

To a solution of alkyne **142** (257 mg, 1.2 mmol) in MeOH (11.8 mL) was added quinoline (154 μL, 1.24 mmol) and Lindlar's catalyst (258 mg, 0.12 mmol, 5% Pd by weight). The solution was stirred under an atmosphere of H_2 for 20 min. A further portion of Lindlar's catalyst (200 mg) was added and the mixture was stirred under an atmosphere of H_2 for 50 min. The mixture was then filtered over Celite® and concentrated in vacuo. The crude product was purified by flash chromatography (eluent: 1:99 → 5:95 v/v EtOAc/petroleum benzines) to yield diene **149** as a colourless oil (230 mg, 88%).

^{1}H-NMR ($CDCl_3$, 400 MHz) δ (ppm) 7.89 (d, J = 16.0 Hz, 1H), 7.60 (d, J = 7.8 Hz, 1H), 7.36–7.21 (m, 3H), 6.58 (dd, J = 11.4, 1.5 Hz, 1H), 6.39 (d, J = 16.0 Hz, 1H), 5.97–5.92 (m, 1H), 4.26 (q, J = 7.1 Hz, 2H), 1.66 (dd, J = 7.0, 1.8 Hz, 3H), 1.33 (t, J = 7.1 Hz, 3H); **^{13}C-NMR** ($CDCl_3$, 100 MHz) δ (ppm) 167.2, 143.1, 138.1, 133.0, 130.2, 129.6, 129.4, 127.9, 127.2, 126.6, 119.2, 60.5, 14.5 14.4; **LRMS** (+ESI) *m/z* 217 $[M+H]^+$; **HRMS** (+ESI) Calc. for $C_{14}H_{16}O_2$ $[M+Na]^+$: 239.1043, Found: 239.1043; **IR** ν_{max} (ATR) 2979, 1708, 1632, 1477, 1445, 1311, 1268, 1165 cm^{-1}.

(*E*)-3-(2-((*Z*)-prop-1-en-1-yl)phenyl)acrylic acid (130)

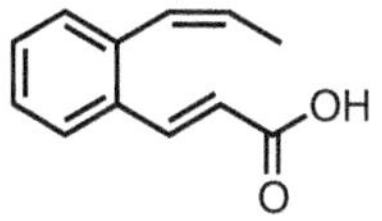

To a solution of diene **149** (189 mg, 0.87 mmol) in ethanol (2.8 mL, 95%) at 0 °C was added LiOH (84 mg, 3.5 mmol). The solution was warmed to room temperature and stirred for 16 h. The mixture was cooled to 0 °C and acidified to pH = 1 with aqueous HCl (1 M). The mixture was extracted with EtOAc (4 × 20 mL). The combined organic layers were dried (Na_2SO_4), filtered and concentrated in vacuo to yield acid **130** as an off-white solid (152 mg, 92%).

^{1}H-NMR ($CDCl_3$, 400 MHz) δ (ppm) 8.01 (d, J = 16.0 Hz, 1H), 7.66 (dd, J = 7.6, 0.7 Hz, 1H), 7.38–7.24 (m, 4H), 6.59 (dd, J = 11.4, 1.5 Hz, 1H), 6.42 (d, J = 16.0, 1H), 5.98 (dq, J = 11.4, 7.0 Hz, 1H), 1.66 (dd, J = 7.0, 1.8 Hz, 3H); **^{13}C-NMR** ($CDCl_3$, 100 MHz) δ (ppm) 172.4, 145.7, 138.5, 132.6, 130.3, 130.2, 129.8, 127.8, 127.3, 126.8, 118.1, 14.6; **LRMS** (+ESI) *m/z* 189 $[M+H]^+$; **HRMS** (+ESI) Calc. for $C_{12}H_{12}O_2$ $[M+Na]^+$: 211.0730, Found: 211.0730; **IR** ν_{max} (ATR) 3014 (broad), 1678, 1623, 1479, 1422, 1332, 1300, 1288, 1224 cm^{-1}; **mp.** 128.7–132.2 °C.

5.5.2 Synthesis of Reduced Cinnamoyl Moiety 131

ethyl 3-(2-iodophenyl)propanoate (204)

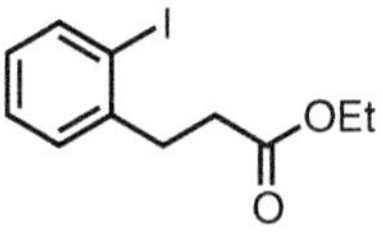

To a solution of ester **140** (500 mg, 1.65 mmol) in ethanol (5.3 mL, 95%) at 0 °C was added LiOH (159 mg, 6.62 mmol). The solution was warmed to room temperature and stirred for 16 h. The mixture was cooled to 0 °C and acidified to pH = 1 with aqueous HCl (0.1 M). The mixture was extracted with EtOAc (3 × 20 mL). The combined organic layers were dried ($MgSO_4$), filtered and concentrated in vacuo to yield acid **205** as a white solid (447 mg) which was used without further purification.

To a solution of the above crude acid **205** (447 mg, 1.63 mmol) in a solution of ethanol (5 mL) and EtOAc (5 mL) was added guanidine nitrate (36.5 mg, 0.163 mmol) followed by hydrazine hydrate (217 μL, 4.89 mmol). The reaction mixture was stirred under an atmosphere of O_2 for 16 h, at which point an aliquot was analysed by ^{1}H-NMR. Starting material was detected, and so the reaction mixture was stirred under O_2 for a further 16 h. The mixture was then concentrated under a stream of N_2. Water (30 mL) was added and then acidified to pH = 1 with aqueous HCl (1 M). The mixture was extracted with EtOAc (3 × 20 mL). The combined organic layers were dried ($MgSO_4$), filtered and concentrated in vacuo to yield reduced acid **206** as a white solid (446 mg) which was used without further purification.

To a solution of the above crude reduced acid **206** (398 mg, 1.44 mmol) in ethanol (6 mL) at 0 °C was added $SOCl_2$ (526 μL, 7.20 mmol) dropwise. The solution was warmed to room temperature and stirred for 16 h. The solution was concentrated in vacuo. The crude product was purified by flash chromatography (eluent: 5:95 v/v EtOAc/petroleum benzines) to yield reduced ethyl ester **204** as a colourless oil (413 mg, 94%).

^{1}H-NMR ($CDCl_3$, 400 MHz) δ (ppm); 7.82 (d, *J* = 8.4 Hz, 1H), 7.30–7.22 (m, 2H), 6.90 (ddd, *J* = 7.9, 6.6, 2.5, 1H), 4.14 (q, *J* = 7.2 Hz, 2H), 3.05 (t, *J* = 7.6 Hz, 2H), 2.62 (t, *J* = 8.2 Hz, 2H), 1.25 (d, *J* = 7.1 Hz, 3H); **LRMS** (+ESI) *m/z* 327 $[M+Na]^+$. These data are in agreement with those previously reported by Tummatorn and Dudley [8].

ethyl 3-(2-(prop-1-yn-1-yl)phenyl)propanoate (209)

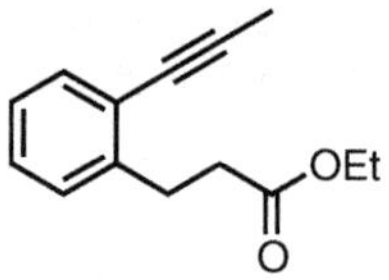

To a solution of 1-bromo-1-propene (0.57 mL, 6.71 mmol) in THF (5 mL) at −78 ° C was added *n*-BuLi (3.58 mL, 8.96 mmol, 2.5 M in hexanes) dropwise. The solution was stirred at this temperature for 45 min. H_2O (161 μL, 8.96 mmol) was added and the resultant solution was warmed to room temperature. After 45 min, a solution of iodide **204** (340 mg, 1.12 mmol) in THF (1.5 mL + 0.3 mL rinse) was added dropwise, followed by CuI (43 mg, 0.22 mmol), *i*Pr_2NH (3.12 mL) and Pd $(PPh_3)_2Cl_2$ (79 mg, 0.112 mmol). The resultant mixture was stirred at room temperature for 16 h. The reaction mixture was then poured onto saturated aqueous NH_4Cl (30 mL), and extracted with Et_2O (3 × 30 mL). The combined organic layers were dried ($MgSO_4$), filtered and concentrated in vacuo. The crude product was purified by flash chromatography (eluent: 5:95 v/v EtOAc/petroleum benzines) to yield alkyne **209** as a colourless oil (202 mg, 83%).

^{1}H-NMR ($CDCl_3$, 500 MHz) δ (ppm) 7.37 (d, *J* = 7.6 Hz, 1H), 7.21–7.11 (m, 3H), 4.13 (q, *J* = 7.2 Hz, 2H), 3.08 (t, *J* = 7.6, 2H), 2.65 (t, *J* = 8.1 Hz, 2H), 2.09 (s, 3H), 1.24 (t, *J* = 7.3 Hz, 3H); **^{13}C-NMR** ($CDCl_3$, 125 MHz) δ (ppm) 173.3, 142.4, 132.4, 128.8, 127.9, 126.3, 123.6, 90.1, 78.1, 60.5, 35.0, 30.1, 14.4, 4.6; **LRMS** (+ESI) *m/z* 239 $[M+Na]^+$; **HRMS** (+ESI) Calc. for $C_{14}H_{16}O_2$ $[M+Na]^+$:

239.1043, Found: 239.1043; **IR** ν_{max} (ATR) 2980, 2917, 2252, 1730, 1485, 1447, 1371, 1293, 1181, 1155 cm^{-1}.

(Z)-ethyl 3-(2-(prop-1-en-1-yl)phenyl)propanoate (210)

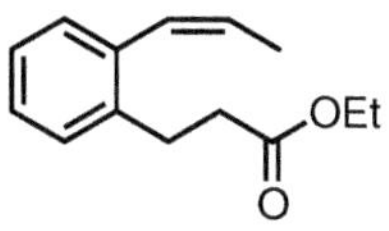

To a solution of alkyne **209** (102 mg, 0.47 mmol) in MeOH (4.7 mL) was added quinoline (58 μL, 0.47 mmol) and Lindlar's catalyst (100 mg, 0.047 mmol, 5% Pd by weight). The solution was stirred under an atmosphere of H_2 for 1 h. A further portion of Lindlar's catalyst (100 mg) was added and the mixture was stirred under an atmosphere of H_2 for 45 min. The reaction was monitored by TLC every 20 min, and if the reaction was not complete a further portion of Lindlar's catalyst (200–300 mg) was added. The reaction was dosed a total of 6 times in order to push it to completion. The mixture was then filtered over Celite® and concentrated in vacuo. The crude product was purified by flash chromatography (eluent: 5:95 v/v EtOAc/petroleum benzines) to yield alkene **210** as a colourless oil (85 mg, 83%).

^{1}H-NMR ($CDCl_3$, 400 MHz) δ (ppm) 7.22–7.17 (m, 4H), 6.53 (ap. dd, J = 11.5, 1.7 Hz, 1H), 5.87 (dq, J = 11.5, 7.0 Hz, 1H), 4.13 (q, J = 7.1 Hz, 2H), 2.93 (t, J = 7.8 Hz, 2H), 2.54 (t, J = 8.3 Hz, 2H), 1.73 (dd, J = 7.0, 1.8 Hz, 3H), 1.24 (t, J = 7.2 Hz, 3H); **^{13}C-NMR** ($CDCl_3$, 100 MHz) δ (ppm) 173.2, 138.8, 136.4, 129.8, 129.0, 128.4, 127.8, 127.1, 126.1, 60.5, 35.1, 28.9, 14.4, 14.3; **LRMS** (+ESI) m/z 241 $[M+Na]^+$; **HRMS** (+ESI) Calc. for $C_{14}H_{18}O_2$ $[M+Na]^+$: 241.1199, Found: 241.1199; **IR** ν_{max} (ATR) 2980, 1730, 1447, 1371, 1288, 1252, 1178, 1157, 1113 cm^{-1}.

(Z)-3-(2-(prop-1-en-1-yl)phenyl)propanoic acid (131)

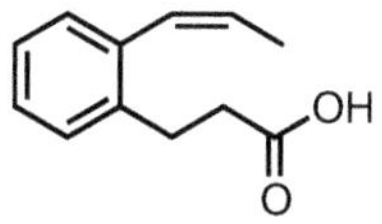

To a solution of alkene **210** (79 mg, 0.36 mmol) in ethanol (4 mL, 95%) at 0 °C was added LiOH (35 mg, 1.45 mmol). The solution was warmed to room temperature and stirred for 16 h. The mixture was cooled to 0 °C and acidified to pH = 1 with aqueous HCl (1 M). The mixture was extracted with EtOAc (3 × 20 mL). The combined organic layers were dried (Na_2SO_4), filtered and concentrated in vacuo to yield acid **131** as an off-white solid (62 mg, 89%).

^{1}H-NMR ($CDCl_3$, 500 MHz) δ (ppm) 7.24–7.14 (m, 4H), 6.53 (dq, J = 11.4, 1.8 Hz, 1H), 5.88 (dq, J = 11.4, 7.0 Hz, 1H), 2.94 (t, J = 7.7 Hz, 2H), 2.61 (t, J = 8.3 Hz, 2H), 1.73 (dd, J = 7.0, 1.8 Hz, 3H); **^{13}C-NMR** ($CDCl_3$, 125 MHz) δ (ppm) 179.6, 138.4, 136.4, 129.9, 129.0, 128.3, 128.0, 127.2, 126.3, 34.8, 28.5, 14.4; **LRMS** (+ESI) m/z 213 $[M+Na]^+$; **HRMS** (+ESI) Calc. for $C_{12}H_{14}O_2$ $[M+Na]^+$: 213.0886, Found: 213.0886; **IR** ν_{max} (ATR) 3014 (broad), 1701, 1435, 1407, 1319, 1275, 1215 cm^{-1}; **mp.** 62.1–65.0 °C.

5.5.3 *Synthesis of Garner's Aldehyde (158) and (159)*

(*R*)-3-*tert*-butyl 4-methyl 2,2-dimethyloxazolidine-3,4-dicarboxylate (164)

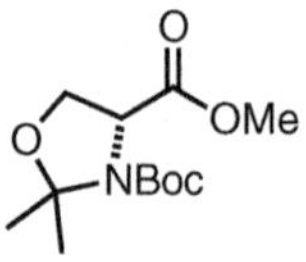

To a solution of D-serine methyl ester hydrochloride **163** (15 g, 96.4 mmol) in a mixture of saturated aqueous $NaHCO_3$ solution (120 mL) and THF (30 mL) was added di-*tert*-butyl dicarbonate (25.8 g, 118 mmol) and the reaction mixture was stirred at room temperature for 16 h. The mixture was then poured onto water (150 mL) and extracted with EtOAc (3 × 150 mL). The combined organic extract was washed with saturated aqueous NaCl (150 mL), dried (Na_2SO_4), filtered and concentrated in vacuo to afford the desired Boc-protected amino acid as a colourless oil which was used without purification (21.1 g, quant).

To a solution of the above crude Boc amino acid (21.1 g, 96.4 mmol) in acetone (314 mL) and 2,2-dimethoxypropane (103.8 mL, 845 mmol) was added $BF_3{\bullet}OEt_2$ (0.67 mL, 5.2 mmol). The reaction mixture was stirred at room temperature for 3 h before being concentrated in vacuo. The residue was dissolved in CH_2Cl_2 (150 mL) and washed with saturated aqueous $NaHCO_3$ (150 mL) and saturated aqueous NaCl (150 mL), dried ($MgSO_4$), filtered and concentrated in vacuo. The crude product was purified by flash chromatography (eluent: 5:95 → 8:92 v/v EtOAc/petroleum benzines) to yield (*R*)-methyl ester **164** as a yellow oil (18.7 g, 75%).

^{1}H-NMR ($CDCl_3$, 500 MHz, rotamers) δ (ppm) 4.48 (dd, *J* = 6.8, 2.7 Hz, 0.4H) and 4.37 (dd, *J* = 7.0, 3.0 Hz, 0.6H), 4.17–4.0 (m, 2H), 3.75 (s, 3H), 1.68–1.62 (m, 3H), 1.55–1.47 (m, 7H), 1.40 (s, 5H); **LRMS** (+ESI) *m/z* 282 $[M+Na]^+$; $[\alpha]_D$ = +64.7 (*c* 0.17, CH_2Cl_2). These data are in agreement with those previously reported by Garner and Park [10].

(*R*)-*tert*-butyl 4-formyl-2,2-dimethyloxazolidine-3-carboxylate (158)

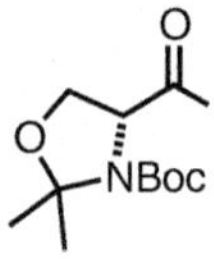

To a solution of (*R*)-methyl ester **164** (10.4 g, 40.1 mmol) in toluene (80 mL) at −78 °C was added a solution of DIBAL-H (70.2 mL, 70.2 mmol, 1 M in hexane) over 2 h via dropping funnel. The reaction mixture was stirred for 30 min at this temperature before being quenched with MeOH (10 mL) and warmed to room temperature. The solution was poured onto cold aqueous HCl (150 mL, 1 M) and extracted with EtOAc (3 × 100 mL). The combined organic phase was washed with cold aqueous HCl (150 mL, 1 M), dried (Na_2SO_4), filtered and concentrated in vacuo. The crude product was purified by flash chromatography (eluent: 15:85 → 25:75 v/v EtOAc/petroleum benzines) to yield (*R*)-Garner's aldehyde **158** as a colourless oil (7.5 g, 83%).

^{1}H-NMR ($CDCl_3$, 300 MHz, roatmers) δ (ppm) 9.58 (s, 0.4H) and 9.52 (s, 0.6H), 4.34–4.26 (m, 0.4H) and 4.21–4.13 (m, 0.6H), 4.11–3.99 (m, 2H), 1.62 (s, 1.7H) and 1.57 (s, 1.3H), 1.52 (s, 1.7H) and 1.47 (s, 1.3H), 1.49 (s, 4.2H) and 1.40 (s, 4.8H); **LRMS** (+ESI) *m/z* 252 $[M+Na]^+$; **$[\alpha]_D$** = +78.3 (*c* 0.95, CH_2Cl_2). These data are in agreement with those previously reported by Garner and Park [10].

(*S*)-*tert*-butyl 4-formyl-2,2-dimethyloxazolidine-3-carboxylate (159)

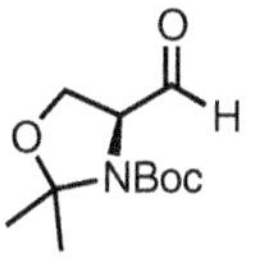

The antipode of Garner's aldehyde synthesized above was synthesized via the same method from the corresponding (*S*)-methyl ester (**165**) (10.0 g, 38.6) to yield (*S*)-Garner's aldehyde (**159**) as a colourless oil (7.37 g, 83%).

The ^{1}H-NMR data and LRMS data are identical to that described for (*R*)-Garner's aldehyde (**158**). **$[\alpha]_D$** = −80.0 (*c* 0.98, CH_2Cl_2). These data are in agreement with those previously reported by Garner and Park [10].

5.5.4 Synthesis of Oxazolidine Protected Fmoc-β-OH-Leu-OH 166

(*S*)-*tert*-butyl 4-((*S*)-1-hydroxy-2-methylpropyl)-2,2-dimethyloxazolidine-3-carboxylate (167)

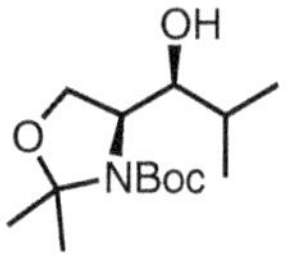

Magnesium granules (3.9 g, 159.5 mmol) were heated under vacuum with a heat gun for 2 min. THF (10 mL) and iodine (10 crystals) were then added and the mixture was stirred vigorously under argon for 15 min. A solution of isopropyl bromide (9 mL, 05.9 mmol) in THF (35 mL) was added dropwise over 20 min by which point the mixture turned a dull grey colour and began to self reflux. It was heated at reflux for 1 h and then cooled to room temperature. The solution was then added dropwise via cannula over 15 min, to a solution of (*S*)-Garner's aldehyde (**159**) (7.3 g, 31.9 mmol) at −78 °C. The solution was stirred for 2 h at this temperature and then warmed to room temperature and stirred for 1 h. The mixture was then diluted with Et_2O (100 mL), and quenched with saturated aqueous NH_4Cl (200 mL). The aqueous phase was extracted with Et_2O (2 × 100 mL) and the combined organic layers were washed with saturated aqueous NH_4Cl (2 × 150 mL), dried ($MgSO_4$), filtered and concentrated in vacuo. The crude product was purified by flash chromatography (eluent: 8:92 → 10:90 v/v EtOAc/petroleum benzines) to yield a single diastereomer of alcohol **167** as a white crystalline solid (3.57 g, 41%).

^{1}H-NMR ($CDCl_3$, 500 MHz) δ (ppm) 4.09–3.96 (br. s, 1H), 3.93 (dd, *J* = 9.3, 5.8 Hz, 1H), 3.76 (d, *J* = 9.1 Hz, 1H), 3.50 (br. d, *J* = 7.4 Hz, 1H), 1.66 (septd, *J* = 6.8, 2.4 Hz, 1H), 1.60 (s, 3H), 1.51 (s, 3H), 1.49 (s, 9H), 1.03 (d, *J* = 6.8 Hz, 3H), 0.90 (d, *J* = 6.7 Hz, 3H); **LRMS** (+ESI) *m/z* 296 $[M+Na]^+$; **$[\alpha]_D$** = −51.1° (*c* 0.45, CH_2Cl_2); **mp**. 84.6–87.3 °C. These data are in agreement with those previously reported by Williams et al. [11].

(4*R*,5*S*)-3-(((9*H*-fluoren-9-yl)methoxy)carbonyl)-5-isopropyl-2,2-dimethyloxazolidine-4-carboxylic acid (166)

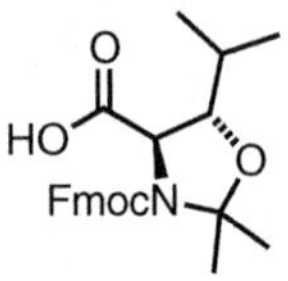

To a solution of alcohol **167** (3.46 g, 12.65 mmol) in THF (400 mL) was added aqueous HCl (17.0 mL, 0.5 M). The solution was stirred at room temperature for 16 h at which point it was dried ($MgSO_4$), filtered and concentrated in vacuo to yield diol **169** as a colourless oil which was used without further purification.

Crude diol **169** (3.49 g, 12.65 mmol) was oxidised by general procedure A to produce carboxylic acid **170** which was used without further purification. Crude acid **170** (2.48 g, 10 mmol) was subject to protecting group manipulation via general procedure B to produce Fmoc amino acid **179** which was used without further purification. Crude Fmoc amino acid **179** (3.20 g, 8.65 mmol) was oxazolidine protected using general procedure C. Crude oxazolidine protected amino acid **166** was purified by flash chromatography (eluent: 10:90 → 60:40 v/v EtOAc/petroleum benzines) to yield oxazolidine protected amino acid **166** as an off-white crystalline solid (2.72 g, 53% from alcohol **167**).

^{1}H-NMR ($CDCl_3$, 500 MHz, rotamers) δ (ppm) 8.52 (br. s, 1H), 7.78–7.67 (m, 2H), 7.58–7.49 (m, 2H), 7.40–7.25 (m, 4H), 4.66 (d, *J* = 4.2 Hz, 1H), 4.42 (m, 1H), 4.22–4.05 (m, 2H), 3.87 (br. t, *J* = 6.1 Hz, 0.5H) and 3.79 (br. t, *J* = 6.4 Hz, 0.5H), 1.91–1.71 (m, 1H), 1.60 (m, 3H), 1.09–0.87 (m, 9H); **^{13}C-NMR** ($CDCl_3$, 125 MHz, rotamers) δ (ppm) 176.6 and 176.1, 153.1 and 152.0, 144.2 and 144.0, 143.8 and 143.7, 141.7 and 141.6, 141.5 and 141.4, 127.8, 127.3 and 127.2, 127.2 and 127.2, 125.0, 124.5 and 124.5, 120.0 and 120.0, 96.0 and 95.0, 83.9 and 83.9, 67.2 and 66.9, 62.6 and 61.8, 47.4, 32.1 and 31.7, 26.9, 24.8 and 24.7, 18.8 and 18.5, 17.7 and 17.6 **LRMS** (+ESI) *m/z* 432 $[M+Na]^+$; **HRMS** (+ESI) Calc. for $C_{24}H_{27}NO_5$ $[M+Na]^+$: 432.1781, Found: 432.1782; **IR** ν_{max} (ATR) 2961, 2924, 1714, 1697, 1411, 1348, 1259 cm^{-1}; **$[\alpha]_D$** = +29.6° (*c* 0.48, CH_2Cl_2); **mp**. 59.5–60.5 °C.

5.5.5 *Synthesis of Fmoc-β-OH-Phe-OH (180)*

(4*R*)-*tert*-butyl 4-(hydroxy(phenyl)methyl)-2,2-dimethyloxazolidine-3- carboxylate (183)

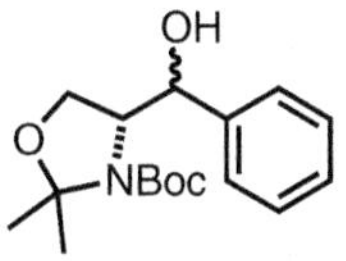

Magnesium granules (545 mg, 22.4 mmol) were heated under vacuum with a heat gun for 2 min. Ether (4 mL) and iodine (2 crystals) were then added and the mixture was stirred vigorously under argon for 20 min. A solution of bromobenzene (1.24 mL, 11.8 mmol) in THF (12 mL) was added dropwise over 10 min by which point the mixture turned a dull grey colour and began to self reflux. The reaction was refluxed for 1 h and then cooled to 0 °C. To the reaction mixture was then added dropwise a solution of (*R*)-Garner's aldehyde (**158**) (2.00 g, 8.72 mmol). The reaction mixture was stirred for 20 min at this temperature and then warmed to room temperature and stirred for 1 h. The mixture was then quenched with saturated aqueous NH_4Cl (100 mL). The aqueous phase was extracted with EtOAc (3 × 70 mL) and the combined organic layers were dried ($MgSO_4$), filtered and concentrated in vacuo. The crude product was purified by flash chromatography (eluent: 20:80 v/v EtOAc/petroleum benzines) to yield alcohol **183** as a white solid (2.12 g, 79%, *syn*/*anti* 2:2.7).

^{1}H-NMR ($CDCl_3$, 400 MHz) δ (ppm) 7.43–7.18 (m, 10H), 5.05 (m, 1H), 4.74 (d, *J* = 8.8 Hz, 1H), 4.40–3.50 (m, 3H), 1.60–1.33 (m, 15H); **LRMS** (+ESI) *m/z* 330 $[M+Na]^+$; $[\alpha]_D$ = +8.0° (*c* 0.5, CH_2Cl_2); **mp.** 79.9–84.0 °C. These data are in agreement with those previously reported by Malins et al. [12].

(*R*)-*tert*-butyl 4-benzoyl-2,2-dimethyloxazolidine-3-carboxylate (184)

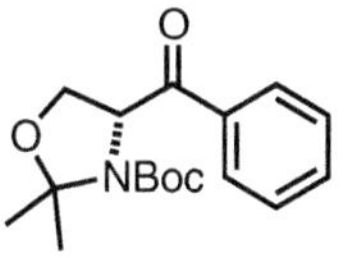

To a solution of oxalyl chloride (0.90 mL, 10.6 mmol) in CH_2Cl_2 (16 mL) at −78 °C was added a solution of dimethylsulfoxide (1.20 mL, 18.1 mmol) in CH_2Cl_2 (4.3 mL) dropwise. The reaction mixture was stirred at this temperature for 10 min. A solution of alcohol **183** (1.63 g, 5.3 mmol) in CH_2Cl_2 (4.3 mL) was added dropwise and the mixture was stirred at −78 °C for 1 h. *i*Pr_2NEt (3.25 mL, 18.6 mmol) was added and the mixture was stirred at this temperature for 10 min and then warmed to room temperature. The reaction mixture was poured onto saturated aqueous NH_4Cl (100 mL). The aqueous phase was extracted with CH_2Cl_2 (3 × 70 mL) and the combined organic layers were dried ($MgSO_4$), filtered and concentrated in vacuo. The crude product was purified by flash chromatography (eluent: 20:80 v/v EtOAc/ petroleum benzines) to yield ketone **184** as a white solid (1.38 g, 85%).

^{1}H-NMR ($CDCl_3$, 400 MHz, rotamers) δ (ppm) 7.93–7.89 (m, 2H), 7.62–7.44 (m, 3H), 5.47 (dd, J = 7.4, 2.9 Hz, 0.4H) and 5.36 (dd, J = 7.7, 3.7 Hz, 0.6H), 4.31 (ddd, J = 9.0, 7.6, 4.8, 1H) and 3.98–3.90 (m, 1H), 1.76 (s, 2H) and 1.73 (s, 1H), 1.60 (s, 2H) and 1.56 (s, 1H), 1.50 (s, 4H) and 1.28 (s, 5H); **LRMS** (+ESI) *m/z* 328 $[M+Na]^+$; **$[\alpha]_D$** = +59.6° (*c* 0.26, CH_2Cl_2); **mp**. 114.6–117.5 °C These data are in agreement with those previously reported by Malins et al. [12].

(*R*)-*tert*-butyl 4-((*S*)-hydroxy(phenyl)methyl)-2,2-dimethyloxazolidine-3-carboxylate (181)

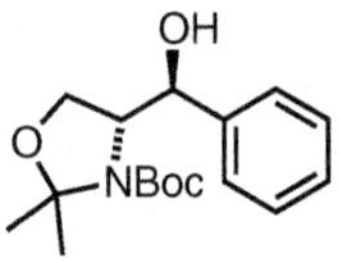

OH

O NBoc

To a solution of ketone **184** (1.38 g, 4.51 mmol) in THF (190 mL) at 0 °C was added diisobutyl aluminium hydride (13.5 mL, 1 M in hexanes, 13.5 mmol) dropwise over 30 min. The solution was stirred at this temperature for 30 min. MeOH (10 mL) was added and the reaction mixture was poured onto ice cold HCl (150 mL, 1 M). The aqueous phase was extracted with EtOAc (3 × 70 mL) and the combined organic layers were dried ($MgSO_4$), filtered and concentrated in vacuo. The crude product was purified by flash chromatography (eluent: 20:80 v/v EtOAc/petroleum benzines) to yield a 14:1 mixture of diastereomers of alcohol **181** as a white solid (1.23 g, 89%).

^{1}H-NMR ($CDCl_3$, 500 MHz, 328 K) δ (ppm) 7.41–7.26 (m, 5H), 5.11 (s, 1H), 4.24–4.21 (m, 1H), 4.07 (dd, J = 9.1, 2.0 Hz, 1H), 3.86–3.84 (m, 1H), 1.55–1.45 (m, 15H); **LRMS** (+ESI) *m/z* 330 $[M+H]^+$; **$[\alpha]_D$** = +22.9° (*c* 0.48, CH_2Cl_2); 96.2–102.0 °C. These data are in agreement with those previously reported by Malins et al. [12].

***tert*-butyl ((1*S*,2*R*)-1,3-dihydroxy-1-phenylpropan-2-yl)carbamate (185)**

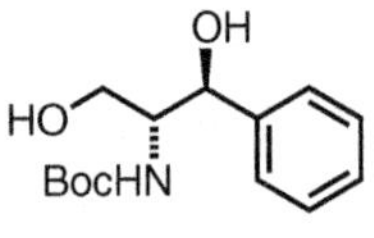

OH

HO

BocHN

To a solution of oxazolidine **181** (800 mg, 2.6 mmol) in MeOH (27 mL) was added *p*-toluene sulfonic acid (198 mg, 0.46 mmol). The mixture was stirred at room temperature for 2.5 h. The reaction mixture was poured onto saturated aqueous $NaHCO_3$ (50 mL). The aqueous phase was extracted with EtOAc (3 × 30 mL) and the combined organic layers were dried ($MgSO_4$), filtered and concentrated in vacuo. The crude product was purified by flash chromatography (eluent: 50:50 v/v EtOAc/petroleum benzines) to yield a 14:1 mixture of diastereomers of diol **185** as a white solid (510 mg, 74%).

^{1}H-NMR ($CDCl_3$, 400 MHz, 328 K) δ (ppm) 7.41–7.24 (m, 5H), 5.30 (d, J = 7.6 Hz, 1H), 5.02–4.96 (m, 1H), 3.83–3.73 (m, 1H), 3.65–3.57 (m, 1H), 3.34 (d, J = 3.7 Hz, 1H), 2.67 (brs, 1H), 1.43 (s, 9H); **LRMS** (+ESI) *m/z* 290 $[M+Na]^+$;

[α]$_D$ = +2.1° (*c* 0.27, CH_2Cl_2); **mp**. 80.8–84.2 °C. These data are in agreement with those previously reported by Williams et al. [11].

(2*S*,3*S*)-2-((((9*H*-fluoren-9-yl)methoxy)carbonyl)amino)-3-hydroxy-3- phenyl propan oic acid (180)

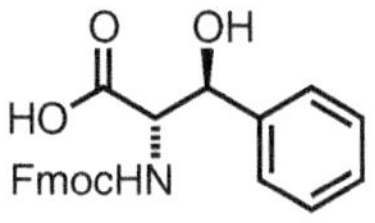

Diol **185** (479 mg, 1.79 mmol) was oxidised by general procedure A to produce carboxylic acid **186** which was used without further purification. Crude acid **186** (371 mg, 1.32 mmol) was subject to protecting group manipulation via general procedure B to afford Fmoc amino acid **180** which was purified by flash chromatography (eluent: 70:30 → 100:0 + 0.1% AcOH v/v EtOAc/petroleum benzines) to yield Fmoc protected amino acid **180** as an off-white crystalline solid (463 mg, 64% from diol **185**).

^{1}H-NMR (MeOD, 400 MHz) δ (ppm) 7.76 (d, *J* = 7.6 Hz, 2H), 7.57–7.50 (m, 2H), 7.46–7.19 (m, 9H), 4.99 (d, *J* = 7.3 Hz, 1H), 4.52 (d, *J* = 7.3 Hz, 1H), 4.30–4.04 (m, 3H); **^{13}C-NMR** (MeOD, 100 MHz) δ (ppm) 174.0, 158.1, 145.2, 145.1, 142.5, 142.4, 142.0, 129.2, 128.9, 128.7, 128.1, 128.0, 126.3, 126.2, 120.9, 75.1, 68.2, 61.3 48.2 **LRMS** (+ESI) *m/z* 426 $[M+Na]^+$; **HRMS** (+ESI) Calc. for $C_{24}H_{21}NO_5$ $[M+Na]^+$: 426.1312, Found: 436.1313; **IR** ν_{max} (ATR) 3408, 3314, 3064, 3083, 1711, 1522, 1450, 1414, 1332, 1260, 1233 cm^{-1}; **[α]$_D$** = +24.1° (*c* 0.19, CH_2Cl_2:MeOH, 0.91:0.9); **mp**. 164.1–173.0 °C

5.5.6 *Synthesis of Oxazolidine Protected Fmoc-β-OH-O-Me-Tyr-OH (187)*

(*R*)-*tert*-butyl 4-((*S*)-hydroxy(4-methoxyphenyl)methyl)-2,2-dimethyloxazolid ine-3- carboxylate (188)

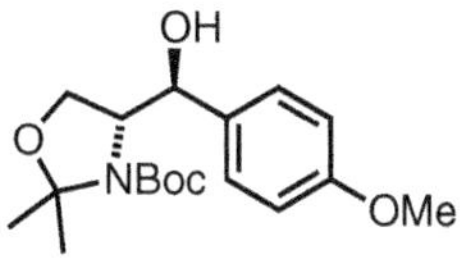

To a solution of dry LiBr (6.66 g, 88.8 mmol) in THF (140 mL) was added *p*-bromoanisole **189** (3.97 mL, 31.53 mmol). The reaction mixture was cooled to −78 °C and *n*-BuLi (13.2 mL, 33.02 mmol, 2.5 M in hexanes) was added dropwise. The reaction mixture was stirred at this temperature for 45 min. To the reaction mixture was added a solution of (*R*)-Garner's aldehyde (**158**) (3.44 g, 15.01 mmol) in THF (16 mL) dropwise. The reaction mixture was stirred at −78 °C for 4 h before being quenched with saturated aqueous NH_4Cl (30 mL). The mixture was warmed to room temperature and poured onto saturated aqueous NH_4Cl (100 mL). The aqueous phase was extracted with EtOAc (3 × 100 mL) and the combined organic layers were dried (Na_2SO_4), filtered and concentrated in vacuo. The crude product was purified by flash chromatography (eluent: 15:85 → 25:75 v/v EtOAc/petroleum benzines) to yield a 5.1:1 mixture of diastereomers of alcohol **188** as a colourless oil (3.80 g, 75%, *anti/syn* 5.1:1).

^{1}H-NMR ($CDCl_3$, 400 MHz, 328 K) *major diastereomer (anti)* δ (ppm) 7.27 (d, *J* = 8.6 Hz, 2H), 6.87 (d, *J* = 8.7 Hz, 2H), 5.04–4.98 (m, 1H), 4.23–4.07 (m, 1H), 4.02 (dd, *J* = 9.4, 2.2 Hz, 1H), 3.88–3.80 (m, 1H), 3.78 (s, 3H), 1.56–1.42 (m, 15H); *minor diastereomer (syn)* δ (ppm) 7.27 (d, *J* = 8.6 Hz, 2H), 6.87 (d, *J* = 8.7 Hz, 2H), 4.70 (dd, *J* = 8.6, 3.5 Hz, 1H), 4.23–4.07 (m, 1H), 3.78 (s, 3H), 3.73–3.61 (m, 2H), 1.56–1.42 (m, 15H); **^{13}C-NMR** ($CDCl_3$, 100 MHz, 328 K) *major diastereomer (anti)* δ (ppm) 159.7, 159.3, 133.6, 127.4, 114.0, 94.8, 80.8, 74.0, 64.0, 63.5, 55.4, 28.6 24.0; *minor diastereomer (syn)* δ (ppm) 159.7, 159.3, 133.6, 128.5, 114.1, 94.8, 80.8, 77.2, 65.0, 63.5, 55.4, 28.6, 26.7 **LRMS** (+ESI) *m/z* 360 $[M+Na]^+$; **HRMS** (+ESI) Calc. for $C_{18}H_{27}NO_5$ $[M+Na]^+$: 360.1781, Found: 360.1782; **IR** ν_{max} (ATR) 3476, 2977, 2933, 1695, 1512, 1457, 1391, 1248, 1173 cm^{-1}; **$[\alpha]_D$** = +28.7° (*c* 1.0, CH_2Cl_2).

(*R*)-*tert*-butyl 4-(4-methoxybenzoyl)-2,2-dimethyloxazolidine-3-carboxylate (191)

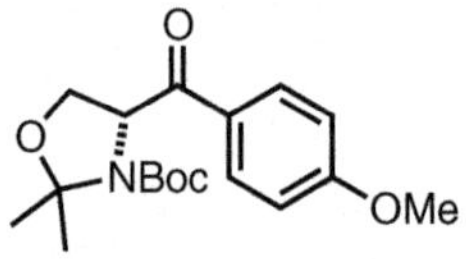

To a solution of alcohol **188** (350 mg, 1.04 mmol) in CH_2Cl_2 (3.2 mL) at 0 °C was added Dess-Martin periodinane (571 mg, 1.35 mmol). The reaction mixture was warmed to room temperature and stirred for 16 h. The solvent was concentrated in vacuo and crude product purified by flash chromatography (eluent: 15:85 → 20:80 v/v EtOAc/petroleum benzines) to yield ketone **191** as a white solid (230 mg, 66%).

^{1}H-NMR ($CDCl_3$, 500 MHz, rotamers) δ (ppm); 7.89 (d, *J* = 8.8 Hz, 2H), 6.95 (d, *J* = 8.8 Hz, 1.2H), 6.92 (d, *J* = 8.9 Hz, 0.8H), 5.43 (dd, *J* = 7.5, 2.9 Hz, 0.4H) and 5.32 (dd, *J* = 7.8, 3.9 Hz, 0.6H), 4.32–4.25 (m, 1H), 3.95–3.88 (m, 1H), 3.87 (s, 1.7H) and 3.85 (s, 1.3H), 1.75 (s, 1.7H) and 1.72 (s, 1.3H), 1.60 (s, 1.7H) and 1.55 (s, 1.3H), 1.49 (s, 4H) and 1.27 (s, 5H); These data are in agreement with those previously reported by Geden et al. [13].

***tert*-butyl ((1*S*,2*R*)-1,3-dihydroxy-1-(4-methoxyphenyl)propan-2-yl)carbamate (192)**

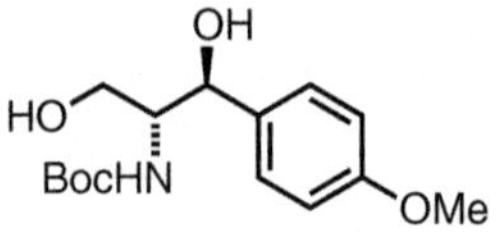

To a solution of oxazolidine **188** (3.64 mg, 10.78 mmol) in methanol (107 mL) was added *p*-toluene sulfonic acid (820 mg, 4.31 mmol). The mixture was stirred at room temperature for 3 h. The reaction mixture was poured onto saturated aqueous

$NaHCO_3$ (150 mL). The aqueous phase was extracted with EtOAc (3 × 100 mL) and the combined organic layers were dried ($MgSO_4$), filtered and concentrated in vacuo. The crude product was purified by flash chromatography (eluent: 50:50 → 60:40 v/v EtOAc/petroleum benzines) to yield a 5.1:1 mixture of diastereomers of diol **192** as a white solid (2.31 g, 73%, *anti*/*syn* 5.1:1).

^{1}H-NMR ($CDCl_3$, 400 MHz, 328 K) *major diastereomer (anti)* δ (ppm) 7.31–7.23 (m, 2H), 6.90–6.83 (m, 2H), 5.24 (d, *J* = 7.8 Hz, 1H), 4.93–4.89 (m, 1H) 3.79 (s, 3H), 3.80–3.70 (m, 2H), 3.64–3.56 (m, 1H), 3.37 (d, *J* = 3.8 Hz, 1H), 2.79 (s, 1H), 1.42 (s, 9H); *minor diastereomer (syn)* δ (ppm) 5.16 (d, *J* = 7.9 Hz, 1H), 4.89–4.86 (m, 1H), 3.78 (s, 3H), 3.26 (s, 1H), 2.73 (s, 1H), 1.36 (s, 9H) (some signals for the minor diastereomer are overlapped with the major diastereomer); **^{13}C-NMR** ($CDCl_3$, 100 MHz, 328 K) *major diastereomer (anti)* δ (ppm) 159.5, 156.3 133.5, 127.4, 114.2, 80.0, 75.7, 62.2, 57.1, 55.5, 28.5; *minor diastereomer (syn)* δ (ppm) 159.6, 156.7, 133.7, 127.5, 114.2, 80.0, 74.1, 64.0, 57.7, 55.5, 28.5; **LRMS** (+ESI) *m/z* 320 $[M+Na]^+$; **HRMS** (+ESI) Calc. for $C_{15}H_{23}NO_5$ $[M+Na]^+$: 320.1468, Found: 320.1469; **IR** ν_{max} (ATR) 3398, 3370, 2975, 2933, 1686, 1612, 1512, 1457, 1392, 1366, 1247, 1171 cm^{-1}; $[\alpha]_D$ = −1.5° (*c* 0.32, CH_2Cl_2); **mp.** 100.6–102.9 °C.

(4*S*,5*S*)-3-(((9*H*-fluoren-9-yl)methoxy)carbonyl)-5-(4-methoxyphenyl)-2,2-dimethyloxazolidine- 4-carboxylic acid (187)

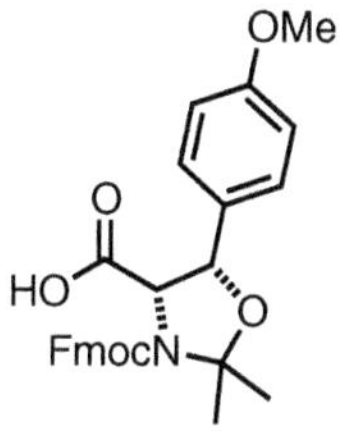

Diol **192** (892 mg, 3.00 mmol) was oxidised by general procedure A to produce carboxylic acid **193** which was used without further purification. Crude acid **193** (720 mg, 2.31 mmol) was subject to protecting group manipulation via general procedure B to afford Fmoc amino acid **194** which was used without further purification. Crude Fmoc amino acid **194** was oxazolidine protected using general procedure C. The product was purified by flash chromatography (eluent: 35:65 → 60:40 v/v EtOAc/petroleum benzines) to yield oxazolidine protected amino acid **187** as an off-white crystalline solid (438 mg, 30% from diol **192**, single diastereomer).

^{1}H-NMR (DMSO-d_6, 500 MHz, rotamers) δ (ppm) 12.47 (brs, 0.5H) and 12.13 (brs, 0.5H), 7.93–7.85 (m, 2H), 7.73–7.62 (m, 2H), 7.45–7.25 (m, 5H), 7.17 (d, *J* = 8.9 Hz, 1H), 6.64 (d, *J* = 8.7 Hz, 1H) and 6.83 (d, *J* = 8.7 Hz, 1H); 5.50 (d, *J* = 6.8 Hz, 0.5H) and 5.22 (d, *J* = 6.8 Hz, 0.5H), 4.79 (dd, *J* = 10.9 and 3.8 Hz, 0.5H) and 4.70 (dd, *J* = 10.7 and 3.5 Hz, 0.5H), 4.66 (d, *J* = 6.70 Hz, 0.5H) and 4.30 (d, *J* = 6.8 Hz, 0.5H), 4.35–4.32 (m, 0.5H) and 4.18–4.13 (m, 0.5H), 4.27–4.18 (m, 1H), 3.76 (s, 1.5H) and 3.70 (s, 1.5H), 1.79 (s, 1.5H) and 1.59 (s, 1.5H), 0.96 (s, 1.5H) and 0.76 (s, 1.5H) **^{13}C-NMR** (DMSO-d_6, 125 MHz, rotamers) δ (ppm) 170.5 and 170.1, 159.2 and 159.1, 151.8 and 151.2, 144.2 and 144.1, 143.6 and 143.6, 141.2 and 141.1, 140.6 and 140.6, 128.0 and 127.8, 127.8 and 127.7, 127.5, 127.2 and 127.2, 127.1, 127.0 and 127.0, 125.4 and 125.3, 124.4 and 124.3, 120.1 and 120.1, 120.0, 113.4 and 113.2, 93.8 and 93.1, 77.1 and 76.2, 67.0 and 66.0, 64.1 and 63.6, 55.0 and 55.0, 46.7 and 46.6, 25.2 and 23.9, 25.1 and 23.9 (extra signals due to further rotational conformations in the Fmoc-region); **LRMS** (+ESI) *m/z* 496 $[M+Na]^+$; **HRMS** (+ESI) Calc. for $C_{28}H_{27}NO_6$ $[M+Na]^+$: 496.1730, Found: 496.1730; **IR** ν_{max} (ATR) 2922, 2853, 1707, 1614, 1515, 1411, 1348, 1249, 1177 cm^{-1}; $[\alpha]_D$ = +30.1° (*c* 0.29, CH_2Cl_2); **mp.** 213.0–214.9 °C.

5.5.7 Synthesis of Fmoc-β-Me-Asp(PhiPr)-OH (212)

(2*S*)-1-benzyl 4-*tert*-butyl 2-(((9*H*-fluoren-9-yl)(phenyl)methyl)amino)succinate (215)

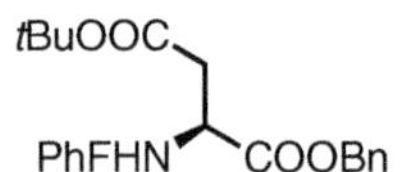

To a solution of H_2N-Asp(*t*Bu)-OH **213** (3.24 g, 17.12 mmol) in CH_2Cl_2 (35 mL) was added TMS-Cl (2.27 ml, 17.91 mmol). The reaction mixture was stirred at room temperature for 2 h. To the mixture was added NEt_3 (5.02 mL, 35.81 mmol), $Pb(NO_3)_2$ (5.16 g, 15.57 mmol) and a solution of phenylfluorenyl bromide (5.00 g, 15.57 mmol) in CH_2Cl_2 (28 mL). The reaction was stirred at room temperature for 4 days. The reaction mixture was then filtered over Celite® and concentrated in vacuo. The crude solid was poured onto EtOAc (100 mL) and saturated NH_4Cl (100 mL). The aqueous phase was extracted with EtOAc (3 × 100 mL) and the combined organic layers were dried ($MgSO_4$), filtered and concentrated in vacuo to yield protected amine **214** (5.9 g).

To a solution of the crude acid **214** (5.90 g, 14.7 mmol) in DMF (62 mL) was added Cs_2CO_3 (5.36 g, 16.5 mmol) and the mixture was stirred for 30 min. To the reaction mixture was added benzyl bromide (1.96 ml, 16.5 mmol) and it was stirred for 16 h at room temperature. The reaction mixture was poured onto water (100 mL) and extracted with EtOAc (3 × 100 mL). The combined organic layers were washed with water (100 mL), saturated aqueous NaCl (2 × 50 mL) dried ($MgSO_4$), filtered and concentrated in vacuo. The crude product was purified by flash chromatography (eluent: 10:90 v/v EtOAc/petroleum benzines) to yield protected amino acid **215** as a pale yellow solid (5.71 g, 64% over two steps).

^{1}H-NMR ($CDCl_3$, 500 MHz) δ (ppm) 7.69–7.62 (m, 2H), 7.41–7.11 (m, 16H), 4.87 (d, *J* = 12.4 Hz, 1H), 4.77 (d, *J* = 12.4 Hz, 1H), 3.00 (t, *J* = 5.4 Hz, 1H), 2.49 (dd, *J* = 15.1, 5.1 Hz, 1H), 2.20 (dd, *J* = 15.1, 5.8 Hz, 1H), 1.44 (s, 9H); **LRMS** (+ESI) *m/z* 542 $[M+Na]^+$; $[\alpha]_D$ = −179.8° (*c* 0.62, CH_2Cl_2); **mp.** 124.0–129.0 °C. These data are in agreement with those previously reported by Schabbert et al. [14].

(2*S*)-1-benzyl 4-*tert*-butyl 2-(((9*H*-fluoren-9-yl)(phenyl)methyl)amino)-3-methylsuccinate (216)

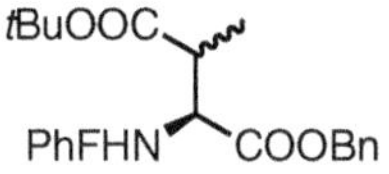

To a solution of protected amino acid **215** (1.0 g, 1.93 mmol) in THF (10 mL) at −78 °C was added dropwise a solution of LiHMDS (3.85 mL, 3.85 mmol, 1 M in THF). The solution was allowed to stir at this temperature for 1 h, before MeI (0.24 mL, 3.85 mmol) was added and the solution was warmed to room temperature and stirred for 16 h. The reaction mixture was poured onto saturated aqueous NH_4Cl (20 mL) and extracted with Et_2O (3 × 30 mL). The combined organic layers were washed with saturated aqueous NaCl (50 mL) dried ($MgSO_4$), filtered and concentrated in vacuo. The crude product was purified by flash chromatography (eluent: 2:3 → 0:1 v/v petroleum benzines/CH_2Cl_2) to yield protected β-methyl amino acid **216** as a yellow oil which solidified upon freezing (0.85 g, 82%, *syn*/*anti* 3:1).

^{1}H-NMR ($CDCl_3$, 500 MHz) *major diastereomer (syn)* δ (ppm) 7.66 (dd, *J* = 7.5, 3.6 Hz, 2H), 7.43 (dd, *J* = 8.0, 1.6 Hz, 2H), 7.37–7.17 (m, 14H), 7.13–7.07 (m, 2H), 4.73 (d, *J* = 12.4 Hz, 1H), 4.55 (d, *J* = 12.2 Hz, 1H), 3.13 (d, *J* = 9.6 Hz, 1H), 2.84 (dd, *J* = 7.3, 9.4 Hz, 1H), 2.56–2.47 (m, 1H), 1.34 (s, 9H), 1.18 (d, *J* = 7.2 Hz, 3H); *minor diastereomer (anti)* δ (ppm) 4.68 (d, *J* = 12.4 Hz, 1H), 4.53 (d, *J* = 12.2 Hz, 1H), 3.00 (apt t, *J* = 8.5 Hz, 1H), 1.42 (s, 9H), 0.89 (d, *J* = 7.0 Hz, 3H); **LRMS** (+ESI) *m/z* 556 $[M+Na]^+$; **$[\alpha]_D$** = −229.3° (*c* 0.16, CH_2Cl_2). These data are in agreement with those previously reported by Schabbert et al. [14].

(2*S*,3*S*)-2-amino-4-(*tert*-butoxy)-3-methyl-4-oxobutanoic acid (217)

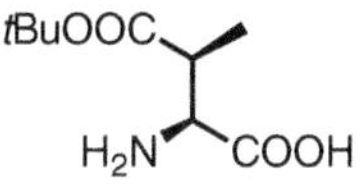

A mixture of β-methyl amino acid **216** (940 mg, 1.76 mmol) and Pearlman's catalyst (169 mg, ~10% Pd by weight) in methanol (35 mL) was stirred under an atmosphere of H_2 for 16 h. The reaction mixture was filtered over Celite® and concentrated in vacuo. The crude product was purified by flash chromatography (eluent: 8:2:1 v/v/v EtOAc/*i*PrOH/H_2O) to yield unprotected β-methyl methyl amino acid **217** as a white solid (123 mg, 35%, *syn* only).

^{1}H-NMR (D_2O, 400 MHz) δ (ppm) 3.95 (d, *J* = 4.2 Hz, 1H), 3.03 (qd, *J* = 7.5, 4.2 Hz, 1H), 1.43 (s, 9H), 1.19 (d, *J* = 7.5 Hz, 3H); **LRMS** (+ESI) *m/z* 226 $[M+Na]^+$; **$[\alpha]_D$** = −8.4° (*c* 0.19, dioxane: H_2O, 1:1); **mp.** 161.0–187.0 °C. These data are in agreement with those previously reported by Schabbert et al. [14].

(2*S*,3*S*)-1-allyl 4-(2-phenylpropan-2-yl) 2-((((9*H*-fluoren-9-yl)methoxy)carbonyl)amino)-3-methylsuccinate (220)

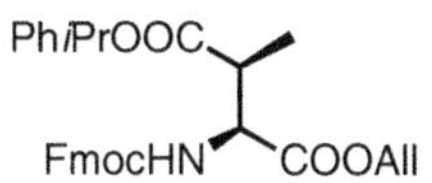

To a solution of amino acid **217** (306 mg, 1.51 mmol) in THF (10 mL) was added saturated aqueous $NaHCO_3$ (5 mL) and Fmoc-OSu (538 mg, 1.59 mmol) and the reaction was stirred overnight at room temperature. The reaction mixture was poured onto water (20 mL) and extracted with Et_2O (2 × 30 mL). The aqueous phase was then acidified with 1 M HCl to pH = 4 and extracted with EtOAc (3 × 40 mL) and the combined organic layers were dried ($MgSO_4$), filtered and concentrated in vacuo to yield the crude acid (562 mg).

To a solution of crude acid (526 mg, 1.32 mmol) in DMF (2.6 mL) at 0 °C was added *i*Pr_2NEt (0.46 mL, 2.64 mmol) followed by allyl bromide (0.23 mL, 2.64 mmol). The reaction mixture was warmed to room temperature and stirred for 16 h before being diluted with EtOAc (50 mL) and washed with HCl (3 × 30 mL, 0.1 M), saturated aqueous NaCl (30 mL) and the organic layer was dried (Na_2SO_4), filtered and concentrated in vacuo to yield allyl ester **218**.

To a solution of crude allyl ester (532 mg, 1.14 mmol) in CH_2Cl_2 (5.5 mL) was added TFA (5.5 mL) at room temperature for 3 h, before CH_2Cl_2 (20 mL) was added and the mixture was concentrated in vacuo. The crude solid was dissolved in CH_2Cl_2 (4 mL) to which a solution of Ph*i*Pr trichloroacetimidate **219** (603 mg, 2.28 mmol) in hexanes (8 mL) was added. The reaction mixture was stirred at room temperature for 16 h at which it was judged incomplete by TLC analysis. A further portion of Ph*i*Pr chloroformate (300 mg, 1.14 mmol) in CH_2Cl_2 (4 mL) was added and stirred for a further 16 h. The reaction mixture was concentrated in vacuo. The crude product was purified by flash chromatography (eluent: 20:80 v/v EtOAc/petroleum benzines) to yield Ph*i*Pr ester **220** as a colourless oil (390 mg, 49% over 4 steps).

^{1}H-NMR ($CDCl_3$, 400 MHz) δ (ppm) 7.78 (dd, *J* = 7.5, 0.6 Hz, 2H), 7.60 (dd, *J* = 7.4, 3.7 Hz, 2H), 7.45–7.29 (m, 8H), 7.27 (m, 1H), 5.96–5.82 (m, 1H), 5.57 (d, *J* = 9.1 Hz, 1H), 5.34 (dd, *J* = 17.1, 1.0 Hz, 1H), 5.26 (dd, *J* = 10.4, 1.0 Hz, 1H), 4.77 (dd, *J* = 9.2, 4.6 Hz, 1H), 4.65 (d, *J* = 5.8 Hz, 2H), 4.50–4.34 (m, 2H), 4.25 (t, *J* = 7.2 Hz, 1H), 3.06 (qd, *J* = 7.3, 4.6 Hz, 1H), 1.80 (s, 3H), 1.78 (s, 3H), 1.27 (d, *J* = 7.3 Hz, 3H); **^{13}C-NMR** ($CDCl_3$, 100 MHz) δ (ppm) 171.4, 170.5, 156.1, 145.4, 143.9, 143.8, 141.4, 131.5, 128.3, 127.8, 127.2, 127.2, 125.2, 125.2, 124.4, 120.1, 119.2, 82.8, 67.4, 66.4, 55.9, 47.2, 42.9, 28.7, 28.1, 13.1 **LRMS** (+ESI) *m/z* 550 $[M+Na]^+$; **HRMS** (+ESI) Calc. for $C_{32}H_{33}NO_6$ $[M+Na]^+$: 550.2200, Found: 550.2209; **IR** ν_{max} (ATR) 3347, 2923, 2853, 1727, 1509, 1449, 1261, 1201, 1136, 1100 cm^{-1}; $[\alpha]_D$ = +7.5° (*c* 0.28, CH_2Cl_2).

(2*S*,3*S*)-2-((((9*H*-fluoren-9-yl)methoxy)carbonyl)amino)-3-methyl-4-oxo-4-((2-phenylpropan-2-yl)oxy)butanoic acid (212)

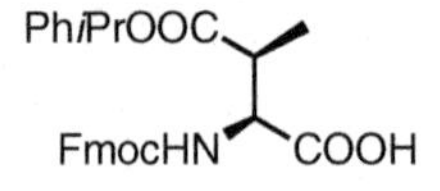

To a solution of allyl ester **220** (108 mg, 0.205 mmol) in THF (2 mL) was added $Pd(PPh_3)_4$ (12 mg, 0.01 mmol) followed by $PhSiH_3$ (0.05 mL, 0.41 mmol). The reaction mixture was stirred at room temperature for 2 h before being concentrated in vacuo. The crude product was purified by flash chromatography (eluent:

30:70 → 100:0 v/v EtOAc/petroleum benzines) to yield acid **212** as a white solid (105 mg, quant).

^{1}H-NMR ($CDCl_3$, 400 MHz) δ (ppm) 7.76 (d, J = 7.4 Hz, 2H), 7.57 (dd, J = 4.5, 6.9 Hz, 2H), 7.44–7.17 (m, 9H), 5.57 (d, J = 8.9 Hz, 1H), 4.71 (dd, J = 4.6, 8.8 Hz, 1H), 4.47–4.34 (m, 2H), 4.22 (ap. t, J = 7.1 Hz, 1H), 3.11–2.99 (m, 1H), 1.76 (s, 6H), 1.28 (d, J = 7.24 Hz, 3H); **^{13}C-NMR** ($CDCl_3$, 100 MHz) δ (ppm) 174.9, 171.8, 156.3, 145.4, 143.9, 143.8, 141.5, 128.4, 127.9, 127.3, 127.3, 125.2, 124.4, 120.1, 83.1, 67.5, 55.8, 47.3, 42.8, 28.6, 28.3, 13.2; **LRMS** (+ESI) *m/z* 510 $[M+Na]^+$; **HRMS** (+ESI) Calc. for $C_{29}H_{29}NO_6$ $[M+Na]^+$: 510.1887, Found: 510.1895; **IR** ν_{max} (ATR) 2923, 2853, 1719, 1517, 1449, 1248, 1219, 1137, 1101 cm^{-1}; **[α]$_D$** = +4.2° (*c* 0.38, CH_2Cl_2); **mp**. 52.5–65.0 °C.

5.5.8 SPPS of Simplified Skyllamycin Analogue 115

Simplified skyllamycin 115

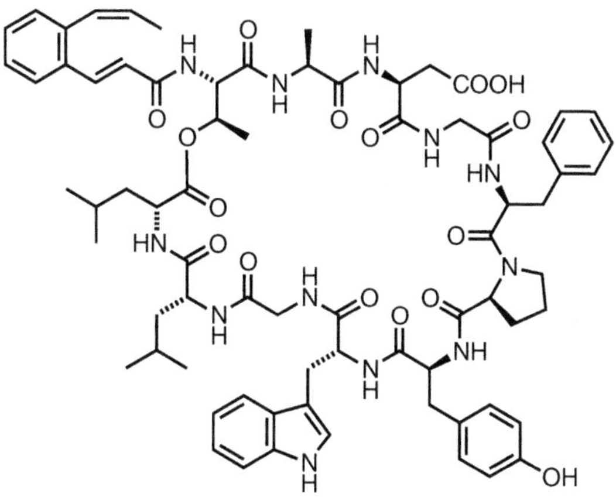

Fmoc-Gly-OH (50 μmol) was loaded to 2-CTC resin as per Sect. 5.5.1. Iterative Fmoc-SPPS was carried out to couple Fmoc-Asp(*t*Bu)-OH, Fmoc-Ala-OH, Fmoc-Thr-OH utilising standard PyBOP coupling conditions. Cinnamoyl moiety **130** was coupled using HATU coupling conditions. Fmoc-D-Leu-OH (141 mg, 8 equiv.) was dissolved in anhydrous CH_2Cl_2 (5 mL) and cooled to 0 °C. *N,N'*-diisopropylcarbodiimide (31 μL, 4 equiv.) was added to this solution which was then warmed to room temperature and stirred for 30 min. The reaction mixture was concentrated under a stream of nitrogen and subsequently redissolved in DMF (0.5 mL). This solution, along with a solution of DMAP (catalytic, 6 crystals) in DMF (0.1 mL), was shaken with the resin for 16 h at room temperature before the solution was expelled and the resin washed with DMF (×5), CH_2Cl_2 (×5) and DMF (×5).

Iterative Fmoc SPPS was continued with the subsequent couplings of Fmoc-D-Leu-OH, Fmoc-Gly-OH, Fmoc-D-Trp(Boc)-OH, Fmoc-Tyr(O*t*Bu)-OH, Fmoc-Pro-OH, Fmoc-Phe-OH all performed using PyBOP coupling conditions. The terminal

Fmoc group was removed, and the resin-bound peptide **155** was washed with DMF (×5), CH_2Cl_2 (×5), DMF (×5), CH_2Cl_2 (×20). The resin was treated with a 30 vol.% solution of HFIP in CH_2Cl_2 (~5 mL, 2 × 30 min) and the solution transferred to a round bottom flask and diluted with CH_2Cl_2 (30 mL) before being concentrated under a stream of N_2 and dried in vacuo to afford **156**. A portion of the crude linear peptide (30% by weight) was then subject to cyclisation (general procedure D) and global deprotection (general procedure E) followed by RP-HPLC purification and lyophilisation to afford the desired cyclic peptide **115** in (1.98 mg, 9.5%).

LRMS (+ESI) m/z calculated mass 1413.7 $[M+Na]^+$: m/z observed 1412.9 $[M+Na]^+$; **HRMS** (+ESI) Calc. for $C_{73}H_{90}N_{12}O_{16}$ $[M+2Na]^{2+}$: 718.3191, Found: 718.3196; **Analytical HPLC** R_t 12.6 min (0–100% MeCN (0.1% formic acid) in H_2O (0.1% formic acid) over 15 min, λ = 230 nm). ^1H-NMR and ^{13}C-NMR data is listed below in Table 5.4.

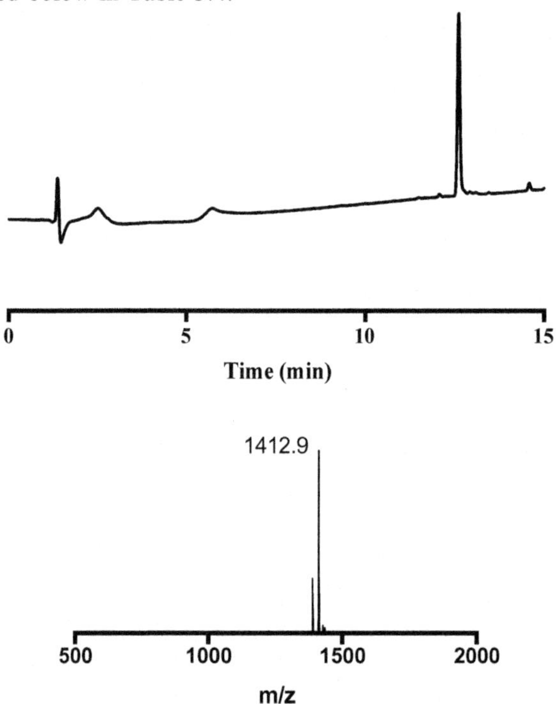

Table 5.4 ^{1}H-NMR (500 MHz) and ^{13}C-NMR (125 MHz) chemical shifts of simplified skyllamycin analogue **115** in CD_3OD referenced at 3.30 ppm (^{1}H) and 49.00 ppm (^{13}C). All assignments were made based on COSY, HSQC and HMBC data in comparison with the reported data for the skyllamycin natural products [15]. ^{13}C-NMR chemical shifts were extracted from the HSQC and HMBC spectra. N-H resonances not included as they slowly exchange with CD_3OD

Residue	Position	δ ^{13}C/ ppm	δ ^{1}H/ ppm	No. H, mult, *J* Hz	Residue	Position	δ ^{13}C/ ppm	δ ^{1}H/ ppm	No. H, mult, *J* Hz
Thr	C=O	176.0			D-Trp	C=O	174.8		
	α	61.8	5.12	1H, brs		7a	137.9		
	β	70.0	5.34	1H, dq, 6.7, 1.6		3a	128.3		
	CH_3	17.2	1.22	3H, d, 6.7		2	125.7	7.07	1H, s
						6	122.3	7.04	1H, t, 7.6
Ala	C=O	174.9				5	119.7	6.90	1H, t, 7.3
	α	52.7	4.13	1H, q, 7.5		4	119.2	7.52	1H, d, 8.1
	CH_3	16.3	1.47	3H, d, 7.5		7	112.2	7.27	1H, m
						3	110.6		
Asp	C=O	172.8				α	54.7	4.88	1H, m
	C=O	173.5				β	29.6	3.31	1H, m
	α	50.7	5.01	1H, dd, 10.3, 4.2		β′		3.11	1H, dd, 10.3, 14.6
	β	37.4	2.89	1H, dd,16.2, 10.1					
	β′		3.28	1H, m	Gly	C=O	171.9		
						α	42.7	4.20	1H, d, 17.5
Gly	C=O	171.5				α′		3.94	1H, d, 16.9
	α	43.1	4.23	1H, m					
	α′		3.48	1H, d, 17.7	D-Leu	C=O	177.1		
						α	55.7	4.25	1H, m
Phe	C=O	173.5				β	41.2	1.66	1H, m
	Aromatic	138.3				β′		1.56	1H, m
	Aromatic	127.5	7.11	2H, m		γ	25.7	1.78	1H, m
	Aromatic	129.3	7.12	1H, m		CH_3	23.0	0.91	3H, d, 6.7

(continued)

Table 5.4 (continued)

Residue	Position	δ ^{13}C/ppm	δ ^{1}H/ppm	No. H, mult, *J* Hz	Residue	Position	δ ^{13}C/ppm	δ ^{1}H/ppm	No. H, mult, *J* Hz
	Aromatic	130.1	7.25	2H, m		CH_3	21.3	0.82	3H, d, 6.6
	α	54.8	4.50	1H, t, 7.2					
	β	37.2	2.60	2H, d, 7.3	D-Leu	C=O	171.1		
						α	51.9	4.37	1H, dd, 11.5, 3.6
Pro	C=O	172.9				β	38.8	1.80	1H, m
	α	612.2	4.25	1H, m		β′		1.69	1H, m
	β	25.2	1.75	1H, m		γ	25.7	1.69	
	β′		1.53	1H, m		CH_3	23.2	0.96	3H, d, 6.0
	γ	29.4	1.86	1H, m		CH_3	20.8	0.86	3H, d, 6.0
	γ′		1.75	1H, m					
	δ	48.0	3.03	1H, m	Cinnamoyl	C=O	171.4		
	δ′		3.51	1H, m		Aromatic	133.7		
						Aromatic	139.1		
Tyr	C=O	171.7				Aromatic	126.9	7.70	1H, m
	1	127.5				Aromatic	130.7	7.14	1H, d, 7.2
	2/6	115.2	6.36	2H, d, 8.4		Aromatic	128.0	727	1H, m
	3/5	131.0	6.29	2H, d, 8.2		Aromatic	130.1	7.30	1H, td, 7.4, 1.2
	4	156.8				*cis*-sp^2	128.7	6.45	1H, dd, 11.3, 1.4
	α	53.8	4.59	1H, t, 6.1		*cis*-sp^2-2	129.7	5.76	1H, dq, 11.3, 7.0
	β	78.2	2.74	1H, dd, 14.3, 5.9		*trans*-sp^2	141.0	7.88	1H, d, 15.9
	β′		2.37	1H, dd, 14.3, 6.6		*trans*-sp^2-2	121.4	7.11	1H, d, 15.8
						CH_3	14.2	1.46	3H, dd, 7.0, 1.9

5.5.9 *Synthesis of Deshydroxy Skyllamycins A–C (116–118)*

Deshydroxy skyllamycin A (116)

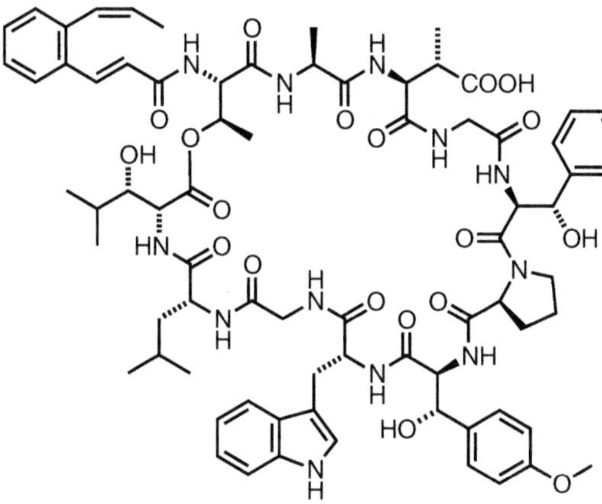

Fmoc-Gly-OH was loaded to 2-CTC as per Sect. 5.5.1 to afford 75 μmol of resin-bound amino acid. Fmoc-β-Me-Asp(Ph*i*Pr)-OH (**212**) was coupled under HATU conditions followed by PyBOP coupling of Fmoc-Ala-OH and Fmoc-Thr-OH before coupling of cinnamoyl moiety **130** using HATU to yield resin-bound peptide **222**. From here, on-resin esterification with protected Fmoc-β-OH-Leu-OH (**166**), followed by microwave assisted coupling of Fmoc-D-Leu-OH as detailed in the general procedures afforded resin-bound depsipeptide. Next, coupling of Fmoc-Gly-OH and Fmoc-D-Trp-OH was carried out using PyBOP coupling conditions followed by coupling of protected Fmoc-β-OH-*O*-Me-Tyr-OH **187** using HATU conditions. Microwave assisted coupling of Fmoc-Pro-OH, followed by coupling of Fmoc-β-OH-Phe-OH (**180**) using HATU, afforded the complete resin-bound linear peptide **224**. Resin-bound linear peptide **224** was cleaved from resin according to the general SPPS protocols (Sect. 5.4.1) to yield crude linear peptide **226** (42.7 mg, 27 μmol, assumed to be 100% pure). This was subject to cyclisation (general procedure D) and global deprotection (general procedure E). Deshydroxy skyllamycin A (**116**) was afforded as a fluffy white solid after RP-HPLC and lyophilisation (5.8 mg, 5.2%).

LRMS (+ESI) m/z calculated mass 1467.7 $[M+H]^+$, 1489.7 $[M+Na]^+$: m/z observed 1467.0 $[M+H]^+$, 1489.0 $[M+Na]^+$; **HRMS** (+ESI) Calc. for $C_{75}H_{94}N_{12}O_{19}$ $[M+Na]^+$: 1489.6650, Found: 1489.6658; **Analytical HPLC** R_t 12.8 min (0–100% MeCN (0.1% formic acid) in H_2O (0.1% formic acid) over 15 min, λ = 230 nm). ^{1}H-NMR and ^{13}C-NMR data is listed below in Table 5.5.

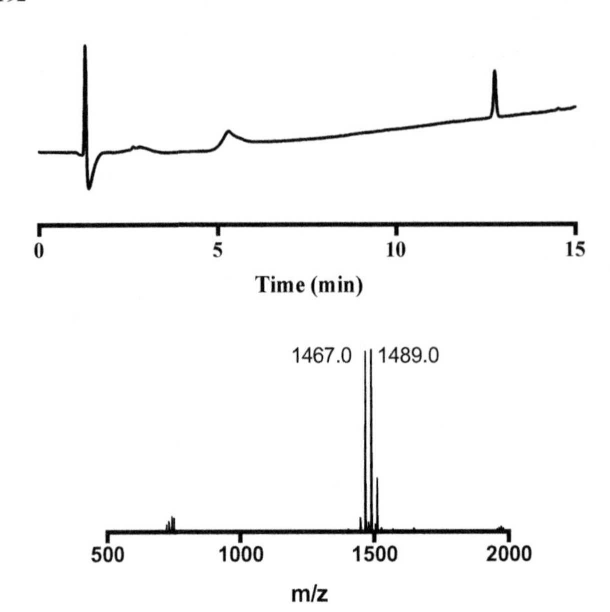

Table 5.5 ^{1}H-NMR (500 MHz) and ^{13}C-NMR (125 MHz) chemical shifts of deshydroxy skyllamycin A (**116**) in CD_3OD referenced at 3.30 ppm (^{1}H) and 49.0 ppm (^{13}C). All assignments were made based on COSY, HSQC and HMBC data in comparison with the skyllamycin natural products [15]. ^{13}C-NMR chemical shifts were extracted from the HSQC and HMBC spectra. N-H resonances not included as they exchange with CD_3OD

Residue	Position	δ ^{13}C/ ppm	δ ^{1}H/ ppm	No. H, mult, *J* Hz	Residue	Position	δ ^{13}C/ ppm	δ ^{1}H/ ppm	(No. H, mult, *J* Hz)
Thr	C=O	176.0			D-Trp	C=O	174.8		
	α	61.8	5.14	1H, m		7a	137.8		
	β	70.3	5.47	1H, q, 6.8		3a	128.3		
	CH_3	17.5	1.33	3H, d, 6.7		2	124.9	7.14	1H, s
						6	122.1	7.04	1H, t, 7.4
Ala	C=O	175.1				5	119.7	6.89	1H, t, 7.4
	α	52.7	4.11	1H, m		4	119.5	7.50	1H, d, 8.1
	CH_3	16.6	1.48	3H, d, 7.3		7	112.0	7.29	1H, m
						3	110.2		

(continued)

Table 5.5 (continued)

Residue	Position	δ ^{13}C/ ppm	δ ^{1}H/ ppm	No. H, mult, *J* Hz	Residue	Position	δ ^{13}C/ ppm	δ ^{1}H/ ppm	(No. H, mult, *J* Hz)
Asp	C=O	172.5				α	54.8	4.81	1H, m
	C=O	177.3				β	28.9	3.27	2H, m
	α	54.5	5.19	1H, t, 7.5		β′			
	β	42.0	3.28	1H, m					
	CH_3	14.2	1.24	3H, d, 7.1	Gly	C=O	171.6		
						α	43.2	4.04	1H, d, 16.6
Gly	C=O	172.3				α′		3.75	1H, m
	α	43.6	4.08	1H, d, 18.2					
	α′		3.44	1H, d, 17.8	D-Leu	C=O	177.0		
						α	56.2	4.26	1H, m
β-OH-Phe	C=O	171.5				β	41.4	1.68	1H, m
	Aromatic	143.0				β′		1.57	1H, m
	Aromatic	127.1	7.39	2H, m		γ	25.8	1.84	1H, m
	Aromatic	128.2	7.23	1H, m		CH_3	22.8	0.96	3H, d, 6.7
	Aromatic	128.9	7.37	2H, m		CH_3	21.7	0.87	3H, d, 6.7
	α	58.0	4.54	1H, m					
	β	73.6	4.60	1H, m	β-OH-Leu	C=O	170.8		
						α	56.1	4.45	1H, d, 1.9
Pro	C=O	173.7				β	76.5	3.79	1H, dd, 9.2, 1.8
	α	61.7	4.25	1H, m		γ	32.3	1.67	1H, m
	β	24.7	1.72	1H, m		CH_3	19.2	1.02	3H, d, 6.7
	β′		1.43	1H, m		CH_3	18.7	0.83	3H, d, 6.7
	γ	30.0	1.95	1H, m					
	γ′		1.43	1H, m	Cinnamoyl	C=O	171.8		
	δ	48.0	3.68	2H, m		Aromatic	133.8		
						Aromatic	138.9		
β-OH-*O*-Me-Tyr	C=O	171.6				Aromatic	127.2	7.92	1H, m
	1	132.8				Aromatic	130.6	7.14	1H, d, 7.3
	2/6	128.9	6.74			Aromatic	128.3	7.41	

(continued)

Table 5.5 (continued)

Residue	Position	δ ^{13}C/ ppm	δ ^{1}H/ ppm	No. H, mult, *J* Hz	Residue	Position	δ ^{13}C/ ppm	δ ^{1}H/ ppm	(No. H, mult, *J* Hz)
				2H, d, 8.6					1H, t, 7.7
	3/5	114.2	6.61	2H, d, 8.8		Aromatic	130.0	7.32	1H, m
	4	160.8				*cis*-sp^2	128.8	6.35	1H, d, 11.4
	OCH_3	55.4	3.67	3H, s		*cis*-sp^2-2	129.8	5.79	1H, dq, 11.4, 6.9
	α	58.9	4.72	1H, m		*trans*-sp^2	139.5	7.82	1H, d, 15.8
	β	75.1	4.46	1H, d, 7.3		*trans*-sp^2-2	122.3	7.24	1H, d, 15.7
						CH_3	14.1	1.48	3H, d, 7.3

Deshydroxy skyllamycin B (117)

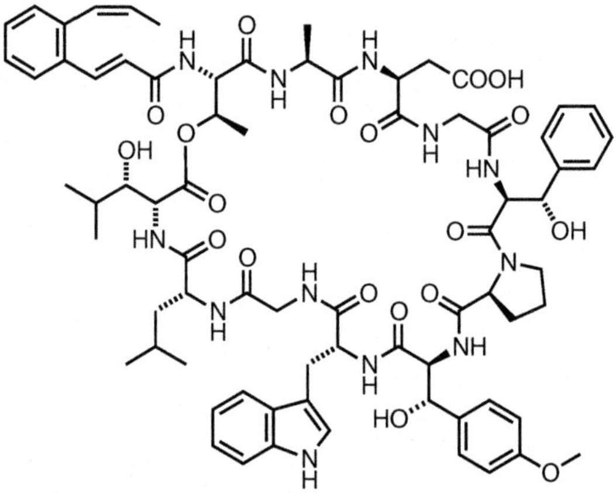

Fmoc-Gly-OH was loaded to 2-CTC as per Sect. 5.5.1 to afford 80 µmol of resin-bound amino acid. Fmoc-Asp(Ph*i*Pr)-OH was coupled under HATU conditions followed by PyBOP coupling of Fmoc-Ala-OH and Fmoc-Thr-OH before coupling of cinnamoyl moiety **130** using HATU to yield resin-bound peptide. From here, on-resin esterification with protected Fmoc-β-OH-Leu-OH (**166**), followed by microwave assisted coupling of Fmoc-D-Leu-OH as detailed in the general procedures afforded resin-bound depsipeptide. Next, coupling of Fmoc-Gly-OH and Fmoc-D-Trp-OH was carried out using PyBOP coupling conditions followed by coupling of protected Fmoc-β-OH-*O*-Me-Tyr-OH **187** using HATU conditions. Microwave assisted coupling of Fmoc-Pro-OH, followed by coupling of Fmoc-β-OH-Phe-OH (**180**) using HATU, afforded the complete resin-bound linear peptide **200**. Resin-bound linear peptide **200** was cleaved from resin according to the general SPPS protocols (Sect. 5.4.1) to yield crude linear peptide **202** (57.8 mg, 36 μmol, assumed to be 100% pure). This was subject to cyclisation (general procedure D) and global deprotection (general procedure E). Deshydroxy skyllamycin B (**117**) was afforded as a fluffy white solid after RP-HPLC and lyophilisation (8.6 mg, 7.4%).

LRMS (+ESI) m/z calculated mass 1453.7 $[M+H]^+$, 1475.7 $[M+Na]^+$: m/z observed 1453.0 $[M+H]^+$, 1475.3 $[M+Na]^+$; **HRMS** (+ESI) Calc. for $C_{74}H_{92}N_{12}O_{19}$ $[M+2Na]^{2+}$: 749.3193, Found: 749.3196; **Analytical HPLC** R_t 12.4 min (0–100% MeCN (0.1% formic acid) in H_2O (0.1% formic acid) over 15 min, λ = 230 nm). ^{1}H-NMR and ^{13}C-NMR data is listed below in Table 5.6.

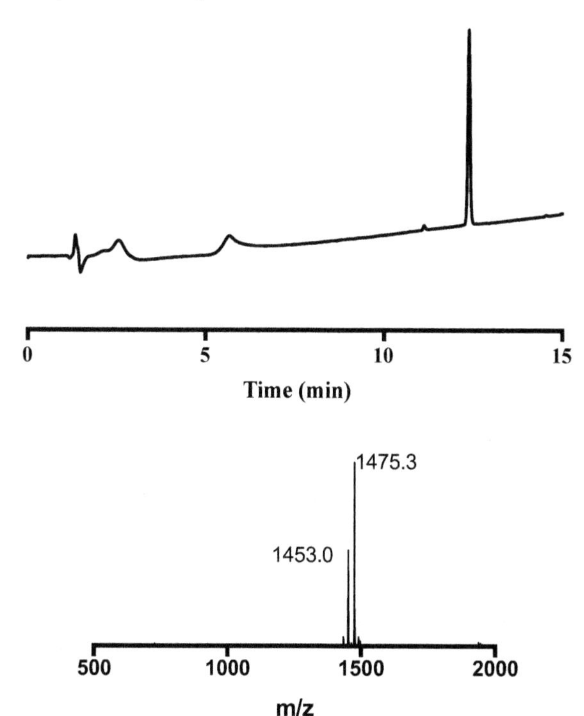

Table 5.6 ^{1}H-NMR (500 MHz) and ^{13}C-NMR (125 MHz) chemical shifts of deshydroxy skyllamycin B (**117**) in CD_3OD referenced at 3.30 ppm (^{1}H) and 49.0 ppm (^{13}C). All assignments were made based on COSY, HSQC and HMBC data in comparison with the skyllamycin natural products [15]. ^{13}C-NMR chemical shifts were extracted from the HSQC and HMBC spectra. N-H resonances not included as they slowly exchange with CD_3OD

Residue	Position	δ ^{13}C/ ppm	δ ^{1}H/ ppm	No. H, mult, *J* Hz	Residue	Position	δ ^{13}C/ ppm	δ ^{1}H/ ppm	No. H, mult, *J* Hz
Thr	C=O	176.7			D-Trp	C=O	175.2		
	α	61.8	5.17	1H, s		7a	138.3		
	β	70.2	5.45	1H, q, 6.8		3a	128.7		
	CH_3	17.5	1.32	3H, d, 7.0		2	125.1	7.13	1H, s
						6	122.2	7.05	1H, t, 7.6
Ala	C=O	175.2				5	119.6	6.89	1H, t, 7.6
	α	52.8	4.08	1H, m		4	119.6	7.50	1H, d, 8.2
	CH_3	16.1	1.45	3H, d, 7.5		7	112.0	7.30	1H, m
						3	110.5		
Asp	C=O	172.8				α	54.8	4.81	1H, m
	C=O	173.8				β	29.0	3.26	2H, m
	α	50.7	5.02	1H, m		β′			
	β	37.9	2.93	1H, dd,16.7, 10.5					
	β′		3.35	1H, dd,16.3, 4.3	Gly	C=O	171.2		
						α	43.2	4.03	1H, d, 16.2
Gly	C=O	172.3				α′		3.75	1H, d, 16.6
	α	43.3	4.09	1H, d, 17.9					
	α′		3.45	1H, d, 17.5	D-Leu	C=O	177.1		
						α	56.2	4.27	1H, m
β-OH-Phe	C=O	171.7				β	41.5	1.67	1H, m
	Aromatic	143.1				β′		1.57	1H, m
	Aromatic	127.2	7.42	2H, m		γ	25.7	1.84	1H, m

(continued)

Table 5.6 (continued)

Residue	Position	δ ^{13}C/ppm	δ ^{1}H/ppm	No. H, mult, *J* Hz	Residue	Position	δ ^{13}C/ppm	δ ^{1}H/ppm	No. H, mult, *J* Hz
	Aromatic	128.2	7.27	1H, m		CH_3	22.8	0.97	3H, d, 6.7
	Aromatic	128.9	7.33	2H, m		CH_3	21.7	0.87	3H, d, 6.6
	α	57.8	4.52	1H, m					
	β	73.9	4.61	1H, m	β-OH-Leu	C=O	170.7		
						α	56.1	4.45	1H, m
Pro	C=O	174.2				β	76.5	3.79	1H, dd, 9.2, 1.8
	α	61.8	4.25	1H, m		γ	32.4	1.65	1H, m
	β	24.8	1.73	1H, m		CH_3	19.3	1.01	3H, d, 6.7
	β′		1.40	1H, m		CH_3	18.7	0.83	3H, d, 6.6
	γ	30.0	1.96	1H, m					
	γ′		1.39	1H, m	Cinnamoyl	C=O	172.5		
	δ	48.0	3.70	2H, m		Aromatic	134.1		
						Aromatic	139.6		
β-OH-*O*-Me-Tyr	C=O	171.6				Aromatic	127.3	7.93	1H, m
	1	132.8				Aromatic	130.9	7.14	1H, d, 7.4
	2/6	129.0	6.78	2H, d, 8.3		Aromatic	128.3	7.40	1H, m
	3/5	114.2	6.65	2H, d, 8.5		Aromatic	130.1	7.31	1H, m
	4	160.7				*cis*-sp^2	128.9	6.45	1H, d, 11.3
	OCH_3	55.4	3.69	3H, s		*cis*-sp^2-2	129.7	5.72	1H, dq, 11.4, 6.9
	α	58.9	4.69	1H, m		*trans*-sp^2	139.8	7.90	1H, d, 15.4
	β	75.1	4.46	1H, m		*trans*-sp^2-2	122.2	7.23	1H, d, 16.0
						CH_3	14.2	1.48	3H, dd, 7.2, 1.8

Deshydroxy skyllamycin C (118)

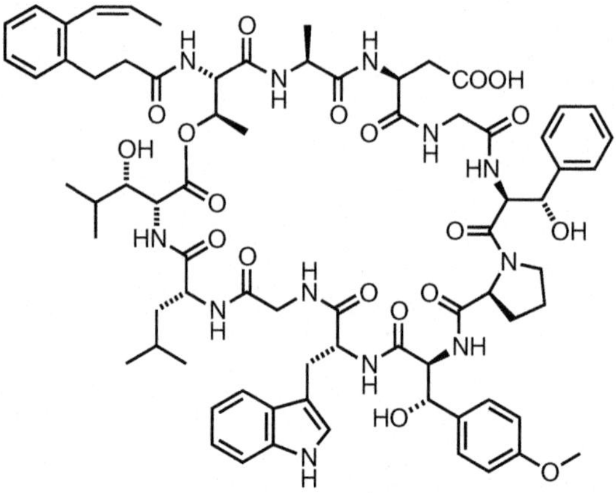

Fmoc-Gly-OH was loaded to 2-CTC as per Sect. 5.5.1 to afford 80 μmol of resin-bound amino acid. Fmoc-Asp(Ph*i*Pr)-OH was coupled under HATU conditions followed by PyBOP coupling of Fmoc-Ala-OH and Fmoc-Thr-OH before coupling of reduced cinnamoyl moiety **131** using HATU to yield resin-bound peptide **223**. From here, on-resin esterification with protected Fmoc-β-OH-Leu-OH (**166**), followed by microwave assisted coupling of Fmoc-D-Leu-OH as detailed in the general procedures. Next, coupling of Fmoc-Gly-OH and Fmoc-D-Trp-OH was carried out using PyBOP coupling conditions followed by coupling of protected Fmoc-β-OH-*O*-Me-Tyr-OH **187** using HATU conditions. Microwave assisted coupling of Fmoc-Pro-OH, followed by coupling of Fmoc-β-OH-Phe-OH (**180**) using HATU, afforded the complete resin-bound linear peptide **225**. Resin-bound linear peptide **225** was cleaved from resin according to the general SPPS protocols (Sect. 5.4.1) to yield crude linear peptide **227** (43.4 mg, 27 μmol, assumed to be 100% pure). This was subject to cyclisation (general procedure D) and global deprotection (general procedure E). Deshydroxy skyllamycin C (**118**) was afforded as a fluffy white solid after RP-HPLC and lyophilisation (4.5 mg, 3.9%).

LRMS (+ESI) m/z calculated mass 1455.7 $[M+H]^+$, 1477.7 $[M+Na]^+$: m/z observed 1455.0 $[M+H]^+$, 1477.3 $[M+Na]^+$; **HRMS** (+ESI) Calc. for $C_{74}H_{94}N_{12}O_{19}$ $[M+2Na]^{2+}$: 750.3271, Found: 750.3284; **Analytical HPLC** R_t 12.5 min (0–100% MeCN (0.1% formic acid) in H_2O (0.1% formic acid) over 15 min, λ = 230 nm). ^{1}H-NMR and ^{13}C-NMR data is listed below in Table 5.7.

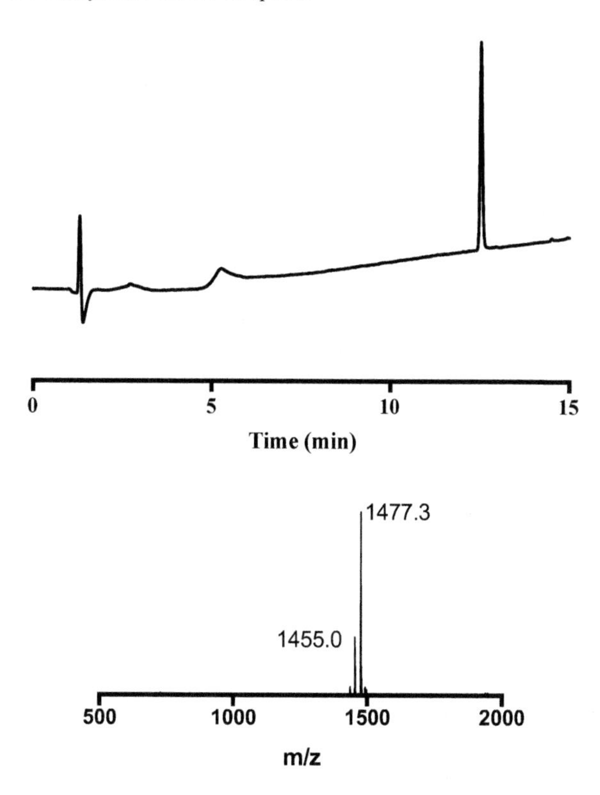

Table 5.7 ^{1}H-NMR (500 MHz) and ^{13}C-NMR (125 MHz) chemical shifts of deshydroxy skyllamycin C (**118**) in CD_3OD referenced at 3.30 ppm (^{1}H) and 49.0 ppm (^{13}C). All assignments were made based on COSY, HSQC and HMBC data in comparison with the skyllamycin natural products [15]. ^{13}C-NMR chemical shifts were extracted from the HSQC and HMBC spectra. N-H resonances not included as they slowly exchange with CD_3OD

Residue	Position	δ ^{13}C/ ppm	δ ^{1}H/ ppm	No. H, mult, *J* Hz	Residue	Position	δ ^{13}C/ ppm	δ ^{1}H/ ppm	No. H, mult, *J* Hz
Thr	C=O	176.6			D-Trp	C=O	174.3		
	α	61.6	5.09	1H, s		7a	138.0		
	β	70.1	5.43	1H, q, 7.0		3a	128.4		
	CH_3	17.2	1.27	3H, d, 6.8		2	125.0	7.17	1H, s
						6	122.3	7.06	1H, t, 7.6
Ala	C=O	174.7				5	119.7	6.93	1H, t, 7.6

(continued)

Table 5.7 (continued)

Residue	Position	δ ^{13}C/ ppm	δ ^{1}H/ ppm	No. H, mult, *J* Hz	Residue	Position	δ ^{13}C/ ppm	δ ^{1}H/ ppm	No. H, mult, *J* Hz
	α	52.7	4.05	1H, m		4	119.5	7.52	1H, d, 8.4
	CH_3	16.4	1.48	3H, d, 7.6		7	112.1	7.30	1H, m
						3	110.5		
Asp	C=O	172.5				α	54.7	4.72	1H, dd, 7.9, 4.8
	C=O	173.2				β	29.1	3.27	1H, dd, 15.1, 5.0
	α	50.5	4.94	1H, m		β′		3.19	1H, dd, 14.7, 7.6
	β	37.6	2.64	1H, dd, 16.2, 9.8					
	β′		3.16	1H, dd,16.2, 4.6	Gly	C=O	171.2		
						α	43.0	3.95	1H, d, 16.4
Gly	C=O	172.0				α′		3.78	1H, d, 16.6
	α	43.2	4.04	1H, d, 17.8					
	α′		3.42	1H, d, 17.7	D-Leu	C=O	177.0		
						α	56.2	4.29	1H, dd, 9.2, 6.1
β-OH-Phe	C=O	171.3				β	41.6	1.68	1H, m
	Aromatic	142.7				β′		1.60	1H, m
	Aromatic	127.3	7.44	2H, m		γ	25.8	1.81	1H, m
	Aromatic	127.4	7.26	1H, m		CH_3	22.8	0.99	3H, d, 6.6
	Aromatic	128.8	7.35	2H, m		CH_3	21.9	0.89	3H, d, 6.6
	α	57.9	4.55	1H, m					
	β	74.1	4.61	1H, m	β-OH-Leu	C=O	170.8		
						α	55.9	4.43	1H, d, 1.8
Pro	C=O	173.8				β	76.5	3.79	1H, m
	α	61.9	4.27	1H, dd, 8.9, 3.5		γ	32.4	1.68	1H, m

(continued)

Table 5.7 (continued)

Residue	Position	δ ^{13}C/ ppm	δ ^{1}H/ ppm	No. H, mult, *J* Hz	Residue	Position	δ ^{13}C/ ppm	δ ^{1}H/ ppm	No. H, mult, *J* Hz
	β	24.8	1.77	1H, m		CH_3	19.3	1.02	3H, d, 6.7
	β′		1.51	1H, m		CH_3	18.5	0.83	3H, d, 7.0
	γ	30.0	1.98	1H, m					
	γ′		1.50	1H, m	Cinnamoyl	C=O	179.7		
	δ	48.1	3.74	2H, m		Aromatic	139.9		
						Aromatic	137.3		
β-OH-*O*-Me-Tyr	C=O	171.7				Aromatic	130.2	7.12	1H, m
	1	132.8				Aromatic	126.3	7.14	1H, m
	2/6	129.0	6.73	2H, d, 8.4		Aromatic	128.4	7.33	1H, m
	3/5	114.2	6.63	2H, d, 8.7		Aromatic	128.3	7.32	1H, m
	4	160.9				*cis*-sp^2	129.5	6.48	1H, d, 11.4
	OCH_3	55.4	3.70	3H, s		*cis*-sp^2-2	128.1	5.59	1H, dq, 11.4, 7.0
	α	59.0	4.75	1H, m		CH_2	36.2	3.04	1H, m
	β	75.0	4.55	1H, m				2.82	1H, m
						CH_2-2	29.5	3.04	2H, m
						CH_3	14.2	1.51	3H, dd, 7.0, 1.6

5.6 General Procedures for Chapter 4

5.6.1 *Fmoc-SPPS Protocols for Chapter 4*

Sieber Amide Loading

Sieber amide resin (550 mg, maximum loading 0.73 mmol/g) was placed in a fritted syringe and swollen in CH_2Cl_2 (10 mL) for 30 min before being washed with DMF (×5). The resin was then treated with a solution of 10 vol.% piperidine in DMF (2 × 3 min) and washed with DMF (×5), CH_2Cl_2 (×5) and DMF (×5) before being shaken with a solution of Fmoc-D-Trp-OH (255 mg, 600 μmol), PyAOP (312 mg, 600 μmol) and (131 μL, 1.2 mmol) in DMF (4 mL) at room temperature for 16 h. The resin was then washed DMF (×5) and CH_2Cl_2 (×5), and then treated with a solution of pyridine: Ac_2O (v/v, 9:1, 5 mL) for 10 min. The resin was then washed DMF (×5), CH_2Cl_2 (×5), and DMF (×5) before resin loading was determined. Loaded resin was treated with a solution of 10 vol.% piperidine in DMF (2 × 3 min) which was then diluted and the absorbance analysed at $\lambda = 301$ nm to determine number of μmol of amino acid loaded to resin.

Sieber Amide Resin Cleavage Conditions
Resin-bound peptide was washed with CH_2Cl_2 (×5) and DMF (×5) before being treated with a solution of 10 vol.% piperidine in DMF (2 × 3 min). The resin was again washed with DMF (×5), CH_2Cl_2 (×5), DMF (×5) and CH_2Cl_2 (×20). The resin was treated with a solution of TFA:*i*Pr_3SiH:CH_2Cl_2 (1:5:94, v/v/v) in CH_2Cl_2 (0.001 M wrt. resin loading) and the solution transferred to a round bottom flask and diluted with CH_2Cl_2 (30 mL) before being concentrated under a stream of N_2 and dried in vacuo. Linear peptides were purified via HPLC using a Waters Sunfire C18 OBD 19 × 150 mm column, using a 0–50 vol.% MeCN in H_2O (0.1% TFA) focussed gradient (0–40 vol.% MeCN over 5 min, 40–50 vol.% over 15 min) at a flow rate flow rate of 16 mL min^{-1} and lyophilised.

5.6.2 General Procedures for the Synthesis of Skyllamycins A–C (101–103)

General Procedure F: Oxidative Cleavage Reaction
To a solution of linear peptide bearing an *N*-terminal serine residue (**250–252**) (6.5–7 μmol) in a mixture of aqueous Na_2HPO_4 (1 M, pH = 8.4) (400 μL) and MeCN (400 μL) was added an aqueous solution of $NaIO_4$ (53–58 μL, 2 equiv., 10 mg in 200 μL of H_2O). The mixture was mixed using a vortex mixer before being allowed to react for 10 min. To the mixture was added an aqueous solution of ethylene glycol (35–38 μL, 10 equiv., 20 μL in 200 μL of H_2O). The mixture was then diluted up to 3.5 mL with a mixture of MeCN and H_2O (with 0.1% TFA) and purified via HPLC using a Waters Sunfire C18 OBD 19 × 150 mm column, using a 0–60 vol.% MeCN in H_2O (0.1% TFA) focussed gradient (0–50 vol.% MeCN over 5 min, 50–60 vol.% over 15 min) at a flow rate flow rate of 16 mL min^{-1} and lyophilised.

General Procedure G: Cyclisation Reaction
Linear peptide aldehyde (**247–249**) (5.8–6.8 mg) was dissolved in MeCN (4 mL) in 2 separate eppendorf tubes. The solutions were incubated at 60 °C and the reaction was monitored by HPLC-MS. After 25 h, each eppendorf was diluted up to 3.5 mL total volume with H_2O and then purified over two runs via HPLC using a Waters Sunfire C18 OBD 19 × 150 mm column, using a 0–60 vol.% MeCN in H_2O (0.1%

TFA) focussed gradient (0–50 vol.% MeCN over 5 min, 50–60 vol.% over 15 min) at a flow rate flow rate of 16 mL min^{-1}. All peaks were collected and lyophilised and sent to the Linington research group at Simon Fraser University for analysis.

5.7 Procedures and Analytical Data Chapter 4

5.7.1 *Synthesis of Oxazolidine Protected Fmoc-β-OH-Phe-OH (256)*

(4*S*,5*S*)-3-(((9*H*-fluoren-9-yl)methoxy)carbonyl)-2,2-dimethyl-5- phenyloxazolidine-4-carboxylic acid (256)

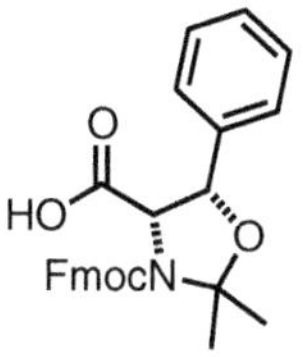

Fmoc-β-OH-Phe-OH (**180**) (450 mg, 1.15 mmol) was protected via general procedure C. Crude oxazolidine protected amino acid **256** was purified by flash chromatography (eluent: 30:70 → 45:55 v/v EtOAc/petroleum benzines) to yield oxazolidine protected amino acid **256** as an off-white crystalline solid (465 mg, 94%).

^{1}H-NMR ($CDCl_3$, 500 MHz, rotamers) δ (ppm) 7.78 (ap. d, *J* = 7.88 Hz, 0.7H) and 7.68 (ap. dd, *J* = 5.7, 7.5 Hz, 1.3H), 7.57 (dd, *J* = 3.5, 7.5 Hz, 0.7H) and 7.47 (ap. t, *J* = 7.0 Hz, 1.3H), 7.44–7.16 (m, 9H), 5.34 (d, *J* = 6.8 Hz, 0.65H) and 5.20 (d, *J* = 6.7 Hz, 0.35H), 4.76 (dd, *J* = 4.2, 10.8 Hz, 0.35H) and 4.39 (dd, *J* = 6.3, 10.8 Hz, 0.65H), 4.69 (dd, *J* = 4.4, 11.0 Hz, 0.35H) and 4.30 (dd, *J* = 6.8, 10.8 Hz), 4.52–4.45 (m, 1H), 4.23 (ap. t, *J* = 4.2 Hz, 0.35H) and 4.02 (ap. t, *J* = 6.6 Hz, 0.65H), 1.87 (s, 2H) and 1.21 (s, 1H), 1.63 (s, 2H) and 1.02 (s, 1H); **^{13}C-NMR** ($CDCl_3$, 125 MHz, rotamers) δ (ppm) 173.6 and 173.4, 152.9 and 151.8, 144.0, 143.9 and 143.6, 141.8 and 141.4, 141.6 and 141.3, 134.2, 129.0 and 128.8, 128.5 and 128.4, 127.9 and 127.9, 127.8 and 127.8, 127.4 and 127.2, 127.3 and 127.2, 126.5 and 126.4, 125.1 and 125.0, 124.6 and 124.5, 120.1 and 120.1, 120.0 and 120.0, 95.5 and 94.7, 78.2 and 77.4, 67.3 and 66.9, 64.6 and 64.0, 47.4 and 47.3, 25.5 and 25.4, 25.0 and 24.3 (extra signals due to further rotational conformations in the Fmoc-region); **LRMS** (+ESI) *m/z* 466 $[M+Na]^+$; **HRMS** (+ESI) Calc. for $C_{27}H_{25}NO_5$ $[M+Na]^+$: 466.1624, Found: 466.1624; **IR** ν_{max} (ATR) 2923, 1744, 1706, 1451, 1410, 1348, 1248, 1216, 1190, 1159 cm^{-1}; **$[\alpha]_D$** = +21.4° (*c* 0.5, CH_2Cl_2); **mp.** 60.6–71.2 °C

5.7.2 *Synthesis of Skyllamycin A (101)*

Skyllamycin A—linear peptide 250

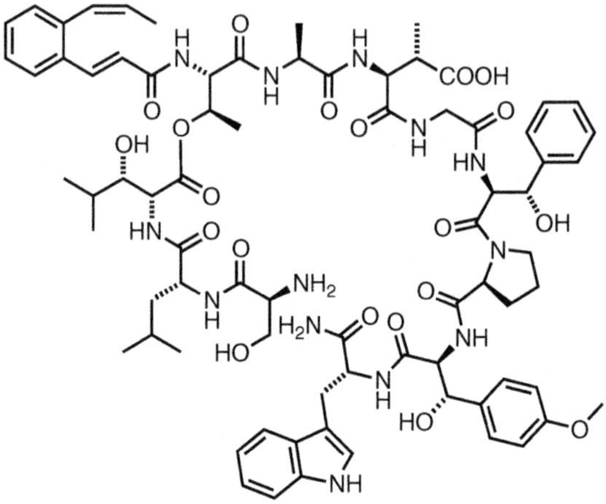

Fmoc-D-Trp-OH was loaded to Sieber amide resin as per the general procedures (Sect. 5.6.1) and after Fmoc-loading it was determined that 120 μmol of amino acid were loaded to resin. Oxazolidine protected Fmoc-β-OH-*O*-Me-Tyr-OH **187** was coupled to the resin under HATU conditions, followed by microwave assisted coupling of Fmoc-Pro-OH. Oxazolidine protected Fmoc-β-OH-Phe-OH **256** was then coupled using HATU conditions, followed by microwave assisted coupling of Fmoc-Gly-OH to yield key resin-bound intermediate **261**. Fmoc-β-Me-Asp(Ph*i*Pr)-OH (**212**) was next coupled using HATU conditions followed by the iterative coupling of Fmoc-Ala-OH and Fmoc-Thr-OH using PyBOP coupling conditions. Cinnamoyl moiety **130** was coupled using HATU conditions followed by on-resin esterification of oxazolidine protected Fmoc-β-OH-Leu-OH (**166**), yielding resin-bound **270**.

Resin-bound peptide **270** was washed with CH_2Cl_2 (×5) and DMF (×5) before being treated with a solution of 10 vol.% piperidine in DMF (2 × 3 min). The resin was again washed with DMF (×5), CH_2Cl_2 (×5) and DMF (×5). The resin was transferred to a Biotage microwave peptide synthesis vessel and treated with a solution of the desired Fmoc-D-Leu-OH (4 equiv.), DIC (4 equiv.) and HOAt (4 equiv.) in DMF (0.1 M in regard to loaded peptide) under microwave irradiation at 50 °C for 1 h. The resin was then transferred to a fritted syringe, the coupling solution was discharged and the resin washed with DMF (×5), CH_2Cl_2 (×5) and DMF (×5). The resin was thoroughly dried, and split and to 60 μmol of resin-bound peptide Fmoc-Ser-OH was coupled PyBOP coupling conditions. The resin was cleaved using the conditions described in Sect. 5.6.1 to afford the linear peptide **250** as a white fluffy solid after lyophilisation (10.1 mg, 11%).

LRMS (+ESI) m/z calculated mass 1514.7 $[M+H]^+$: m/z observed 1513.8 $[M+H]^+$; **HRMS** (+ESI) Calc. for $C_{76}H_{99}N_{13}O_{20}$ $[M+2H]^{2+}$: 757.8637, Found:

757.8645; **Analytical HPLC** R_t 9.8 min (0–100% MeCN (0.1% formic acid) in H_2O (0.1% formic acid) over 15 min, λ = 230 nm).

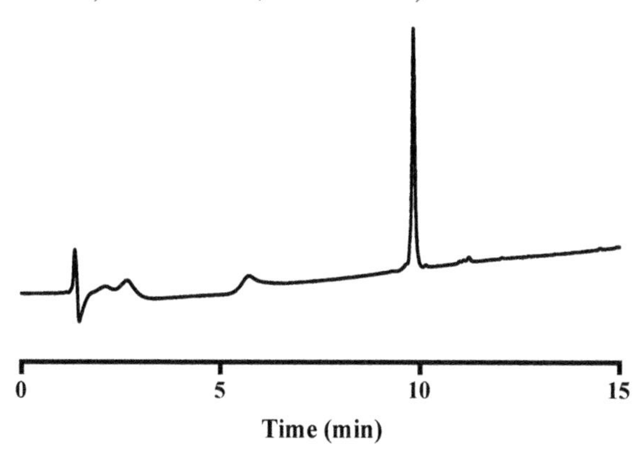

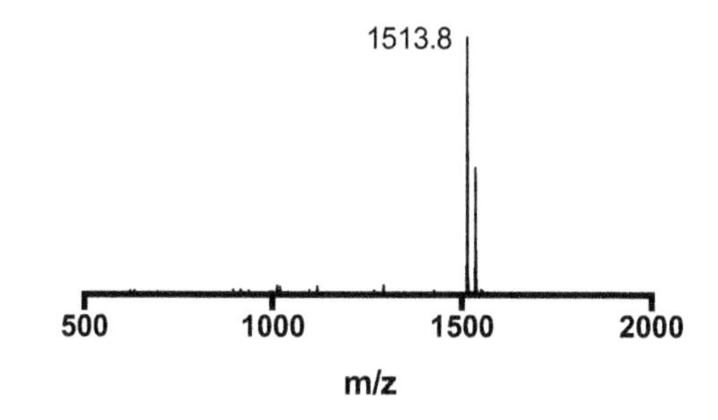

Skyllamycin A—aldehyde 247

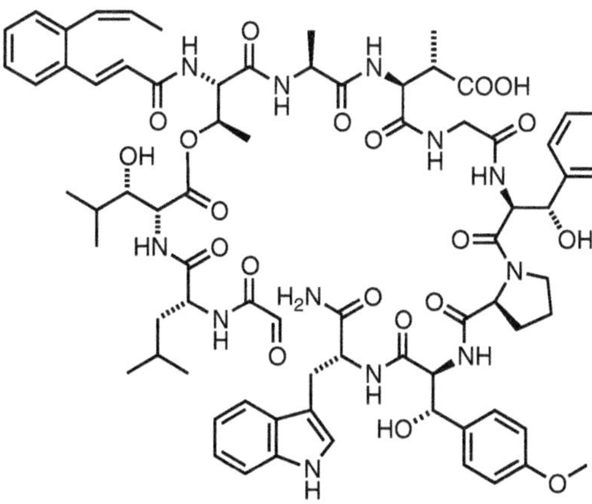

Linear peptide **250** (10.1 mg, 6.2 μmol) was subject to oxidative cleavage conditions (general procedure F) as described in Sect. 5.6.2. Aldehyde **247** was obtained as a white fluffy solid after RP-HPLC and lyophilisation (5.8 mg, 62%).

LRMS (+ESI) m/z calculated mass 1501.7 $[M+H_2O+H]^+$, 1523.7 $[M+H_2O+Na]^+$: m/z observed 1501.0 $[M+H_2O+H]^+$, 1524.1 $[M+H_2O+Na]^+$; **HRMS** (+ESI) Calc. for $C_{75}H_{94}N_{12}O_{20}$ $[M+Na]^+$: 1505.6600, Found: 1505.6624; **Analytical HPLC** R_t 12.2 min (0–100% MeCN (0.1% formic acid) in H_2O (0.1% formic acid) over 15 min, λ = 230 nm).

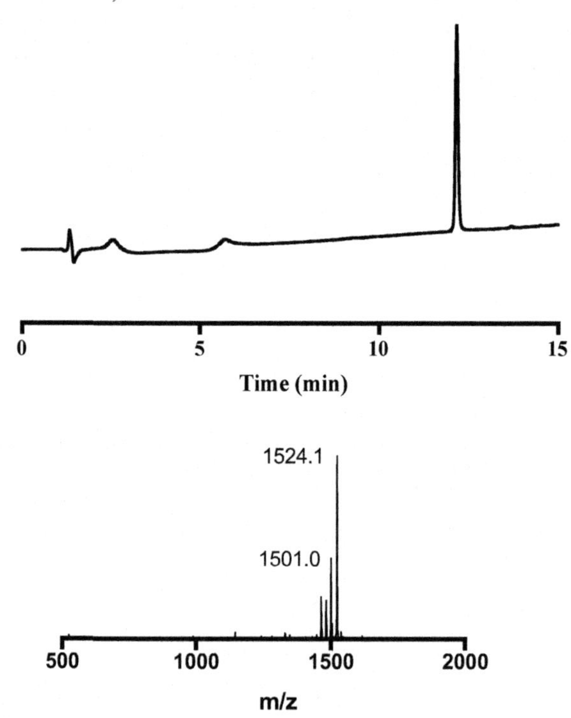

Skyllamycin A (101)

Aldehyde **247** (5.8 mg, 3.8 μmol) was subject to cyclisation conditions (general procedure G) described in Sect. 5.6.1. Skyllamycin A (**101**) was isolated as a white fluffy solid after RP-HPLC and lypohilisation (1.98 mg, 32%).

LRMS (+ESI) m/z calculated mass 1483.7 $[M+H]^+$, 1505.7 $[M+Na]^+$: m/z observed 1483.0 $[M+H]^+$, 1505.1 $[M+Na]^+$; **HRMS** (+ESI) Calc. for $C_{75}H_{94}N_{12}O_{20}$ $[M+H]^+$: 1483.6780, Found: 1483.6768; **Analytical HPLC** R_t 12.9 min (0–100% MeCN (0.1% formic acid) in H_2O (0.1% formic acid) over 15 min, λ = 230 nm). ^{1}H-NMR and ^{13}C-NMR data is listed in Table 5.8.

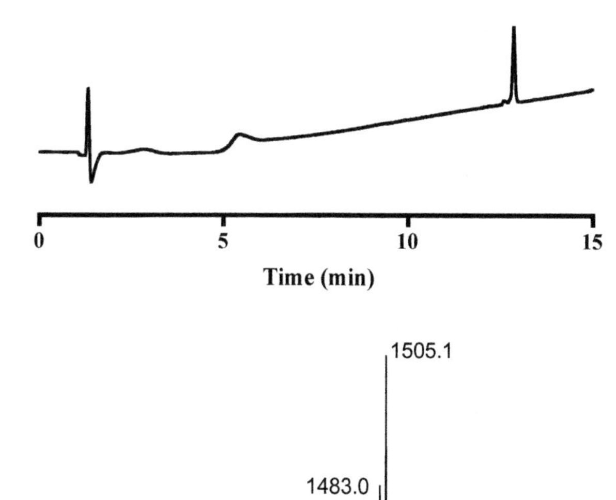

Table 5.8 ^{1}H-NMR (600 MHz) and ^{13}C-NMR (150 MHz) chemical shifts of synthetic skyllamycin A (**101**) in CD_3OD referenced at 3.30 ppm (^{1}H) and 49.0 ppm (^{13}C). All assignments were made based on COSY, HSQC and HMBC data in comparison with the skyllamycin natural products [15]. N-H resonances not included as they exchange with CD_3OD

Residue	Position	δ ^{13}C/ ppm	δ ^{1}H/ ppm	No. H, mult, *J* Hz	Residue	Position	δ ^{13}C/ ppm	δ ^{1}H/ ppm	(No. H, mult, *J* Hz)
Thr	C=O	176.3			D-Trp	C=O	175.0		
	α	61.9	5.07	1H, s		7a	138.2		
	β	70.8	5.47	1H, q, 6.2		3a	128.9		
	CH_3	17.7	1.34	3H, d, 6.7		2	125.2	7.17	1H, s
						6	122.5	7.06	1H, t, 7.7
Ala	C=O	175.5				5	120.0	6.93	1H, t, 7.7
	α	53.1	4.05	1H, q, 7.1		4	119.8	7.59	1H, d, 8.1
	CH_3	16.9	1.47	3H, d, 7.3		7	112.4	7.28	1H, m
						3	110.9		
Asp	C=O	179.4				α	55.6	4.72	1H, dd, 9.2, 4.3
	C=O	177.3				β	29.3	3.35	1H, m
	α	55.6	5.120	1H, d, 5.3		β′		3.16	1H, m
	β	43.1	3.25	1H, m					
	CH_3	17.7	1.28	3H, d, 5.0	α-OH-Gly	C=O	173.3		
						α	74.5	5.43	1H, s
Gly	C=O	172.1							
	α	43.6	4.13	1H, d, 18.2					
	α′		3.46	1H, d, 18.0	D-Leu	C=O	177.5		
						α	56.4	4.34	1H, m
β-OH-Phe	C=O	172.1				β	41.2	1.82	1H, m
	Aromatic	143.4				β′		1.63	1H, m
	Aromatic	127.4	7.42	2H, d, 7.5		γ	26.2	1.85	1H, m
	Aromatic	128.5	7.24	1H, t, 7.5		CH_3	23.3	0.98	3H, d, 6.7

(continued)

Table 5.8 (continued)

Residue	Position	δ ^{13}C/ppm	δ ^{1}H/ppm	No. H, mult, *J* Hz	Residue	Position	δ ^{13}C/ppm	δ ^{1}H/ppm	(No. H, mult, *J* Hz)
	Aromatic	129.2	7.35	2H, t, 7.5		CH_3	21.9	0.91	3H, d, 6.7
	α	58.6	4.53	1H, d, 9.9					
	β	73.8	4.62	1H, d, 9.9	β-OH-Leu	C=O	171.2		
						α	56.7	4.45	1H, m
Pro	C=O	174.1				β	77.0	3.80	1H, dd, 9.5, 1.5
	α	61.8	4.34	1H, m		γ	32.6	1.71	1H, m
	β	24.9	1.77	1H, m		CH_3	19.7	1.03	3H, d, 6.7
	β′		1.43	1H, m		CH_3	19.0	0.84	3H, d, 6.7
	γ	30.2	1.95	1H, m					
	γ′		1.66	1H, m	Cinnamoyl	C=O	171.9		
	δ	48.2	3.68	2H, m		Aromatic	134.6		
						Aromatic	139.3		
β-OH-*O*-Me-Tyr	C=O	171.7				Aromatic	127.7	7.89	1H, m
	1	133.1				Aromatic	131.0	7.14	1H, m
	2/6	129.1	6.70	2H, d, 8.3		Aromatic	128.4	7.29	1H, m
	3/5	114.6	6.58	2H, d, 8.3		Aromatic	130.3	7.16	1H, m
	4	161.2				*cis*-sp^2	129.1	6.37	1H, d, 11.4
	OCH_3	55.7	3.66	3H, s		*cis*-sp^2-2	130.2	5.76	1H, m
	α	58.9	4.75	1H, d, 7.3		*trans*-sp^2	140.2	7.83	1H, d, 15.7
	β	75.0	4.49	1H, d, 7.2		*trans*-sp^2-2	122.7	7.21	1H, d, 15.7
						CH_3	14.5	1.50	3H, dd, 7.0, 1.7

5.7.3 *Synthesis of Skyllamycin B (102)*

Skyllamycin B—linear peptide 251

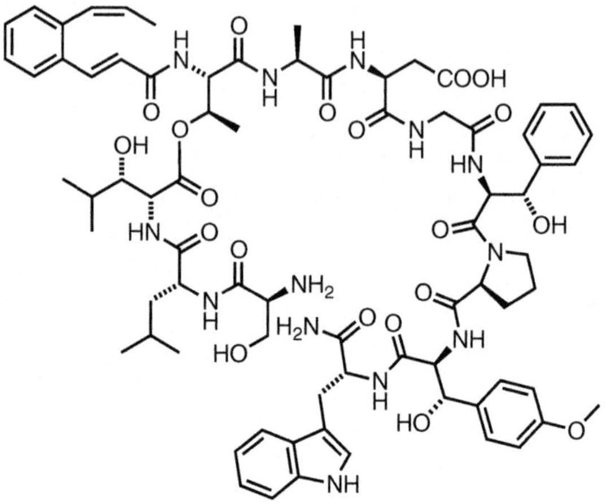

Fmoc-D-Trp-OH was loaded to Sieber amide resin as per the general procedures (Sect. 5.6.1) and after Fmoc-loading it was determined that 120 μmol of amino acid were loaded to resin. Oxazolidine protected Fmoc-β-OH-*O*-Me-Tyr-OH **187** was coupled to the resin under HATU conditions, followed by microwave assisted coupling of Fmoc-Pro-OH. Oxazolidine protected Fmoc-β-OH-Phe-OH **256** was then coupled using HATU conditions, followed by microwave assisted coupling of Fmoc-Gly-OH to yield key resin-bound intermediate **261**. Fmoc-Asp(Ph*i*Pr)-OH (**212**) was next coupled using HATU conditions followed by the iterative coupling of Fmoc-Ala-OH and Fmoc-Thr-OH using PyBOP coupling conditions. Cinnamoyl moiety **130** was coupled using HATU conditions followed by on-resin esterification of oxazolidine protected Fmoc-β-OH-Leu-OH **166**. Fmoc-D-Leu-OH was coupled using microwave assisted coupling conditions. The resin was thoroughly dried and split, and to 60 μmol of resin-bound peptide Fmoc-Ser-OH was coupled using PyBOP coupling conditions. The resin was cleaved using the conditions described in Sect. 5.6.1 to afford the linear peptide **251** as a white fluffy solid after lyophilisation (10.8 mg, 12%).

LRMS (+ESI) m/z calculated mass 1500.7 $[M+H]^+$: m/z observed 1499.8 $[M+H]^+$; **HRMS** (+ESI) Calc. for $C_{75}H_{97}N_{13}O_{20}$ $[M+2H]^{2+}$: 750.8559, Found: 750.8567; **Analytical HPLC** R_t 9.8 min (0–100% MeCN (0.1% formic acid) in H_2O (0.1% formic acid) over 15 min, λ = 230 nm).

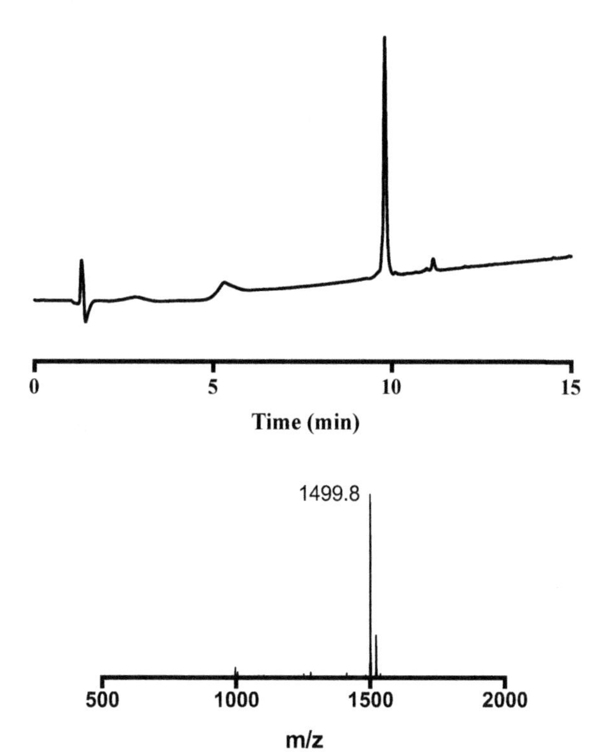

Skyllamycin B—aldehyde 248

Linear peptide **251** (10.8 mg, 6.7 μmol) was subject to oxidative cleavage conditions (general procedure F) as described in Sect. 5.6.2. Aldehyde **248** was obtained as a white fluffy solid after RP-HPLC and lyophilisation (6.2 mg, 63%).

LRMS (+ESI) m/z calculated mass 1469.7 $[M+H]^+$, 1509.7 $[M+H_2O+Na]^+$: m/z observed 1468.9 $[M+X]^+$, 1509.1 $[M+H_2O+Na]^+$; **HRMS** (+ESI) Calc. for $C_{74}H_{92}N_{12}O_{20}$ $[M+Na]^+$: 1491.6443, Found: 1491.6471; **Analytical HPLC** R_t 12.1 min (0–100% MeCN (0.1% formic acid) in H_2O (0.1% formic acid) over 15 min, λ = 230 nm).

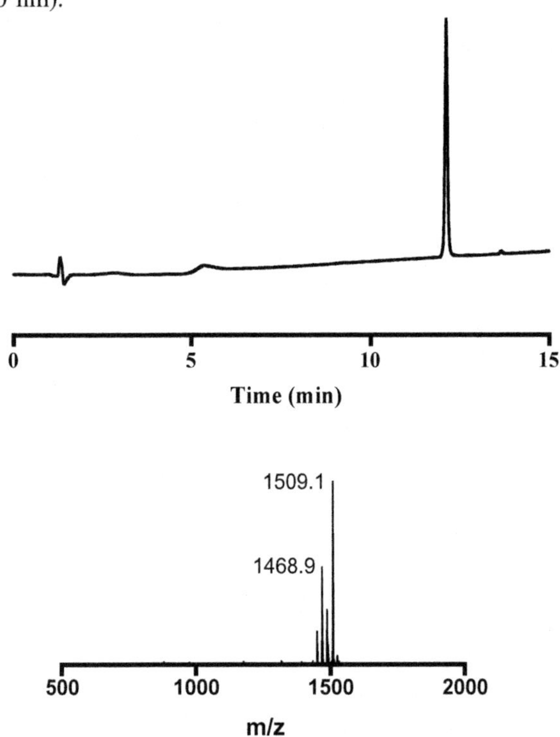

Skyllamycin B (102)

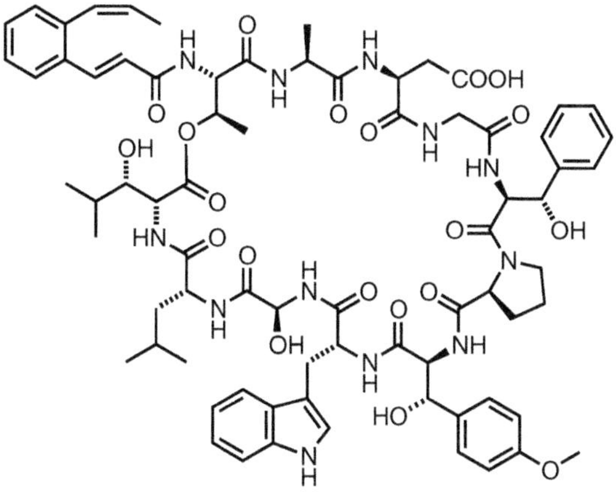

Aldehyde **248** (6.2 mg, 4.2 μmol) was subject to cyclisation conditions (general procedure G) described in Sect. 5.6.2. A mixture of skyllamycin B (**102**) and its epimer were isolated as a white fluffy solid after RP-HPLC and lypohilisation (2.60 mg, 42%). This mixture was subject to further RP-HPLC purification by the Linington research group. Specifically the mixture of epimers was subject to HPLC using a Phenomenex Kinetix XB-C18 4.6 × 250 mm column, using a gradient of 45% MeOH, 21% MeCN, 34% H_2O ramped to 48.7% MeOH, 21% MeCN, 30.3% H_2O over 20 min, then ramped to 69% MeOH, 21% MeCN, 10% H_2O over 2 min, then held for 3 min min with a flow rate of 1.5 mL min^{-1} to yield skyllamycin B (**102**) as a white fluffy solid (1.20 mg, 46% from mixture of epimers, 19% overall).

LRMS (+ESI) m/z calculated mass 1469.7 $[M+H]^+$, 1491.6 $[M+Na]^+$: m/z observed 1468.8 $[M+H]^+$, 1490.7 $[M+Na]^+$; **HRMS** (+ESI) Calc. for $C_{74}H_{92}N_{12}O_{20}$ $[M+H]^+$: 1469.6624, Found: 1469.6600; **Analytical HPLC** R_t 12.4 min (0–100% MeCN (0.1% formic acid) in H_2O (0.1% formic acid) over 15 min, λ = 230 nm). ^{1}H-NMR and ^{13}C-NMR data is listed in Table 5.9.

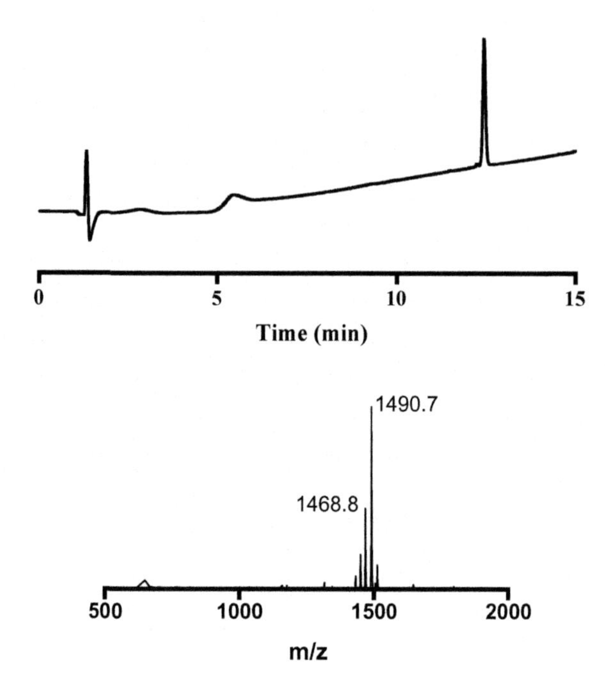

Table 5.9 ^{1}H-NMR (600 MHz) and ^{13}C-NMR (150 MHz) chemical shifts of synthetic skyllamycin B (**102**) in CD_3OD referenced at 3.30 ppm (^{1}H) and 49.0 ppm (^{13}C). All assignments were made based on COSY, HSQC and HMBC data in comparison with the skyllamycin natural products [15]. N-H resonances not included as they slowly exchange with CD_3OD

Residue	Position	δ ^{13}C/ ppm	δ ^{1}H/ ppm	No. H, mult, *J* Hz	Residue	Position	δ ^{13}C/ ppm	δ ^{1}H/ ppm	No. H, mult, *J* Hz
Thr	C=O	176.8			D-Trp	C=O	175.3		
	α	61.9	5.08	1H, s		7a	138.5		
	β	70.7	5.48	1H, q, 7.0		3a	129.2		
	CH_3	17.7	1.34	3H, d, 6.8		2	125.3	7.16	1H, s
						6	122.5	7.06	1H, t, 7.3
Ala	C=O	175.5				5	120.1	6.93	1H, t, 7.3
	α	53.0	4.07	1H, q, 7.3		4	119.7	7.58	1H, d, 8.6
	CH_3	16.5	1.45	3H, d, 7.5		7	112.4	7.28	1H, m
						3	111.3		

(continued)

Table 5.9 (continued)

Residue	Position	δ ^{13}C/ppm	δ ^{1}H/ppm	No. H, mult, *J* Hz	Residue	Position	δ ^{13}C/ppm	δ ^{1}H/ppm	No. H, mult, *J* Hz
Asp	C=O	175.7				α	55.5	4.72	1H, dd, 8.7, 4.7
	C=O	173.8				β	29.4	3.33	1H, under CD_3OD
	α	51.6	5.00	1H, dd, 10.4, 3.7		β′		3.16	1H, dd, 14.9, 9.1
	β	39.4	2.88	1H, dd,15.6, 10.5					
	β′		3.28	1H, under CD_3OD	α-OH-Gly	C=O	173.8		
						α	74.5	5.45	1H, s
Gly	C=O	172.1							
	α	43.4	4.09	1H, d, 18.1					
	α′		3.46	1H, d, 17.8	D-Leu	C=O	177.5		
						α	56.5	4.34	1H, m
β-OH-Phe	C=O	172.2				β	41.2	1.82	1H, m
	Aromatic	143.1				β′		1.62	1H, m
	Aromatic	127.5	7.43	2H, d, 7.4		γ	26.1	1.85	1H, m
	Aromatic	129.2	7.25	1H, m		CH_3	23.3	0.98	3H, d, 6.7
	Aromatic	128.7	7.33	2H, t, 7.5		CH_3	21.9	0.92	3H, d, 6.6
	α	58.4	4.51	1H, m					
	β	74.0	4.62	1H, m	β-OH-Leu	C=O	171.7		
						α	56.6	4.48	1H, m
Pro	C=O	174.4				β	77.0	3.80	1H, dd, 9.2, 1.5
	α	61.9	4.34	1H, m		γ	32.5	1.71	1H, m
	β	24.9	1.76	1H, m		CH_3	19.7	1.02	3H, d, 6.7
	β′		1.38	1H, m		CH_3	19.0	0.84	3H, d, 6.6
	γ	30.2	1.93	1H, m					
	γ′		1.63	1H, m	Cinnamoyl	C=O	172.4		
	δ	48.3	3.71	2H, m		Aromatic	134.9		
						Aromatic	140.1		
	C=O	172.1				Aromatic	127.9	7.90	1H, m

(continued)

Table 5.9 (continued)

Residue	Position	δ ^{13}C/ ppm	δ ^{1}H/ ppm	No. H, mult, *J* Hz	Residue	Position	δ ^{13}C/ ppm	δ ^{1}H/ ppm	No. H, mult, *J* Hz
β-OH-*O*-Me-Tyr									
	1	133.1				Aromatic	131.0	7.13	1H, m
	2/6	129.2	6.74	2H, d, 8.4		Aromatic	128.7	7.29	1H, m
	3/5	114.5	6.61	2H, d, 8.5		Aromatic	128.5	7.27	1H, m
	4	161.0				*cis*-sp^2	129.3	6.48	1H, d, 11.6, 1.5
	OCH_3	55.7	3.67	3H, s		*cis*-sp^2-2	130.0	5.70	1H, dq, 11.5, 7.0
	α	59.2	4.72	1H, d, 7.3		*trans*-sp^2	140.6	7.91	1H, d, 16.0
	β	74.9	4.49	1H, d, 7.3		*trans*-sp^2-2	122.5	7.20	1H, d, 15.9
						CH_3	14.5	1.44	3H, dd, 7.0, 1.6

5.7.4 Synthesis of Skyllamycin C (103)

Skyllamycin C—linear peptide 252

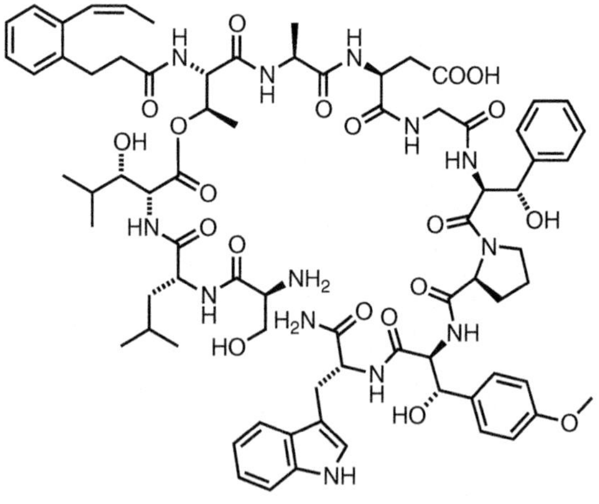

Fmoc-D-Trp-OH was loaded to Sieber amide resin as per the general procedures (Sect. 5.6.1) and after Fmoc-loading it was determined that 120 μmol of amino acid were loaded to resin. Oxazolidine protected Fmoc-β-OH-*O*-Me-Tyr-OH **187** was coupled to the resin under HATU conditions, followed by microwave assisted coupling of Fmoc-Pro-OH. Oxazolidine protected Fmoc-β-OH-Phe-OH **256** was then coupled using HATU conditions, followed by microwave assisted coupling of Fmoc-Gly-OH to yield key resin-bound intermediate **261**. Fmoc-Asp(Ph*i*Pr)-OH (**212**) was next coupled using HATU conditions followed by the iterative coupling of Fmoc-Ala-OH and Fmoc-Thr-OH using PyBOP coupling conditions. Reduced cinnamoyl moiety **131** was coupled using HATU conditions followed by on-resin esterification of oxazolidine protected Fmoc-β-OH-Leu-OH **166**. Fmoc-D-Leu-OH was coupled using microwave assisted coupling conditions. The resin was thoroughly dried and split, and to 60 μmol of resin-bound peptide Fmoc-Ser-OH was then coupled using PyBOP coupling conditions. The resin was cleaved using the conditions described in Sect. 5.6.1 to afford the linear peptide **252** as a white fluffy solid after lyophilisation (11.0 mg, 12%).

LRMS (+ESI) m/z calculated mass 1502.7 $[M+H]^+$: m/z observed 1501.7 $[M+H]^+$; **HRMS** (+ESI) Calc. for $C_{75}H_{99}N_{13}O_{20}$ $[M+2H]^{2+}$: 751.8637, Found: 751.8644; **Analytical HPLC** R_t 9.9 min (0–100% MeCN (0.1% formic acid) in H_2O (0.1% formic acid) over 15 min, λ = 230 nm).

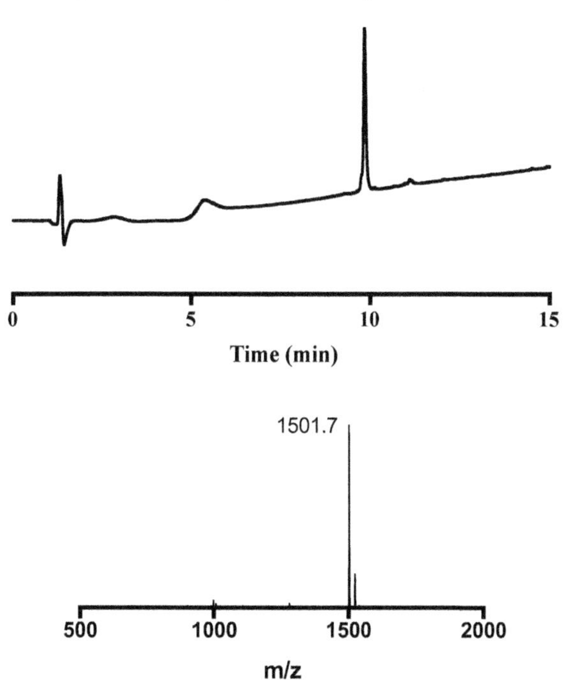

Skyllamycin C—aldehyde 249

Linear peptide **252** (11.0 mg, 6.8 μmol) was subject to oxidative cleavage conditions (general procedure F) as described in Sect. 5.6.2. Aldehyde **249** was obtained as a white fluffy solid after RP-HPLC and lyophilisation (6.8 mg, 68%).

LRMS (+ESI) m/z calculated mass 1489.7 $[M+H_2O+H]^+$, 1511.7 $[M+H_2O+Na]^+$: m/z observed 1489.0 $[M+H_2O+H]^+$, 1510.9 $[M+H_2O+Na]^+$; **HRMS** (+ESI) Calc. for $C_{74}H_{94}N_{12}O_{20}$ $[M+2Na]^{2+}$: 758.3246, Found: 758.3250; **Analytical HPLC** R_t 12.1 min (0–100% MeCN (0.1% formic acid) in H_2O (0.1% formic acid) over 15 min, λ = 230 nm).

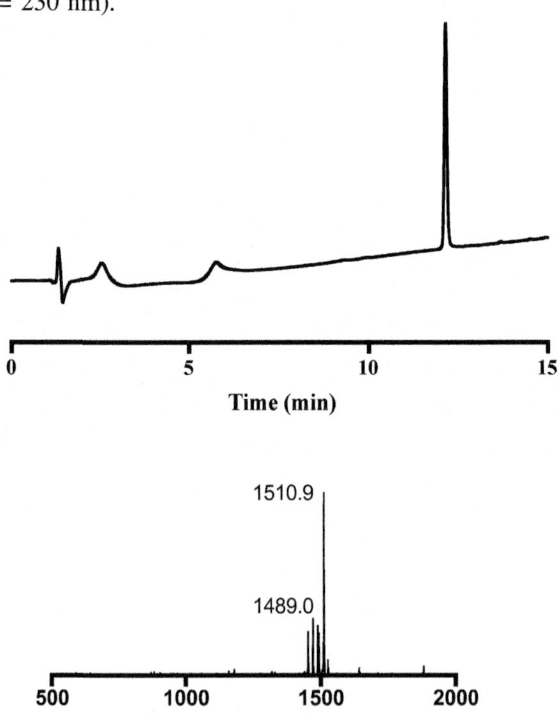

Skyllamycin C (103)

Aldehyde **249** (6.8 mg, 4.6 μmol) was subject to cyclisation conditions (general procedure G) described in Sect. 5.6.2. Skyllamycin C (**103**) was isolated as a white solid after RP-HPLC and lypohilisation (2.16 mg, 33%).

LRMS (+ESI) m/z calculated mass 1471.7 $[M+H]^+$, 1593.7 $[M+Na]^+$: m/z observed 1471.0 $[M+H]^+$, 1592.7 $[M+Na]^+$; **HRMS** (+ESI) Calc. for $C_{74}H_{94}N_{12}O_{20}$ $[M+H]^+$: 1471.6780, Found: 1471.6770; **Analytical HPLC** R_t 12.6 min (0–100% MeCN (0.1% formic acid) in H_2O (0.1% formic acid) over 15 min, λ = 230 nm). ^{1}H-NMR and ^{13}C-NMR data is listed in Table 5.10.

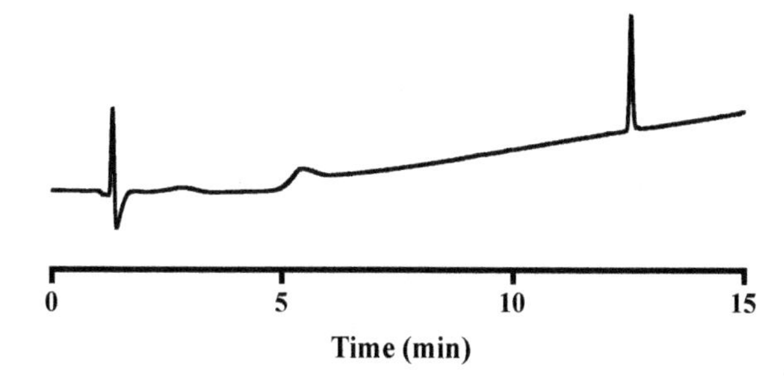

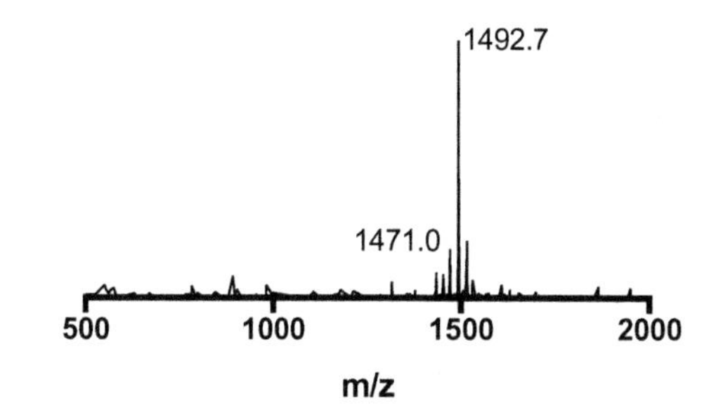

Table 5.10 ^{1}H-NMR (600 MHz) and ^{13}C-NMR (150 MHz) chemical shifts of synthetic skyllamycin C (**103**) in CD_3OD referenced at 3.30 ppm (^{1}H) and 49.0 ppm (^{13}C). All assignments were made based on COSY, HSQC and HMBC data in comparison with the skyllamycin natural products[15]. N-H resonances not included as they slowly exchange with CD_3OD

Residue	Position	δ ^{13}C/ ppm	δ ^{1}H/ ppm	No. H, mult, *J* Hz	Residue	Position	δ ^{13}C/ ppm	δ ^{1}H/ ppm	No. H, mult, *J* Hz
Thr	C=O	171.4			D-Trp	C=O	174.2		
	α	60.6	4.98	1H, s		7a	138.0		
	β	70.9	5.46	1H, q, 7.2		3a	128.7		
	CH_3	16.6	1.23	3H, d, 6.8		2	125.1	7.16	1H, s
						6	122.5	7.07	1H, t, 7.4
Ala	C=O	174.8				5	120.0	6.97	1H, t, 7.4
	α	52.7	4.10	1H, q, 7.5		4	119.6	7.60	1H, d, 8.0
	CH_3	16.8	1.45	3H, d, 7.4		7	112.4	7.30	1H, d, 8.0
						3	110.5		
Asp	C=O	177.7				α	55.4	4.62	1H, dd, 9.3, 4.8
	C=O	173.2				β	29.3	3.27	1H, m
	α	52.7	4.88	1H, m		β′		3.09	1H, dd, 15.3, 9.4
	β	40.8	3.10	1H, m					
	β′		2.76	1H, m	α-OH-Gly	C=O	172.7		
						α	73.9	5.63	1H, s
Gly	C=O	171.8							
	α	43.4	4.05	1H, d, 17.7					
	α′		3.47	1H, d, 17.8	D-Leu	C=O	176.0		
						α	56.1	4.36	1H, m
β-OH-Phe	C=O	173.2				β	41.3	1.78	1H, m
	Aromatic	143.0				β′		1.66	1H, m
	Aromatic	127.8	7.48	2H, d, 7.8		γ	26.2	1.77	1H, m
	Aromatic	128.7	7.25	1H, t, 7.3		CH_3	23.2	0.99	3H, d, 6.6
	Aromatic	129.2	7.35	2H, d, 7.4		CH_3	22.3	0.91	3H, d, 6.6
	α	58.5	4.55	1H, d, 10.2					

(continued)

Table 5.10 (continued)

Residue	Position	δ ^{13}C/ ppm	δ ^{1}H/ ppm	No. H, mult, *J* Hz	Residue	Position	δ ^{13}C/ ppm	δ ^{1}H/ ppm	No. H, mult, *J* Hz
	β	74.1	4.69	1H, d, 10.2	β-OH-Leu	C=O	171.4		
						α	56.5	4.50	1H, s
Pro	C=O	173.5				β	77.3	3.79	1H, d, 9.5
	α	61.9	4.38	1H, dd, 8.0, 3.0		γ	32.4	1.71	1H, m
	β	25.0	1.82	1H, m		CH_3	19.7	1.02	3H, d, 6.7
	β′		1.51	1H, m		CH_3	19.1	0.85	3H, d, 6.7
	γ	30.1	1.97	1H, m					
	γ′		1.78	1H, m	Cinnamoyl	C=O	178.7		
	δ	48.5	3.75	2H, m		Aromatic	140.5		
						Aromatic	137.8		
β-OH-*O*-Me-Tyr	C=O	171.4				Aromatic	130.5	7.11	1H, m
	1	132.8				Aromatic	126.7	7.12	1H, m
	2/6	129.2	6.74	2H, d, 8.0		Aromatic	128.8	7.29	1H, d, 6.8
	3/5	114.5	6.63	2H, d, 8.0		Aromatic	128.2	7.13	1H, m
	4	160.8				*cis*-sp^2	129.8	6.47	1H, d, 11.4
	OCH_3	55.7	3.71	3H, s		*cis*-sp^2-2	128.4	5.52	1H, m
	α	58.9	4.79	1H, d, 6.9		CH_2	37.1	2.94	1H, m
	β	74.9	4.59	1H, d, 6.7				2.78	1H, m
						CH_2-2	30.0	3.07	2H, m
						CH_3	14.5	1.49	3H, dd, 6.9

5.7.5 Circular Dichroism Analysis of Skyllamycins A–C (101–103)

Each of the skyllamycins were diluted in MeOH to a concentration of 0.5 mg/mL (with the exception of synthetic skyllamycin B which was 0.4 mg/mL). From each sample, 600 μL was transferred to quartz cuvettes (path length 2.0 mm) for circular dichroism (CD) analysis. The spectrum was acquired on a Chirascan qCD from 600 to 200 nm at a 1 nm resolution at a scan rate of 0.5 scans/s. The methanol blank scans were subtracted and the average of three scans is reported below as graphs.

5.8 Biofilm Inbition Activity of Skyllamycins A–C (101–103) and Analogues (115–118)

Screening was performed according to the original protocol [15] with the exception that compounds were pinned into the test plate three times instead of one. After overnight LB liquid culture growth, cultures were diluted 1:25 into 50 ml of salt-adjusted 5% LB and allowed to acclimate to nutrient-poor media for approximately 30 min before dispensing into 384-well plates (Corning; catalog no. 3712). A 40 μL volume of the liquid culture was dispensed into each well. Due to the heterogeneous properties of biofilm-state bacteria, it is critical to maintain a light

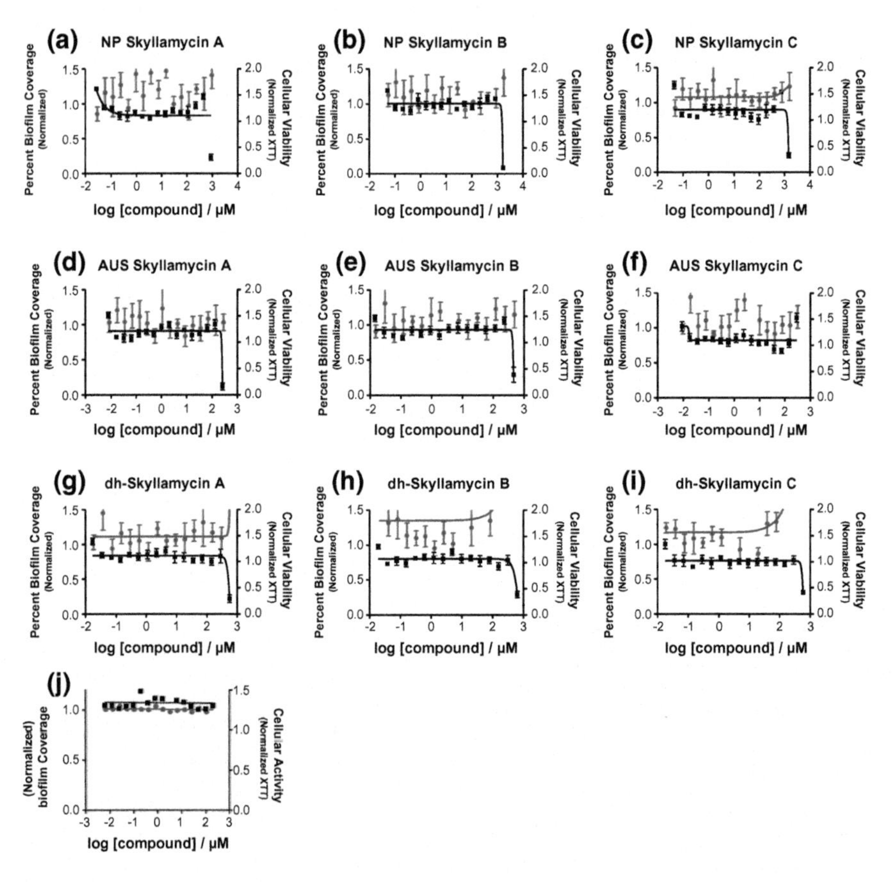

Fig. 5.2 Biofilm inhibition activity (black) and cellular viability (blue) of skyllamycin analogues tested against *Pseudomonas aeruginosa* biofilms. **a** Isolated skyllamycin A (**101**) **b** isolated skyllamycin B (**102**) **c** isolated skyllamycin C (**103**) **d** synthetic skyllamycin A (**101**) **e** synthetic skyllamycin B (**102**) **f** synthetic skyllamycin C (**103**) **g** deshydroxy skyllamycin A (**116**) **h** deshydroxy skyllamycin B (**117**) **i** deshydroxy skyllamycin C (**118**) **j** simplified skyllamycin analogue **115**

swirling motion to keep bacteria evenly distributed in liquid media while dispensing. Compounds were triple-pinned immediately after dispensing of the culture. Assay plates were incubated for 6 h at 30 °C, then 30 μL of 0.5 mg/ml of XTT–phosphate-buffered saline (PBS) buffer with 200 μM menadione (MEN) (CAS; catalog no. 58-27-5) was added. Measurements of the background absorption at 490 nm were taken immediately after dispensing XTT/MEN and again after 2 h of incubation at 30 °C. Immediately following final XTT/MEN reading, each plate was washed with 1X PBS using an automated platewasher (BioTek). For biofilm measurements, each well was quantified by the acquisition of 4 images at 20X magnification. The combined 20X images account for ~20% of the total well surface area, which is critical for reproducible and accurate quantification of biofilm coverage. The images were analyzed using MetaXpress image software (Molecular Devices) to quantify biofilm coverage. Compounds were screened twice (two biological replicates) with each screen consisting of two technical replicates (total of four sets of data) with Z′ scores between 0.54 and 0.65 indicating a successful and reliable screen (Fig. 5.2).

References

1. K. Feichtinger, H.L. Sings, T.J. Baker, K. Matthews, M. Goodman, J. Org. Chem. **63**, 8432–8439 (1998)
2. J. Rudolph, F. Hannig, H. Theis, R. Wischnat, Org. Lett. **3**, 3153–3155 (2001)
3. Y.C. Huang, Y.M. Li, Y. Chen, M. Pan, Y.T. Li, L. Yu, Q.X. Guo, L. Liu, Angew. Chem. Int. Ed. **52**, 4858–4862 (2013)
4. Y.E. Jad, G.A. Acosta, T. Naicker, M. Ramtahal, A. El-Faham, T. Govender, H.G. Kruger, B. G. de la Torre, F. Albericio, Org. Lett. **17**, 6182–6185 (2015)
5. L.L. Ling, T. Schneider, A.J. Peoples, A.L. Spoering, I. Engels, B.P. Conlon, A. Mueller, T.F. Schaberle, D.E. Hughes, S. Epstein, M. Jones, L. Lazarides, V.A. Steadman, D.R. Cohen, C.R. Felix, K.A. Fetterman, W.P. Millett, A.G. Nitti, A.M. Zullo, C. Chen, K. Lewis, Nature **517**, 455–459 (2015)
6. N.K. Taneja, J.S. Tyagi, J. Antimicrob. Chemother. **60**, 288–293 (2007)
7. I. Wiegand, K. Hilpert, R.E.W. Hancock, Nat. Protoc. **3**, 163–175 (2008)
8. J. Tummatorn, G.B. Dudley, Org. Lett. **13**, 1572–1575 (2011)
9. D. Sun, P. Lai, W. Xie, J. Deng, Y. Jiang, Synth. Commun. **37**, 2989–2994 (2007)
10. P. Garner, J.M. Park, J. Org. Chem. **52**, 2361–2364 (1987)
11. L. Williams, Z.D. Zhang, F. Shao, P.J. Carroll, M.M. Joullié, Tetrahedron **52**, 11673–11694 (1996)
12. L.R. Malins, A.M. Giltrap, L.J. Dowman, R.J. Payne, Org. Lett. **17**, 2070–2073 (2015)
13. J.V. Geden, A.J. Clark, S.R. Coles, C.S. Guy, F. Ghelfi, S. Thom, Tetrahedron Lett. **57**, 3109–3112 (2016)
14. S. Schabbert, M.D. Pierschbacher, R.H. Mattern, M. Goodman, Bioorg. Med. Chem. **10**, 3331–3337 (2002)
15. G. Navarro, A.T. Cheng, K.C. Peach, W.M. Bray, V.S. Bernan, F.H. Yildiz, R.G. Linington, Antimicrob. Agents Chemother. **58**, 1092–1099 (2014)

Appendix A
Exemplar NMR Spectra

A. Giltrap, *Total Synthesis of Natural Products with Antimicrobial Activity*,
Springer Theses, https://doi.org/10.1007/978-981-10-8806-3

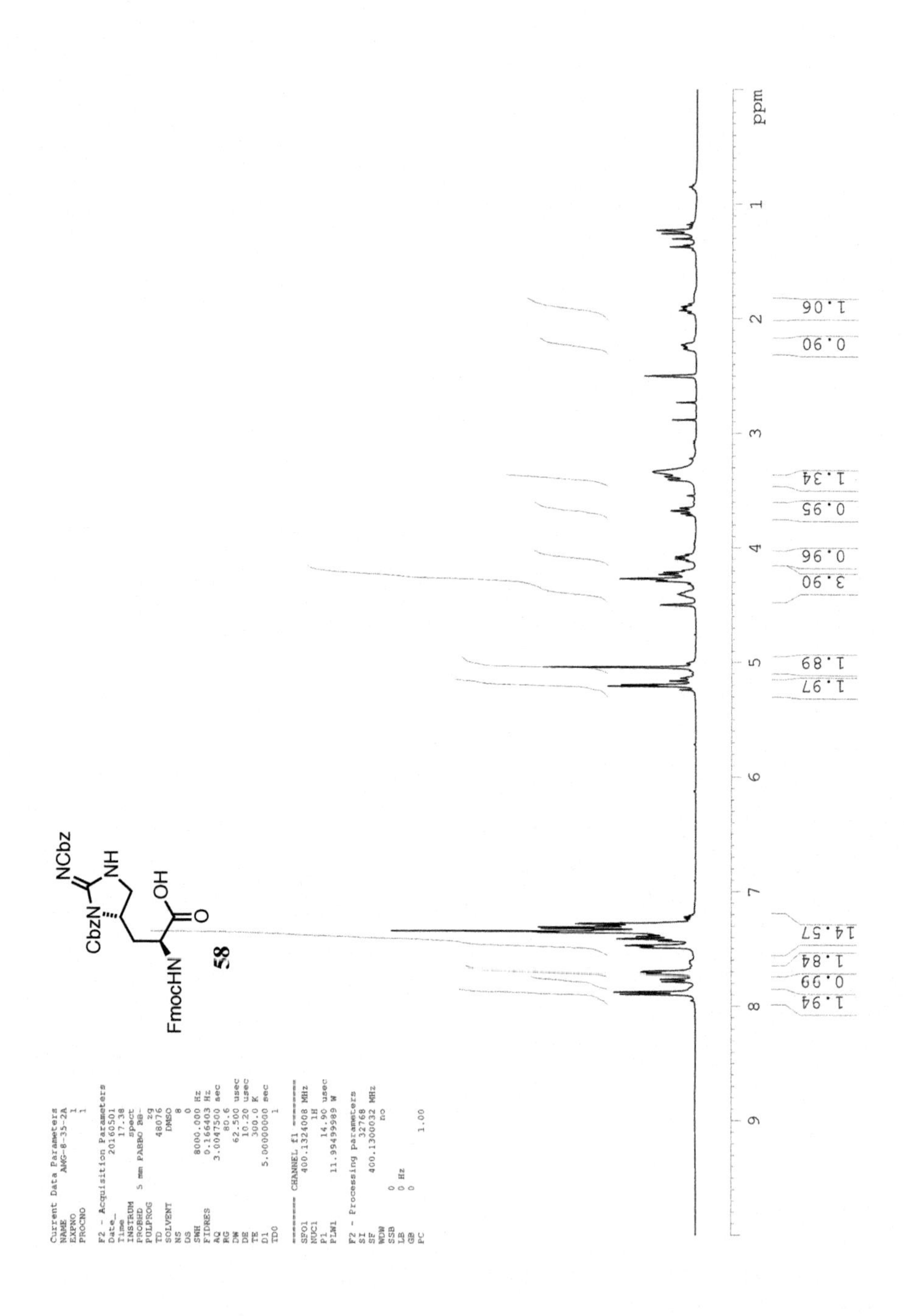

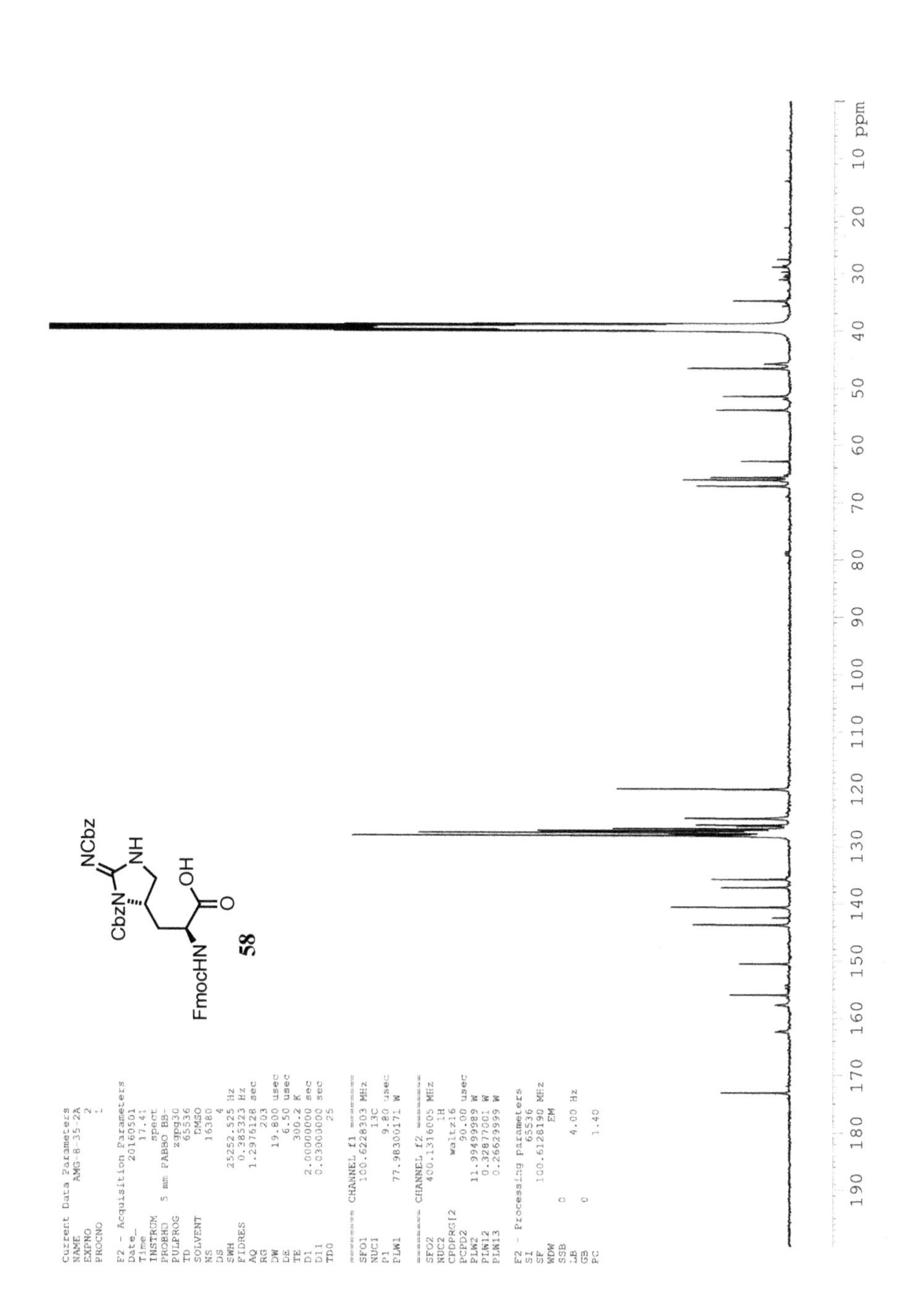

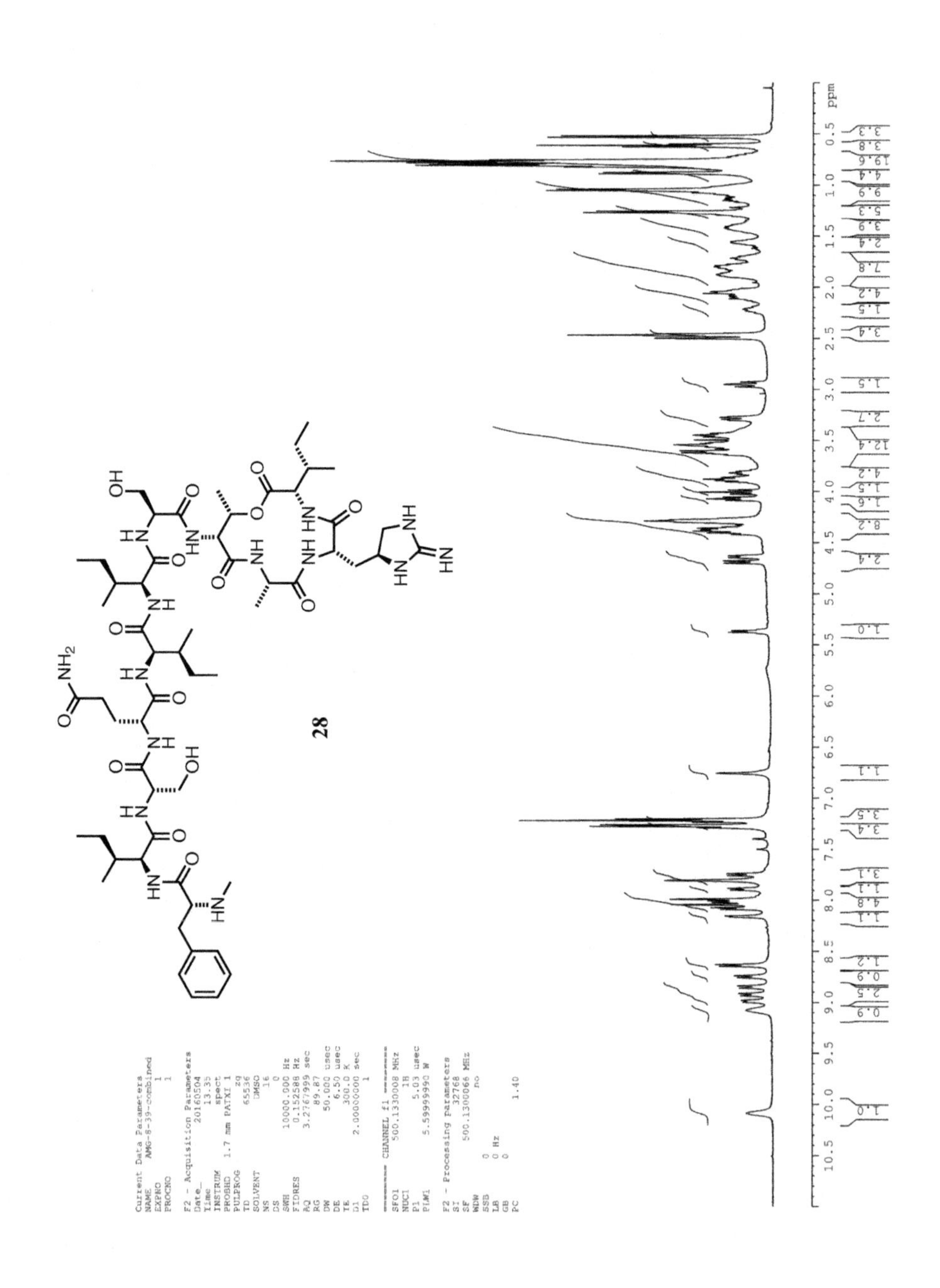
Current Data Parameters
NAME AMG-8-39-combined
EXPNO 1
PROCNO 1
F2 - Acquisition Parameters
Date_ 20160504
Time 13.35
INSTRUM spect
PROBHD 1.7 mm PATXI 1
PULPROG zg
TD 65536
SOLVENT DMSO
NS 16
DS 0
SWH 10000.000 Hz
FIDRES 0.152588 Hz
AQ 3.2767999 sec
RG 89.87
DW 50.000 usec
DE 6.50 usec
TE 300.0 K
D1 2.00000000 sec
TD0 1
CHANNEL f1
SFO1 500.1330008 MHz
NUC1 1H
P1 5.03 usec
PLW1 5.59999990 W
F2 - Processing parameters
SI 32768
SF 500.1300066 MHz
WDW no
SSB 0
LB 0 Hz
GB 0
PC 1.40
28
ppm

28 - COSY

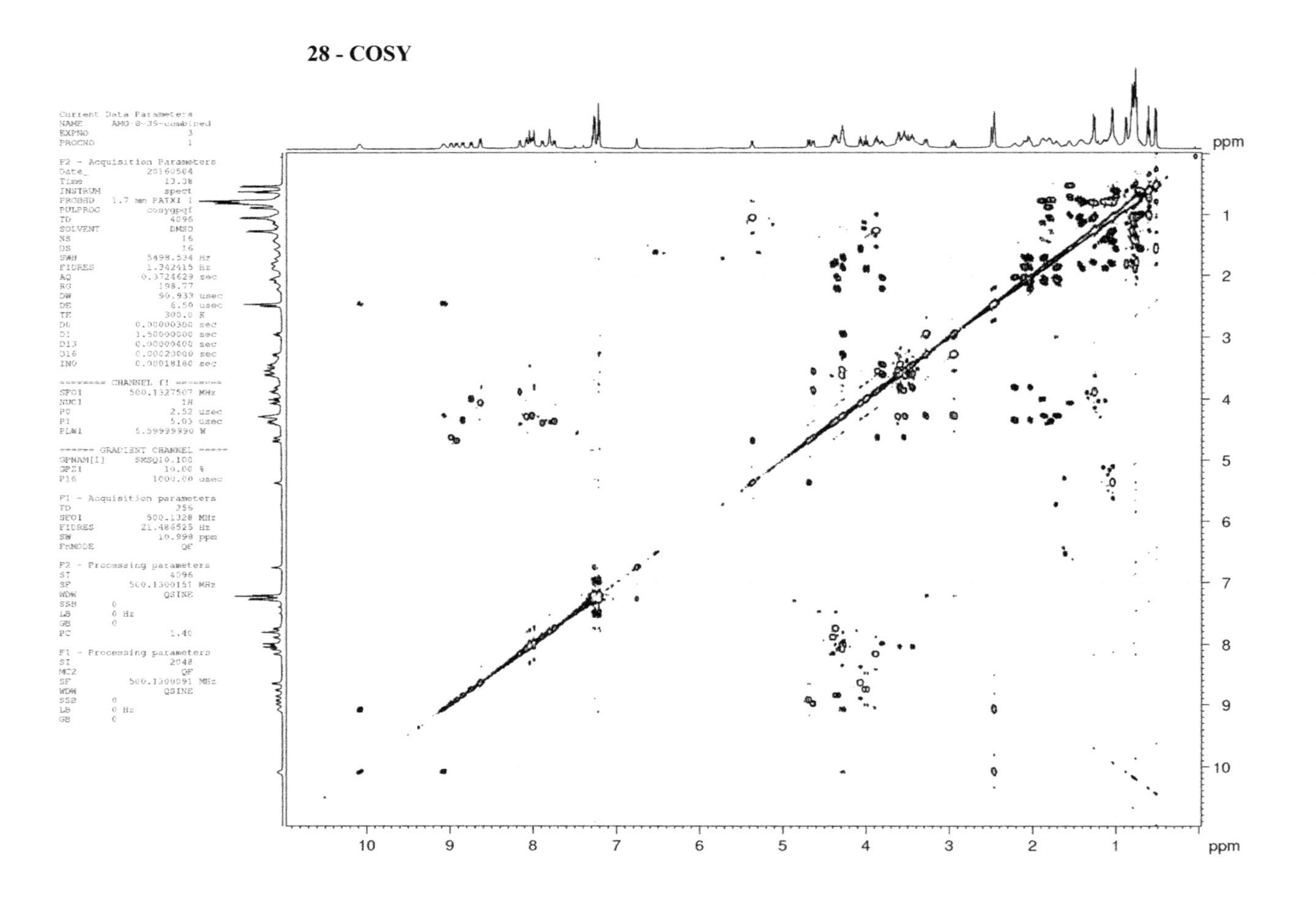

Current Data Parameters
NAME AMG-8-35-combined
EXPNO 3
PROCNO 1
F2 - Acquisition Parameters
Date_ 20160504
Time 13.38
INSTRUM spect
PROBHD 1.7 mm PATXI 1
PULPROG cosygpqf
TD 4096
SOLVENT DMSO
NS 16
DS 16
SWH 5498.534 Hz
FIDRES 1.342415 Hz
AQ 0.3724629 sec
RG 198.77
DW 90.933 usec
DE 6.50 usec
TE 300.0 K
D0 0.00000300 sec
D1 1.50000000 sec
D13 0.00000400 sec
D16 0.00020000 sec
IN0 0.00018180 sec
======== CHANNEL f1 ========
SFO1 500.1327507 MHz
NUC1 1H
P0 2.52 usec
P1 5.03 usec
PLW1 5.59999990 W
====== GRADIENT CHANNEL =====
GPNAM[1] SMSQ10.100
GPZ1 10.00 %
P16 1000.00 usec
F1 - Acquisition parameters
TD 256
SFO1 500.1328 MHz
FIDRES 21.486525 Hz
SW 10.998 ppm
FnMODE QF
F2 - Processing parameters
SI 4096
SF 500.1300151 MHz
WDW QSINE
SSB 0
LB 0 Hz
GB 0
PC 1.40
F1 - Processing parameters
SI 2048
MC2 QF
SF 500.1300091 MHz
WDW QSINE
SSB 0
LB 0 Hz
GB 0
ppm
ppm

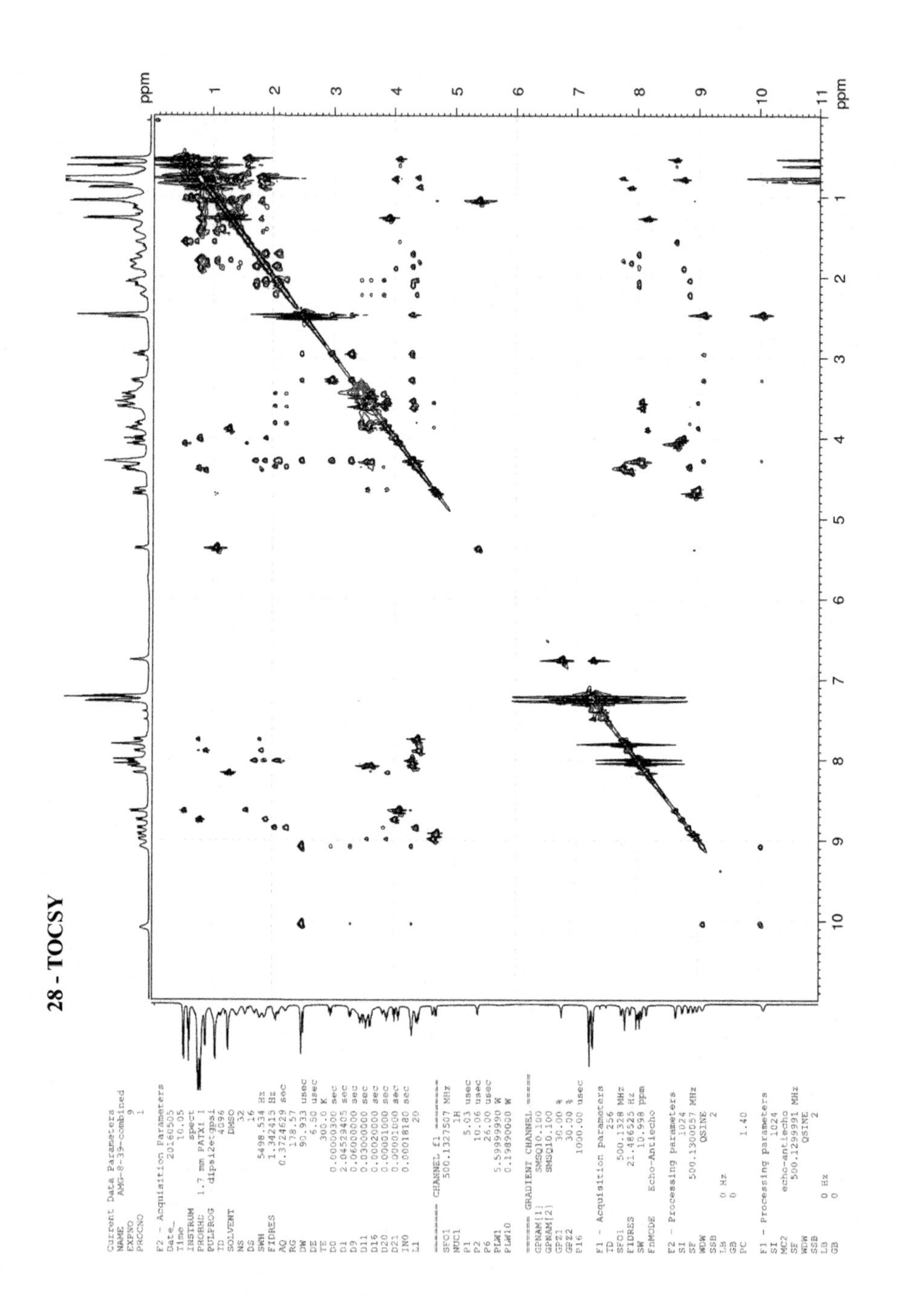
28 - TOCSY
ppm
ppm

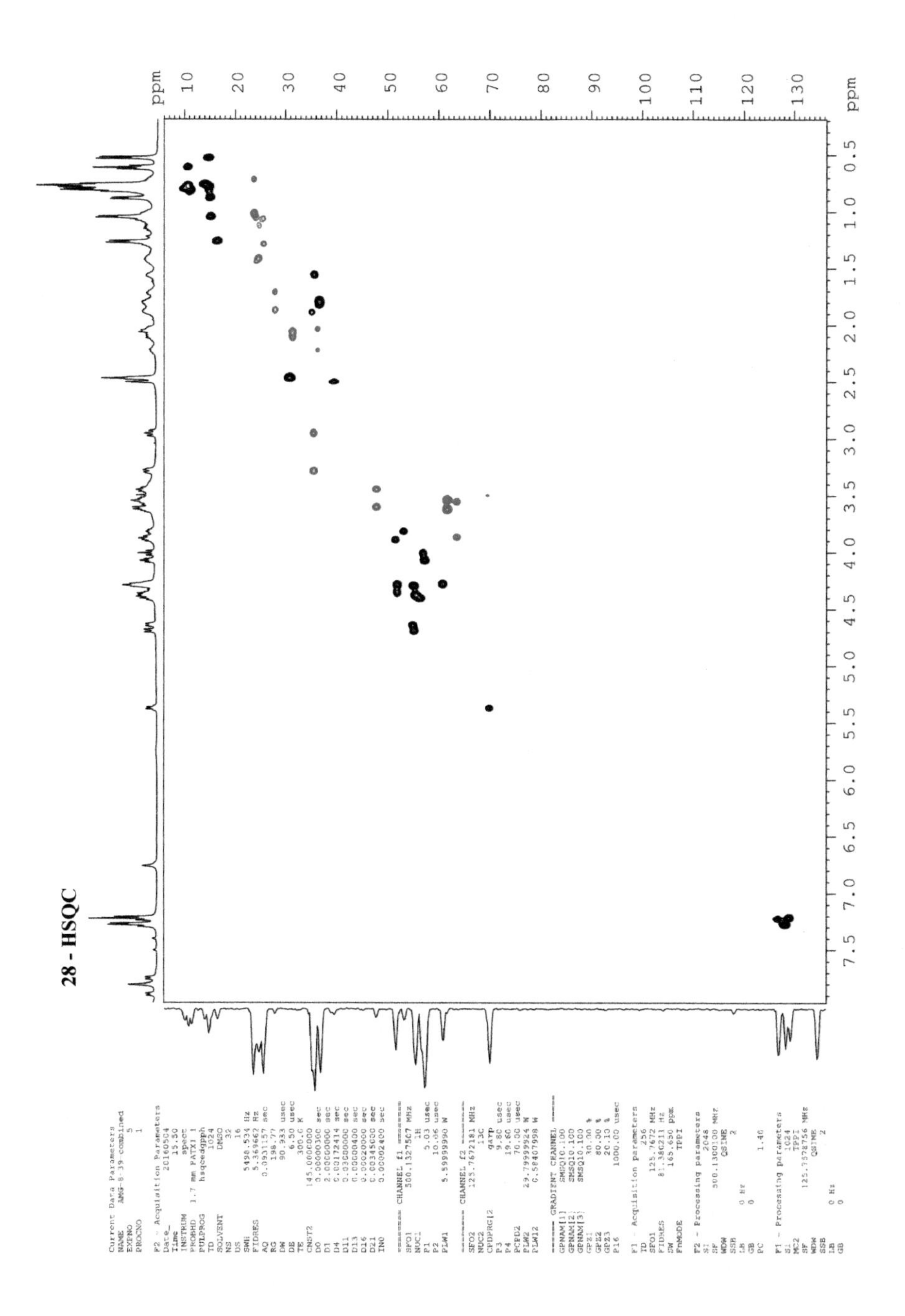
28 - HSQC
ppm
ppm

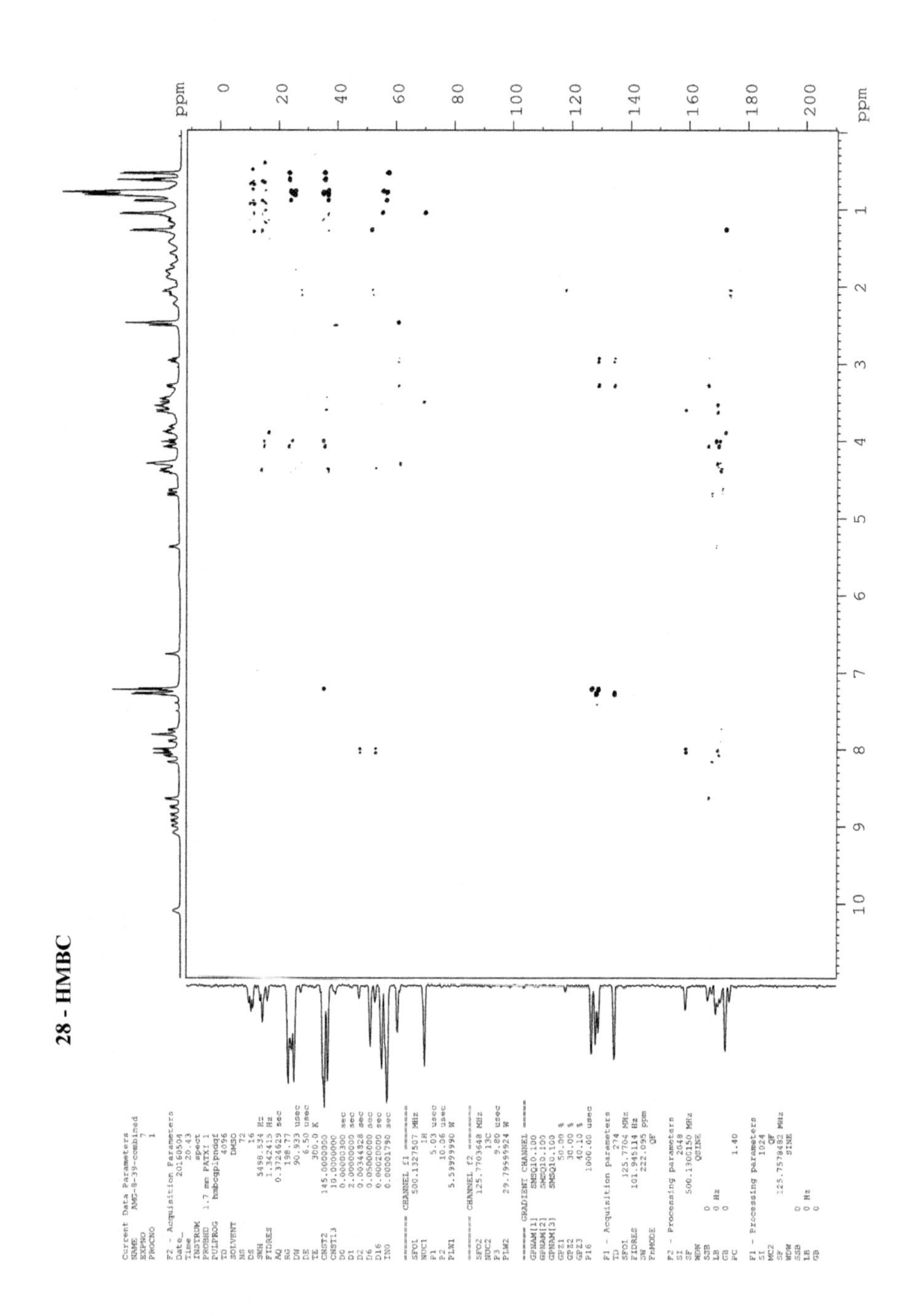
28 - HMBC
ppm

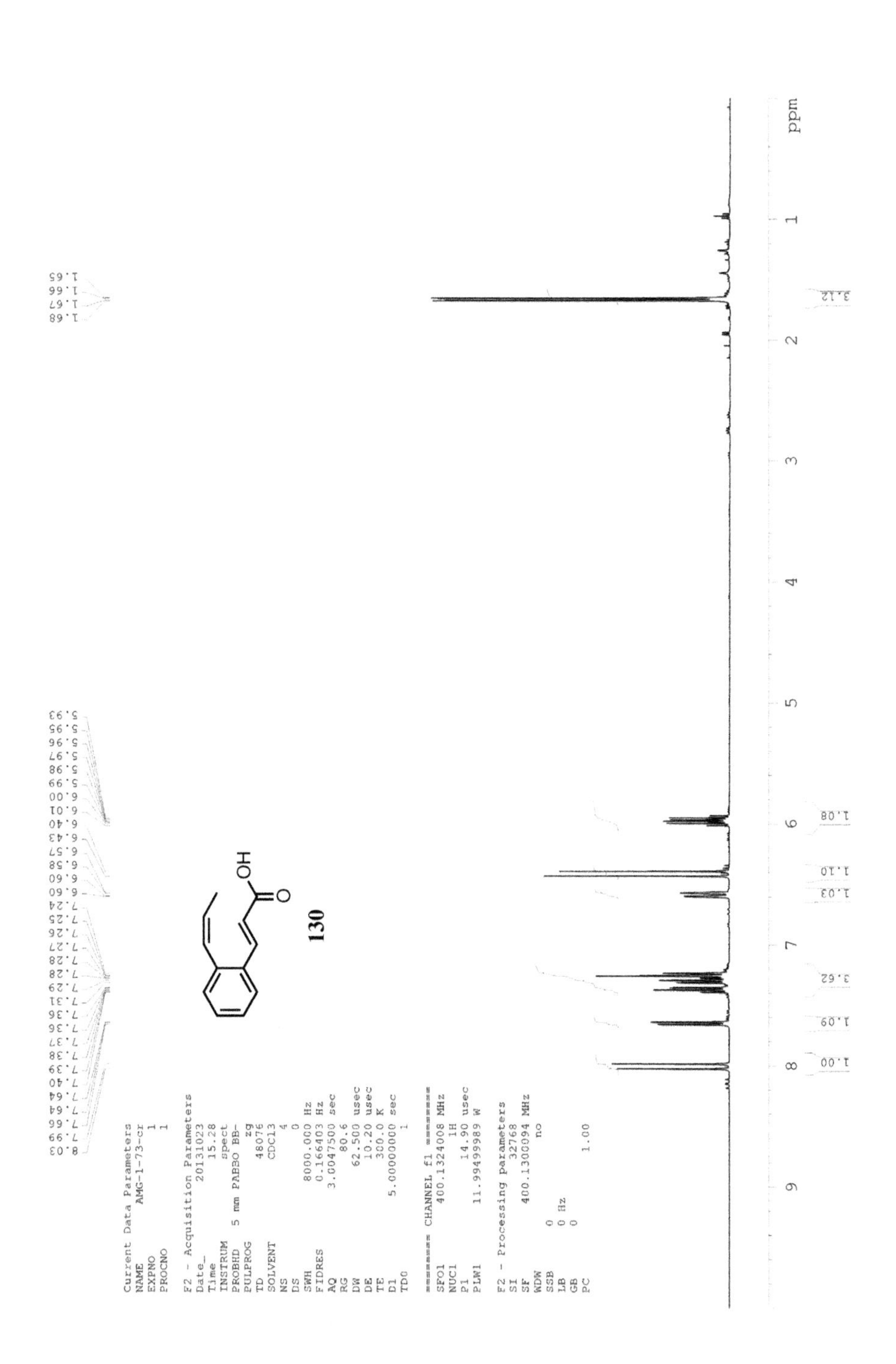

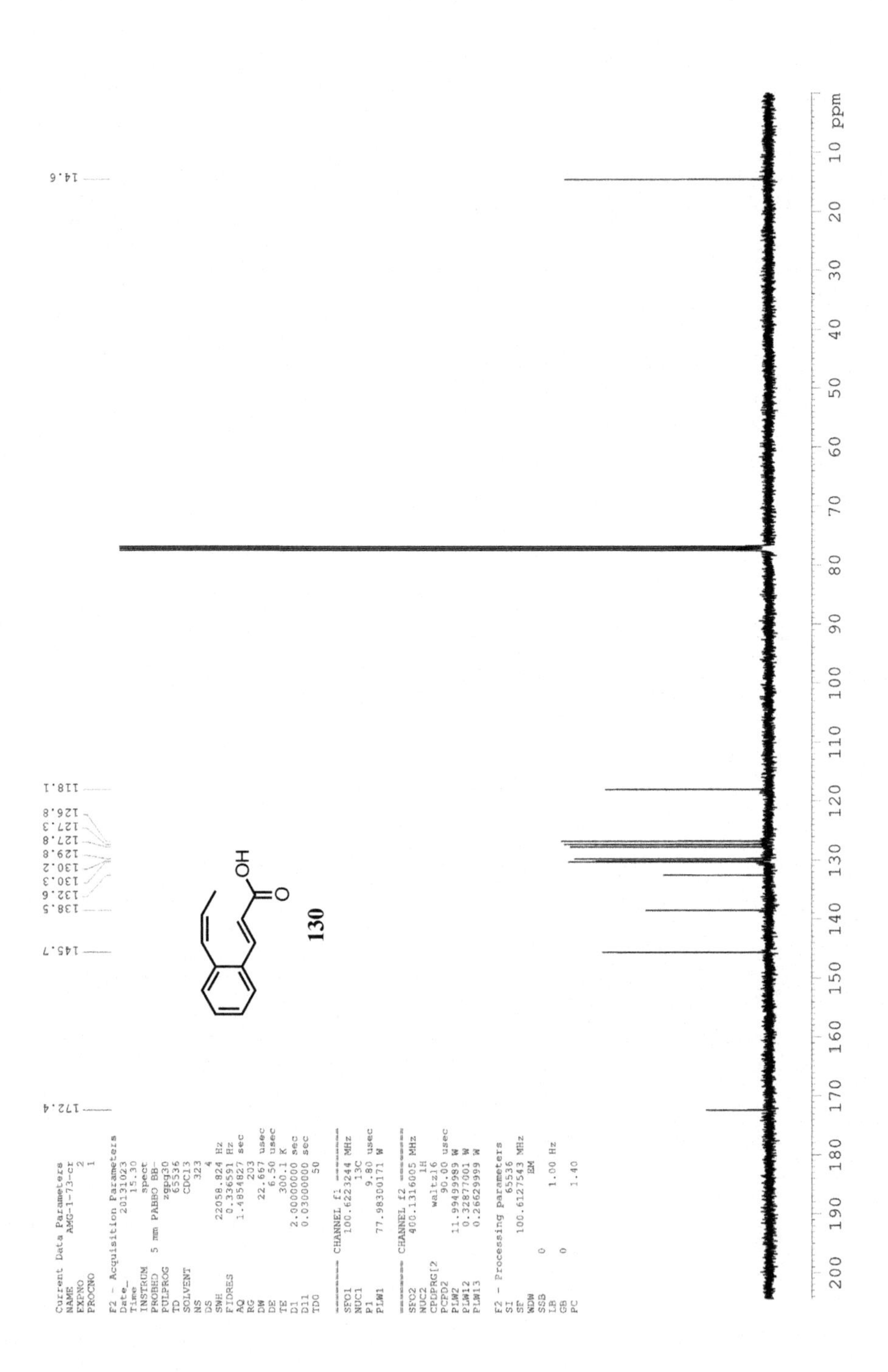

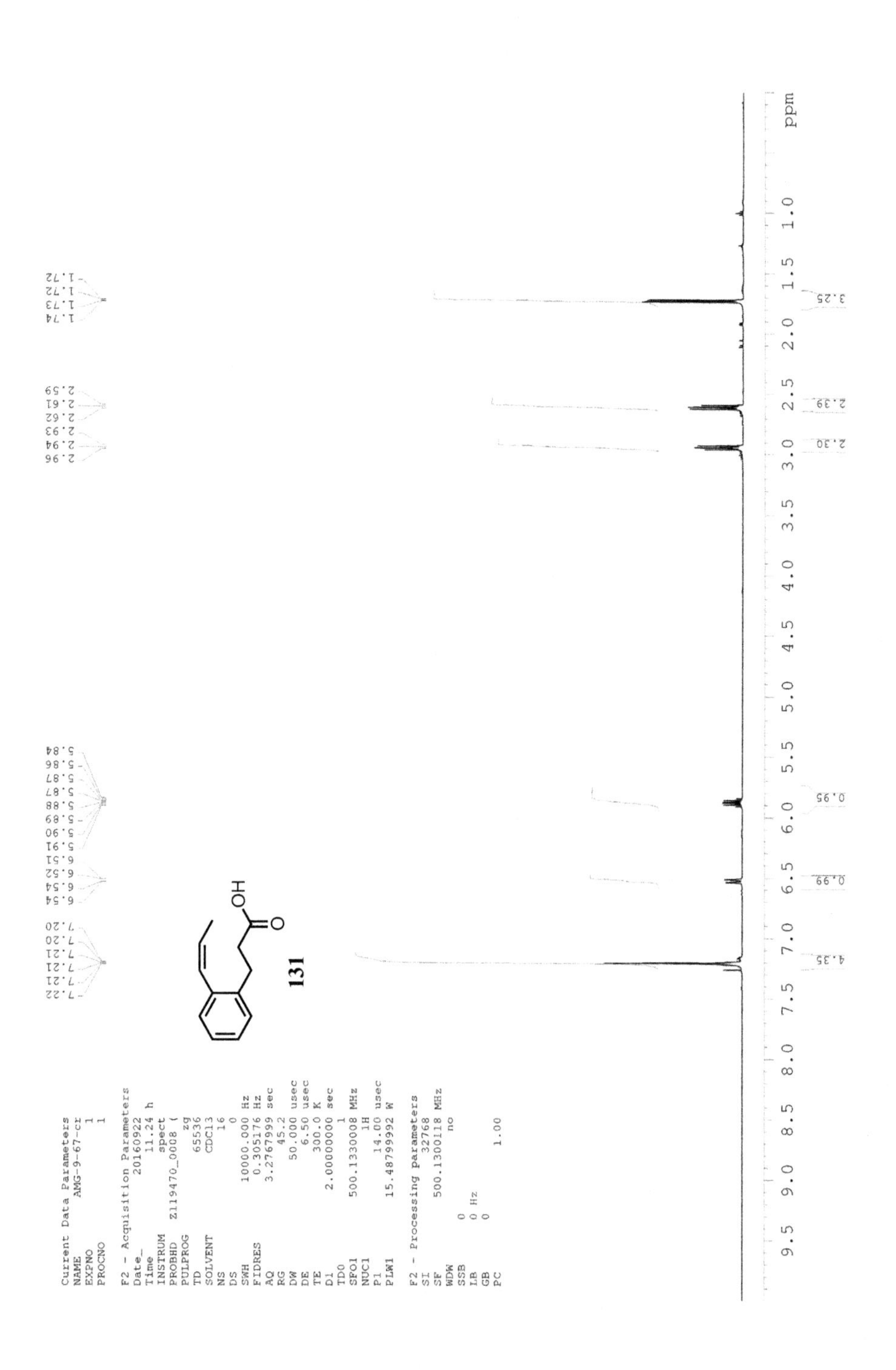

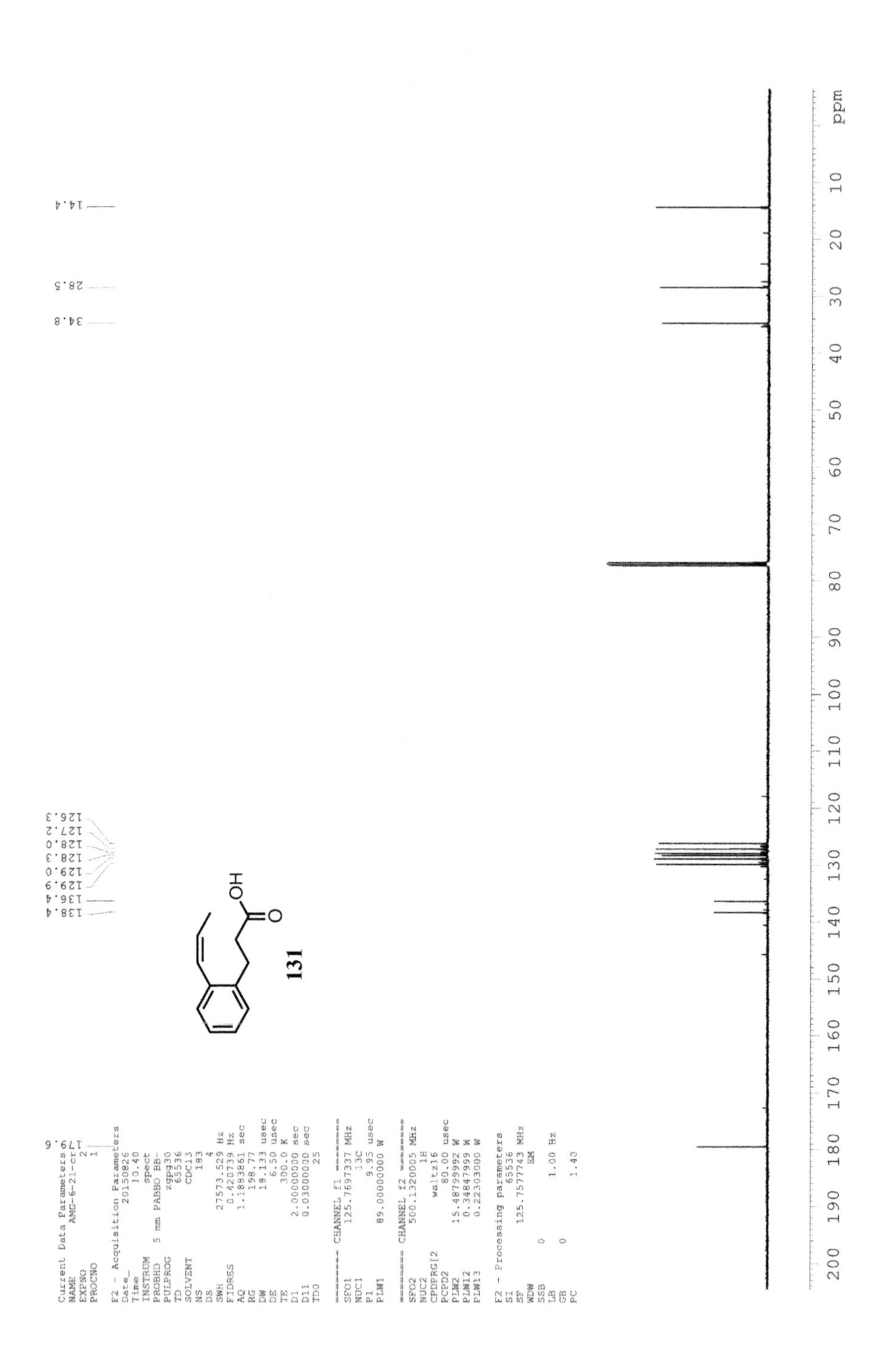
179.6
138.4
136.4
129.9
129.0
128.3
128.0
127.2
126.3
34.8
28.5
14.4
Current Data Parameters
NAME AMG-6-21-cr
EXPNO 2
PROCNO 1
F2 - Acquisition Parameters
Date_ 20150826
Time 10.40
INSTRUM spect
PROBHD 5 mm PABBO BB-
PULPROG zgpg30
TD 65536
SOLVENT CDCl3
NS 183
DS 4
SWH 27573.529 Hz
FIDRES 0.420739 Hz
AQ 1.1883861 sec
RG 198.77
DW 18.133 usec
DE 6.50 usec
TE 300.0 K
D1 2.00000000 sec
D11 0.03000000 sec
TD0 25
======== CHANNEL f1 ========
SFO1 125.7697337 MHz
NUC1 13C
P1 9.95 usec
PLW1 89.00000000 W
======== CHANNEL f2 ========
SFO2 500.1320005 MHz
NUC2 1H
CPDPRG[2 waltz16
PCPD2 80.00 usec
PLW2 15.48799992 W
PLW12 0.34847999 W
PLW13 0.22303000 W
F2 - Processing parameters
SI 65536
SF 125.7577743 MHz
WDW EM
SSB 0
LB 1.00 Hz
GB 0
PC 1.40
OH
O
131
200 190 180 170 160 150 140 130 120 110 100 90 80 70 60 50 40 30 20 10 ppm

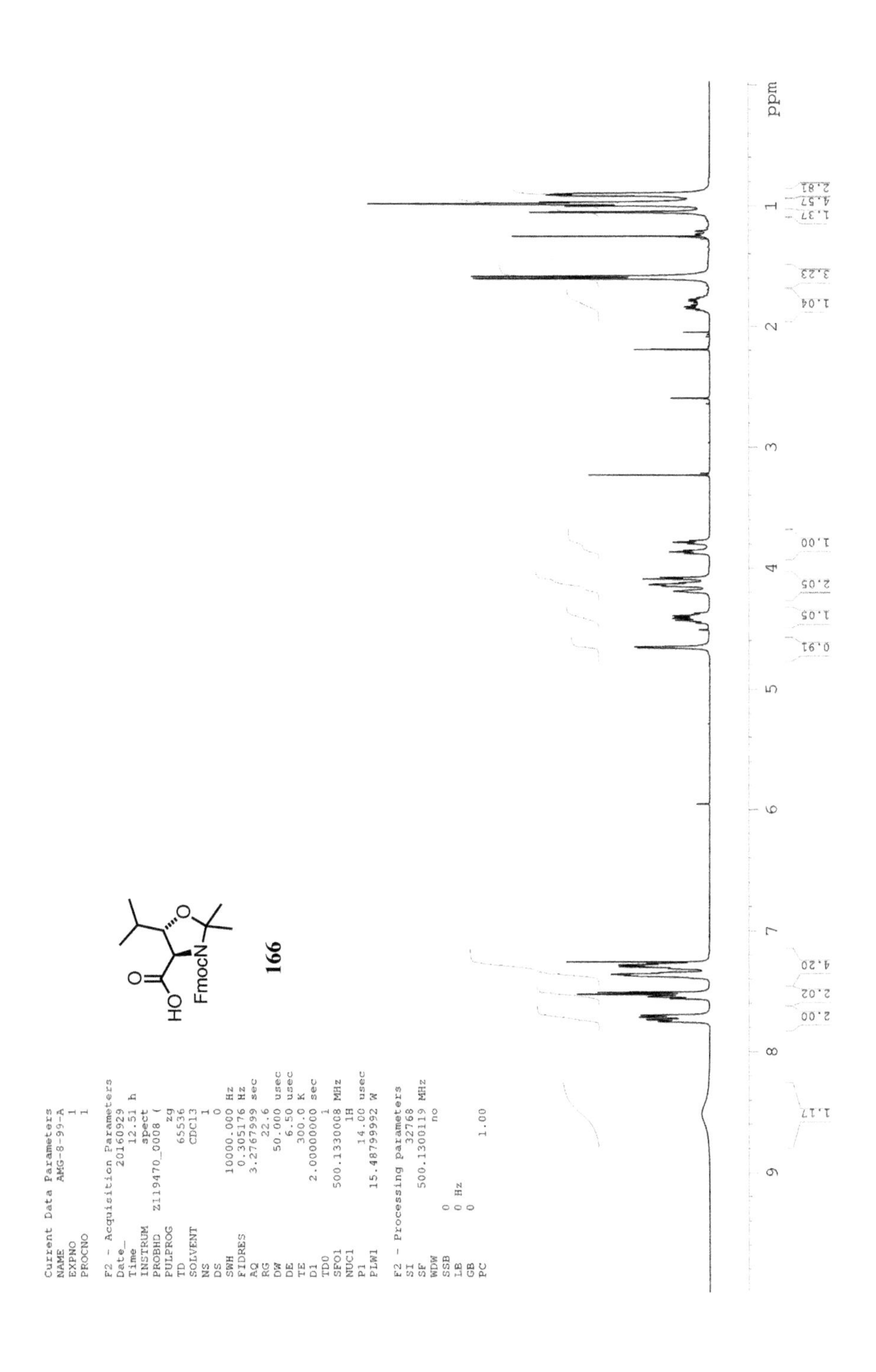
Current Data Parameters
NAME AMG-8-99-A
EXPNO 1
PROCNO 1
F2 - Acquisition Parameters
Date_ 20160929
Time 12.51 h
INSTRUM spect
PROBHD Z119470_0008 (
PULPROG zg
TD 65536
SOLVENT CDCl3
NS 1
DS 0
SWH 10000.000 Hz
FIDRES 0.305176 Hz
AQ 3.2767999 sec
RG 22.6
DW 50.000 usec
DE 6.50 usec
TE 300.0 K
D1 2.00000000 sec
TD0 1
SFO1 500.1330008 MHz
NUC1 1H
P1 14.00 usec
PLW1 15.48799992 W
F2 - Processing parameters
SI 32768
SF 500.1300119 MHz
WDW no
SSB 0
LB 0 Hz
GB 0
PC 1.00
HO
O
FmocN
166
ppm

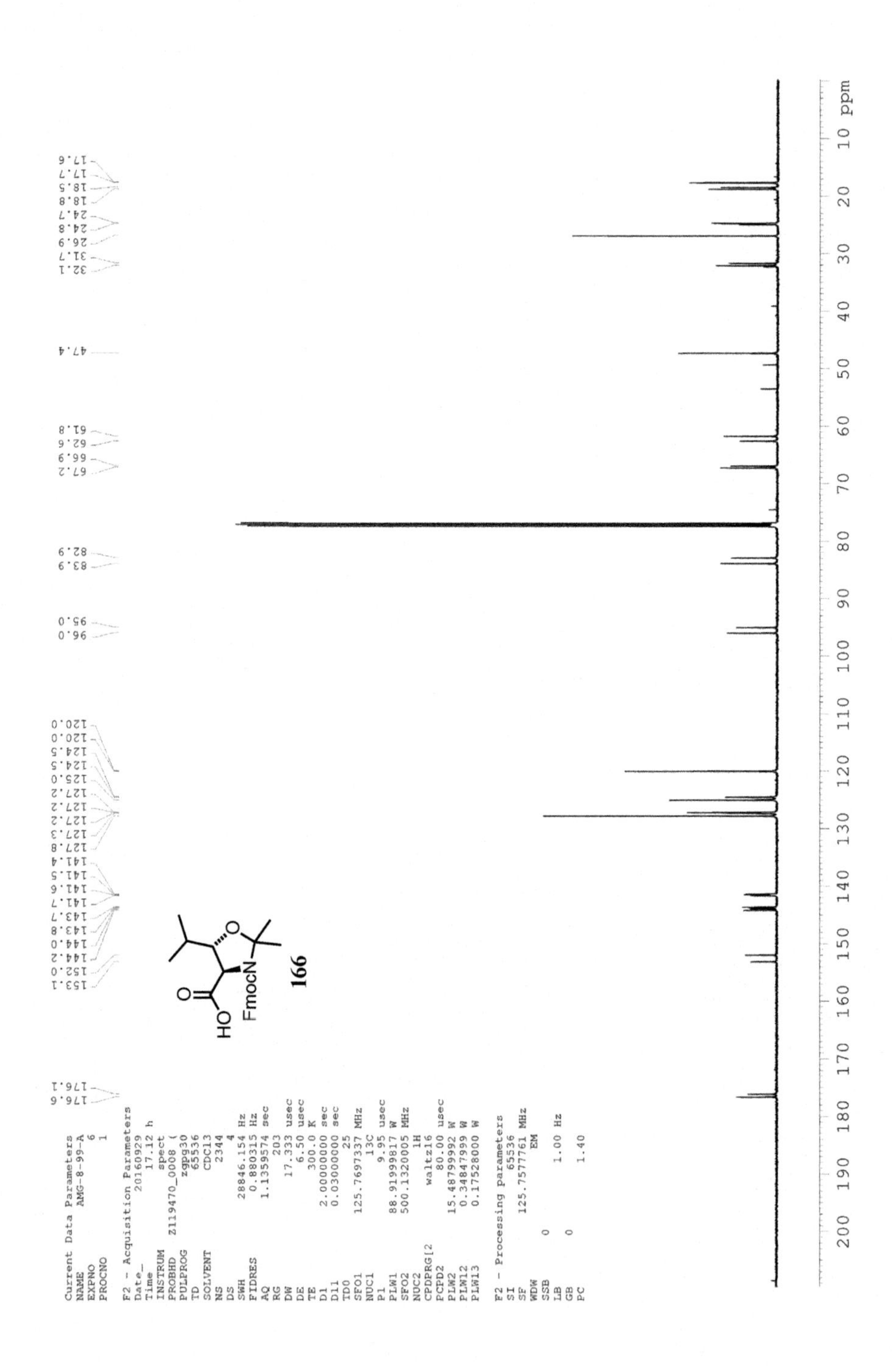
HO
O
FmocN
O
166
ppm

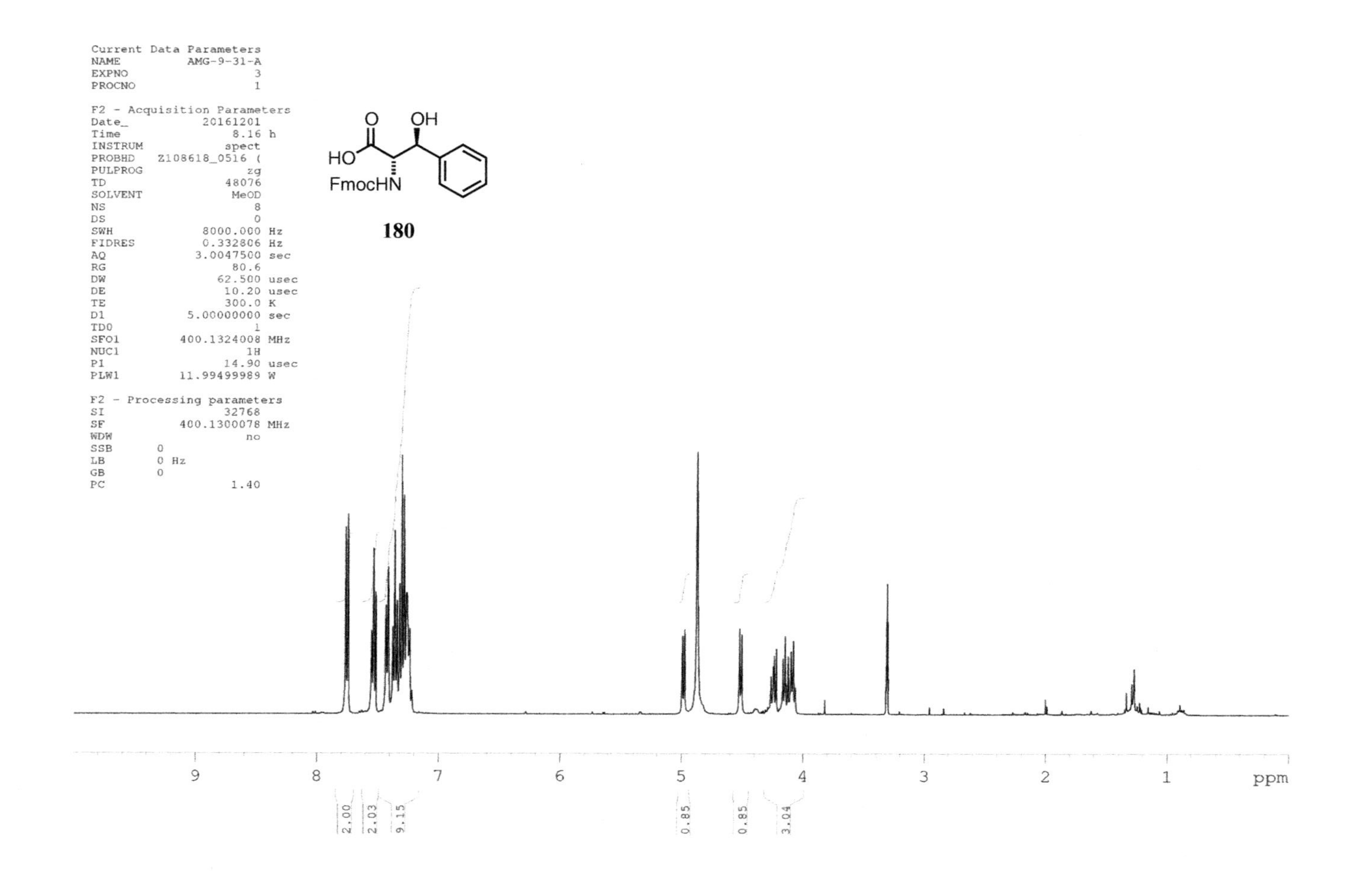
Current Data Parameters
NAME AMG-9-31-A
EXPNO 3
PROCNO 1
F2 - Acquisition Parameters
Date_ 20161201
Time 8.16 h
INSTRUM spect
PROBHD Z108618_0516 (
PULPROG zg
TD 48076
SOLVENT MeOD
NS 8
DS 0
SWH 8000.000 Hz
FIDRES 0.332806 Hz
AQ 3.0047500 sec
RG 80.6
DW 62.500 usec
DE 10.20 usec
TE 300.0 K
D1 5.00000000 sec
TD0 1
SFO1 400.1324008 MHz
NUC1 1H
P1 14.90 usec
PLW1 11.99499989 W
F2 - Processing parameters
SI 32768
SF 400.1300078 MHz
WDW no
SSB 0
LB 0 Hz
GB 0
PC 1.40
O
OH
HO
FmocHN
180
9
8
7
6
5
4
3
2
1
ppm
2.00
2.03
9.15
0.85
0.85
3.04

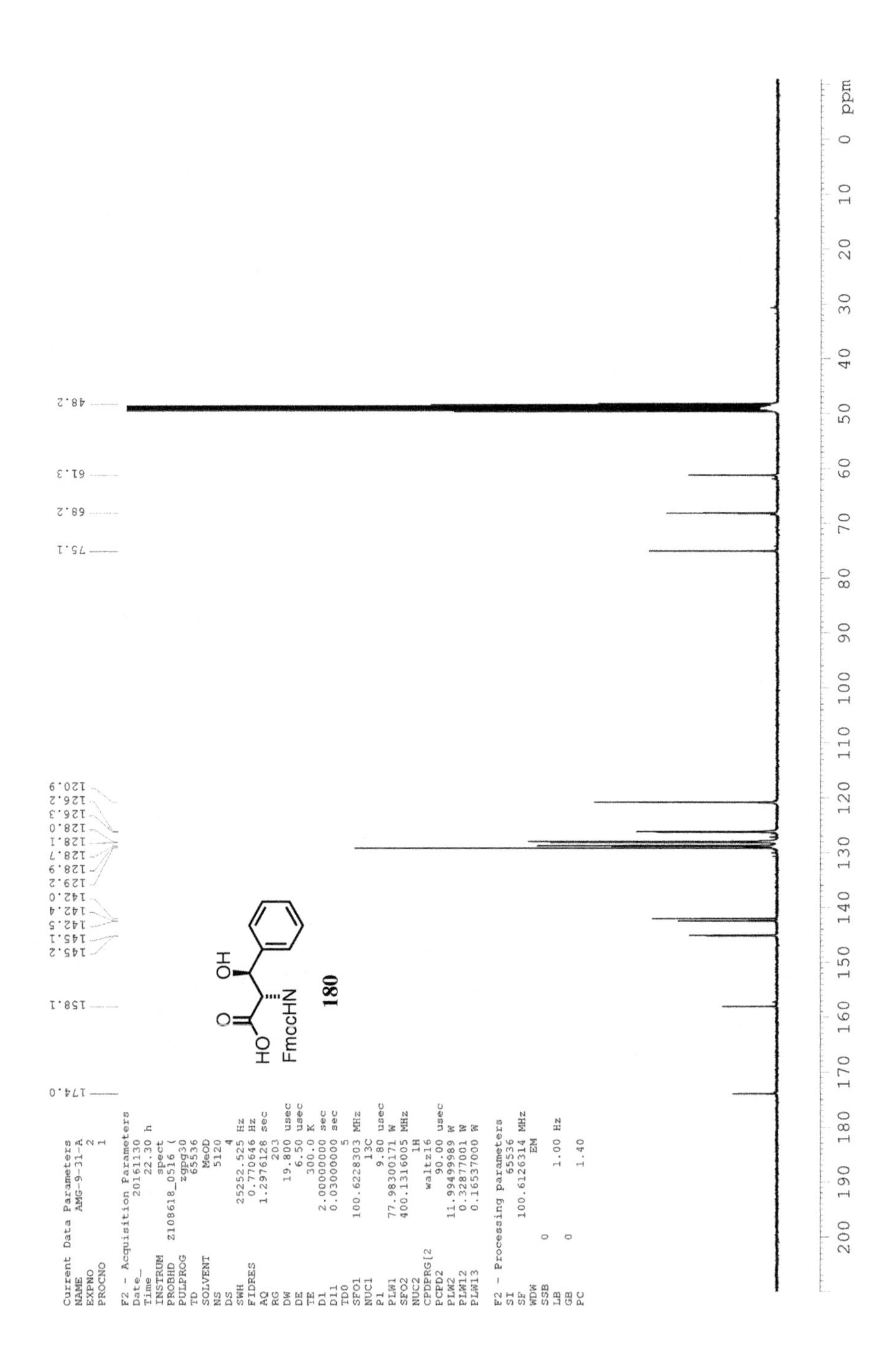

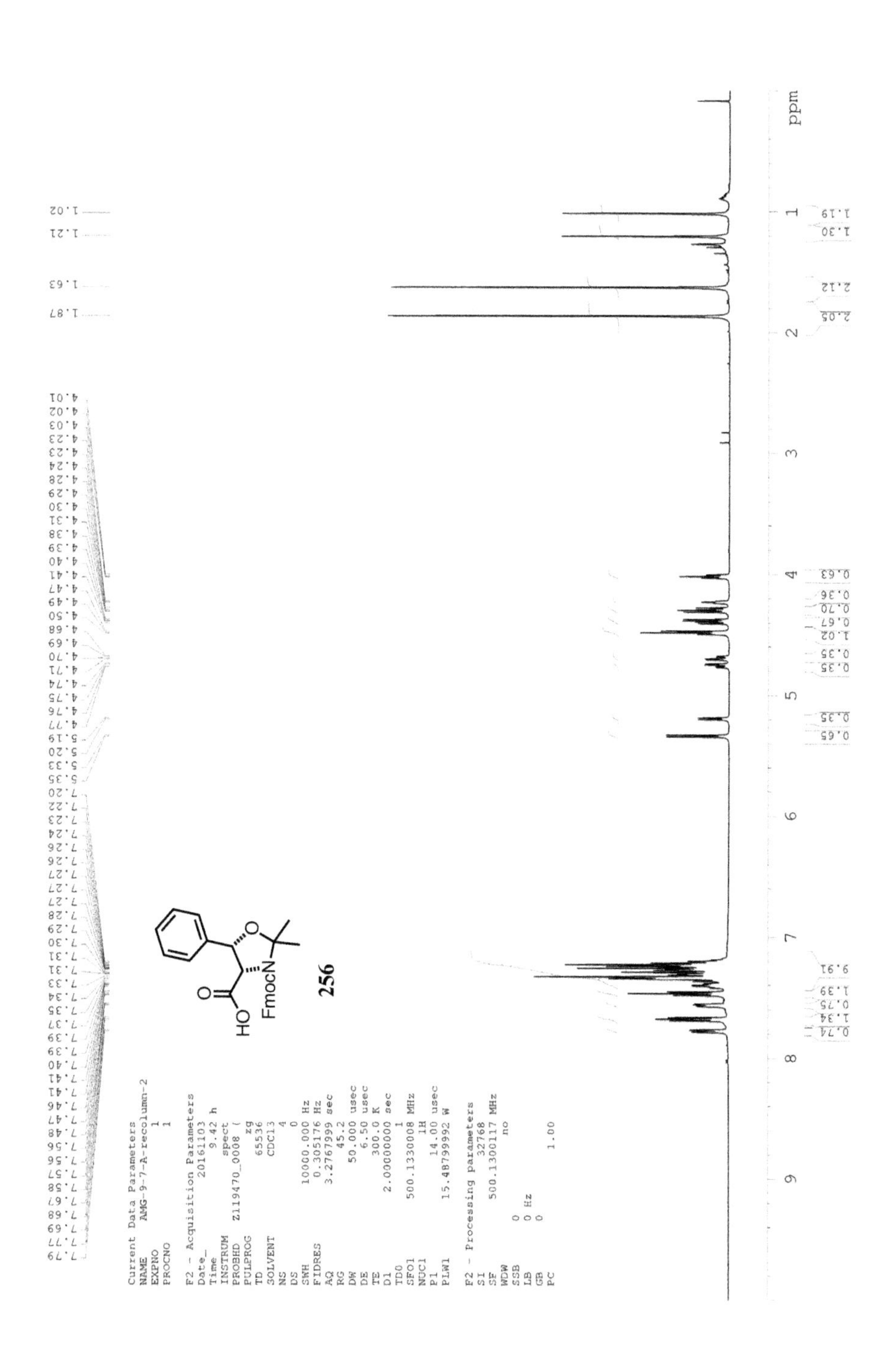

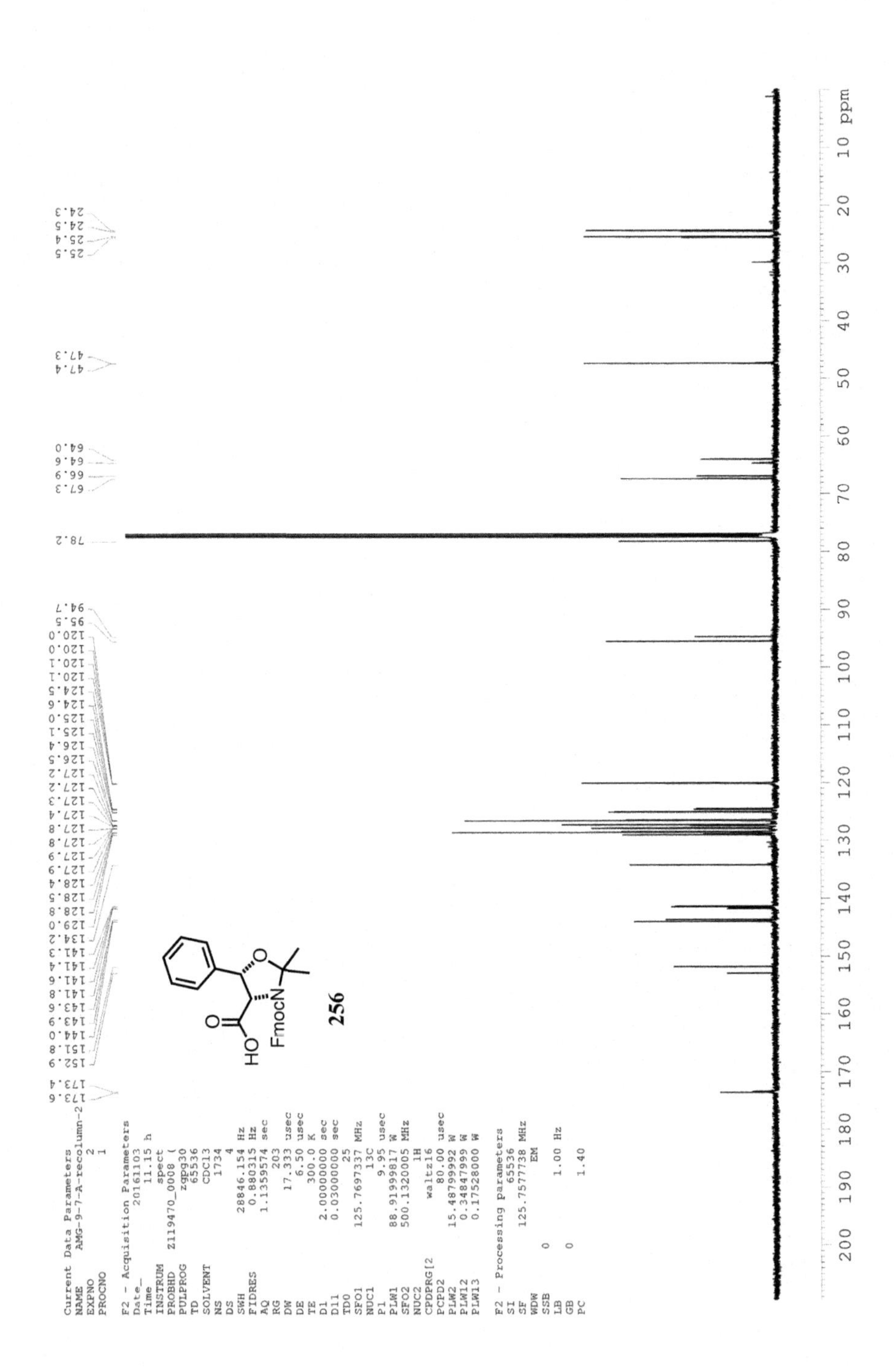

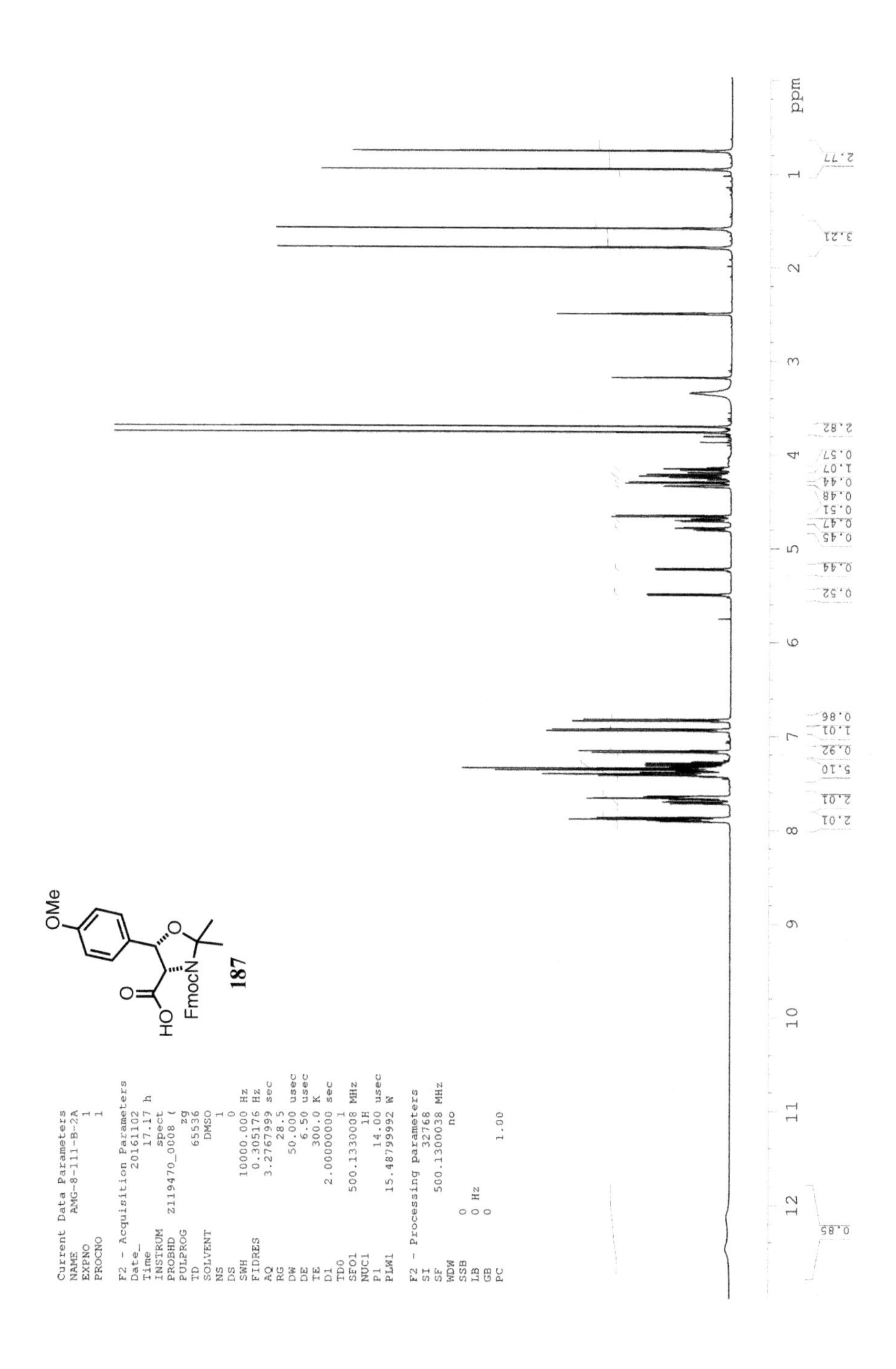

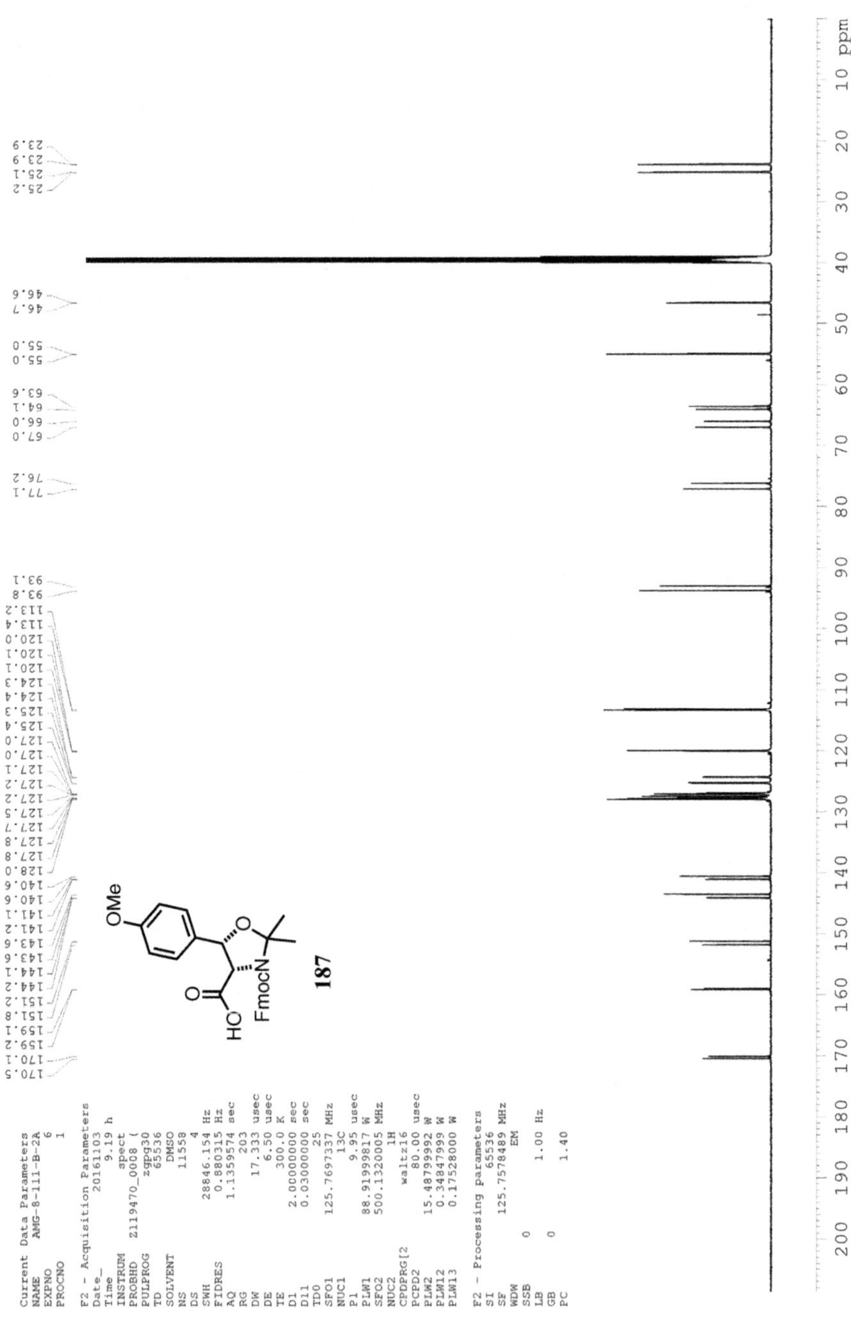
Current Data Parameters
NAME AMG-8-111-B-2A
EXPNO 6
PROCNO 1
F2 - Acquisition Parameters
Date_ 20161103
Time 9.19 h
INSTRUM spect
PROBHD Z119470_0008 (
PULPROG zgpg30
TD 65536
SOLVENT DMSO
NS 11558
DS 4
SWH 28846.154 Hz
FIDRES 0.880315 Hz
AQ 1.1359574 sec
RG 203
DW 17.333 usec
DE 6.50 usec
TE 300.0 K
D1 2.00000000 sec
D11 0.03000000 sec
TD0 25
SFO1 125.7697337 MHz
NUC1 13C
P1 9.95 usec
PLW1 88.91999817 W
SFO2 500.1320005 MHz
NUC2 1H
CPDPRG[2 waltz16
PCPD2 80.00 usec
PLW2 15.48799992 W
PLW12 0.34847999 W
PLW13 0.17528000 W
F2 - Processing parameters
SI 65536
SF 125.7578489 MHz
WDW EM
SSB 0
LB 1.00 Hz
GB 0
PC 1.40
OMe
HO
O
FmocN
O
187
ppm

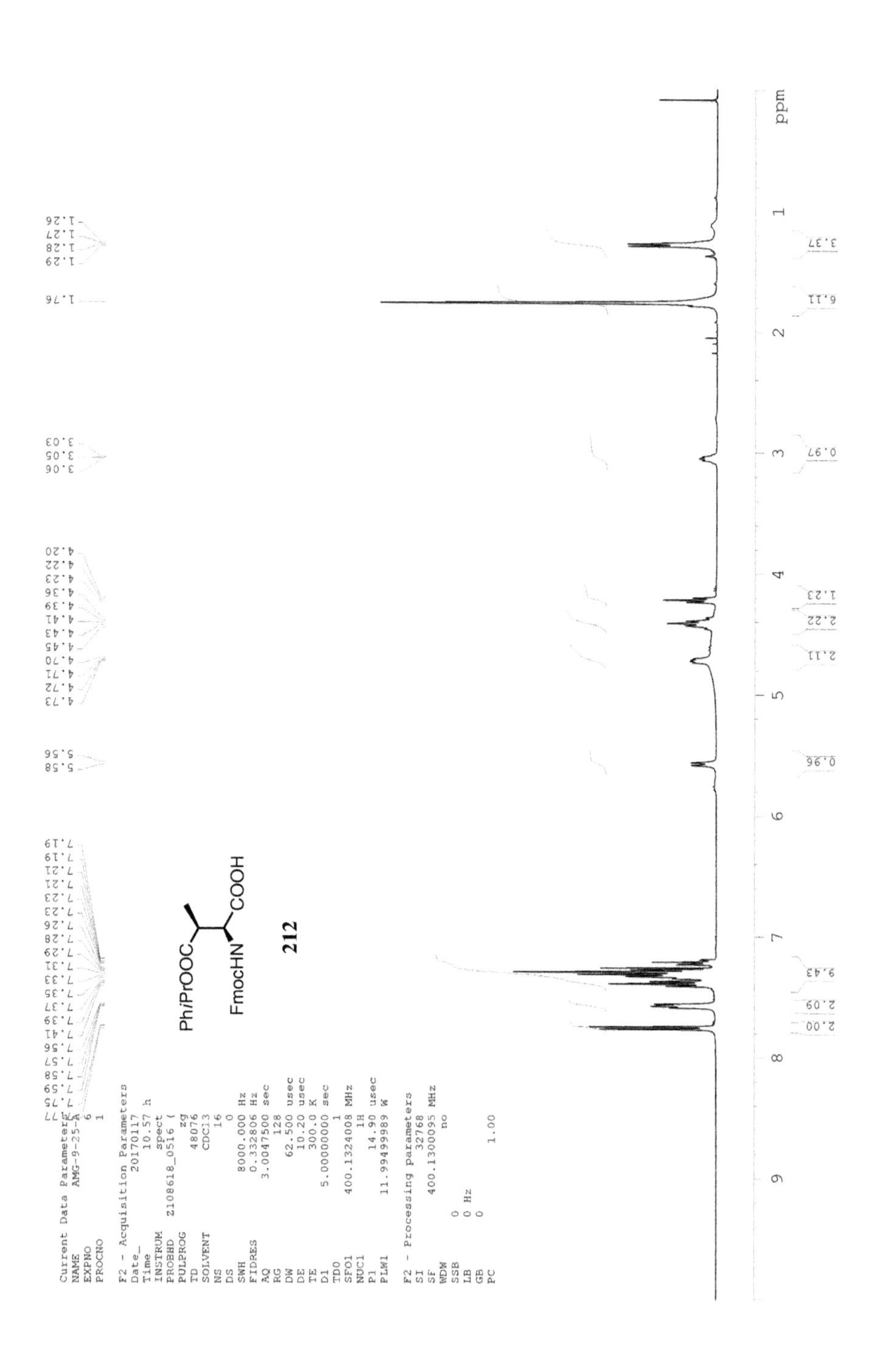

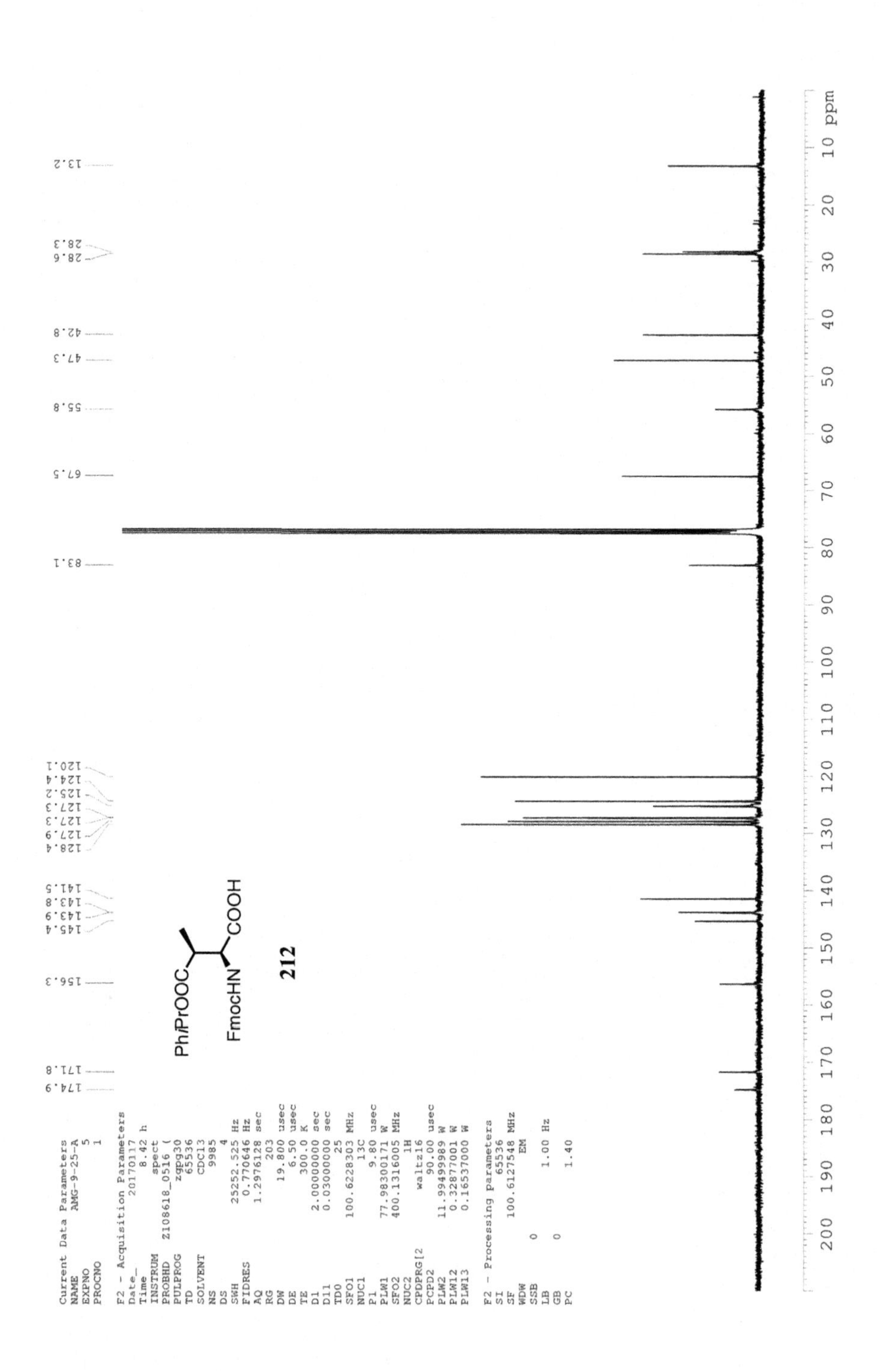
174.9
171.8
156.3
145.4
143.9
143.8
141.5
128.4
127.9
127.3
127.3
125.2
124.4
120.1
83.1
67.5
55.8
47.3
42.8
28.6
28.3
13.2
Ph*i*PrOOC
FmocHN
COOH
212
ppm

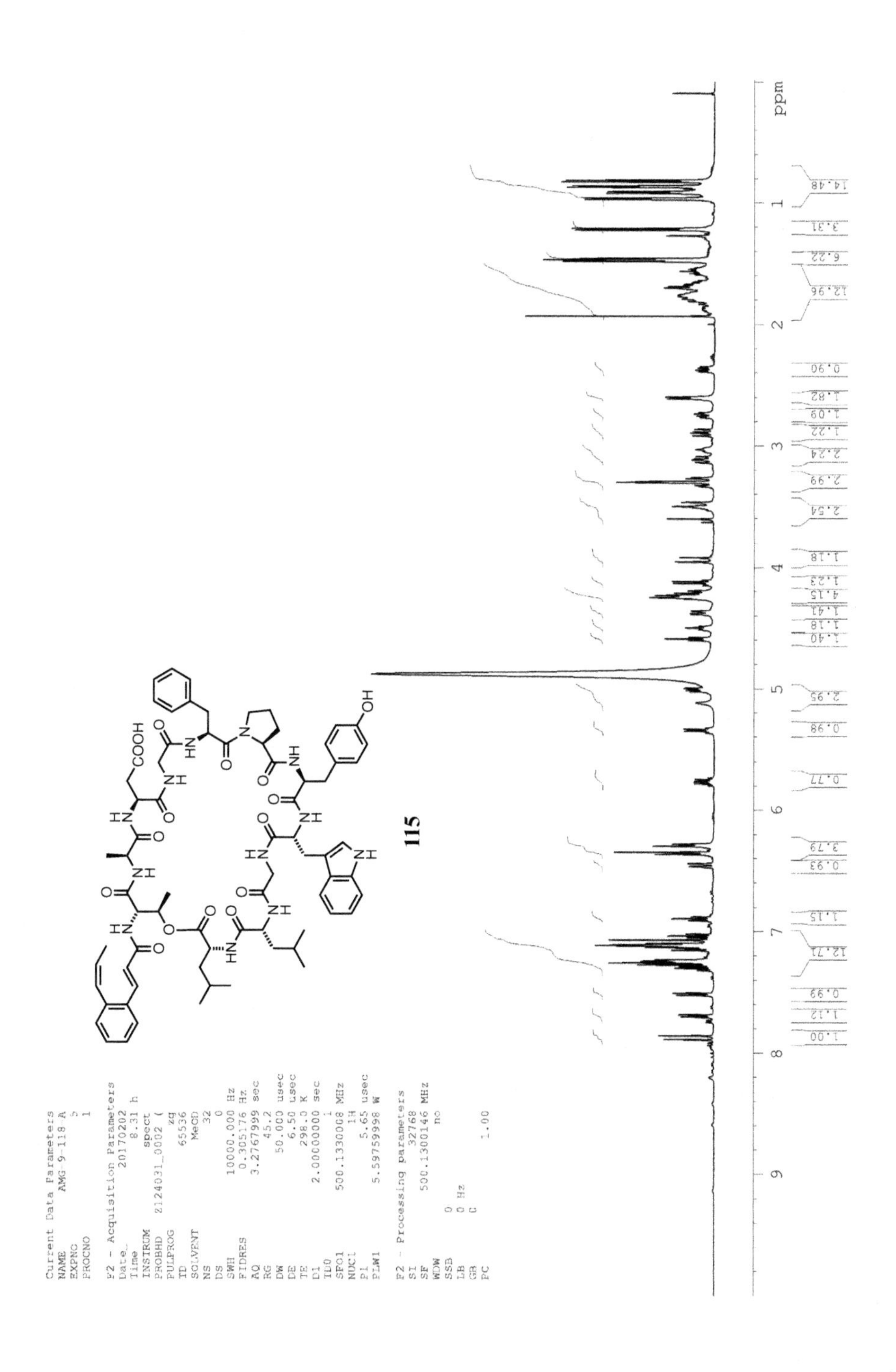
115
COOH
OH
ppm

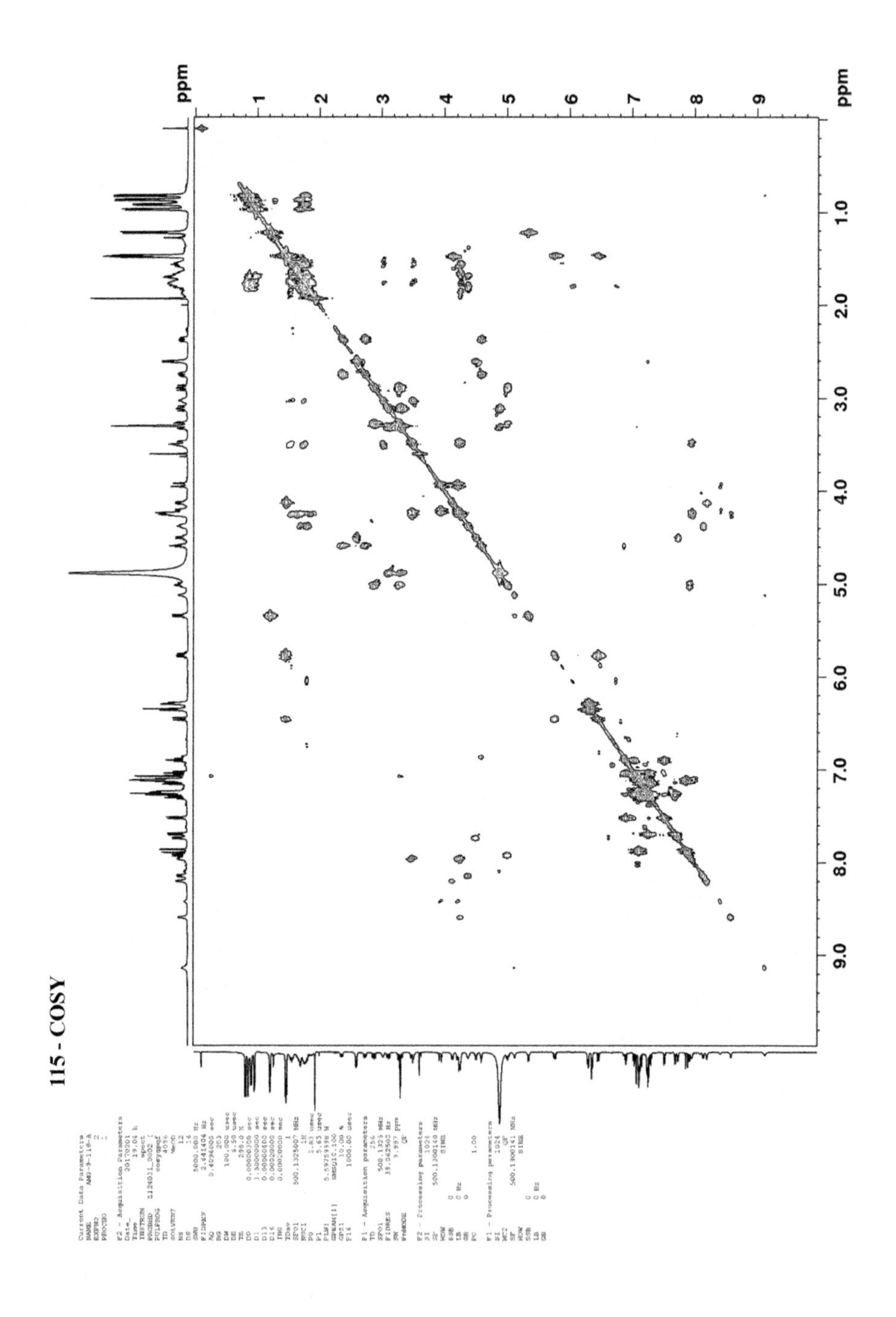
115 - COSY
ppm
1.0
2.0
3.0
4.0
5.0
6.0
7.0
8.0
9.0

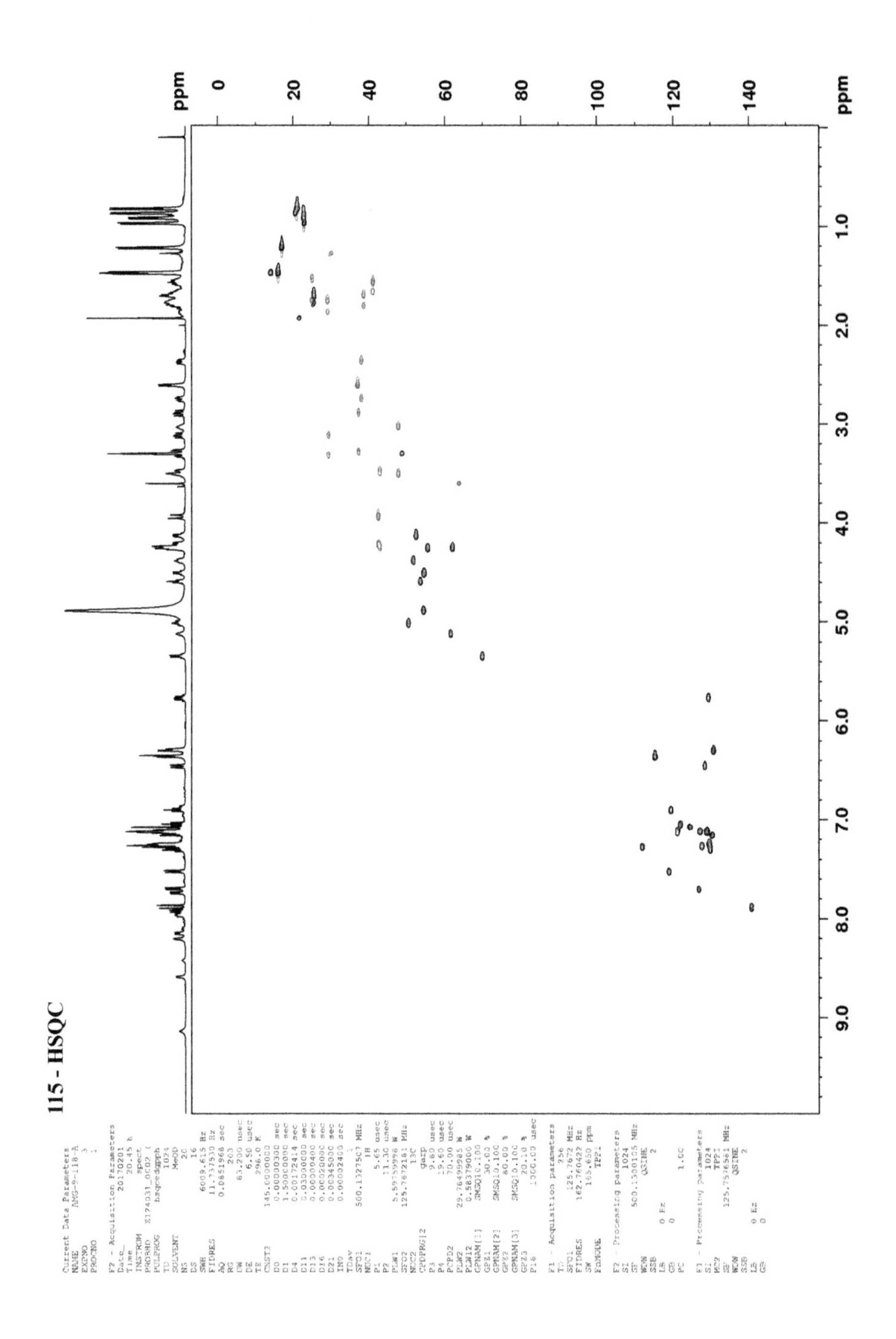
115 - HSQC
ppm
0
20
40
60
80
100
120
140
ppm
1.0
2.0
3.0
4.0
5.0
6.0
7.0
8.0
9.0

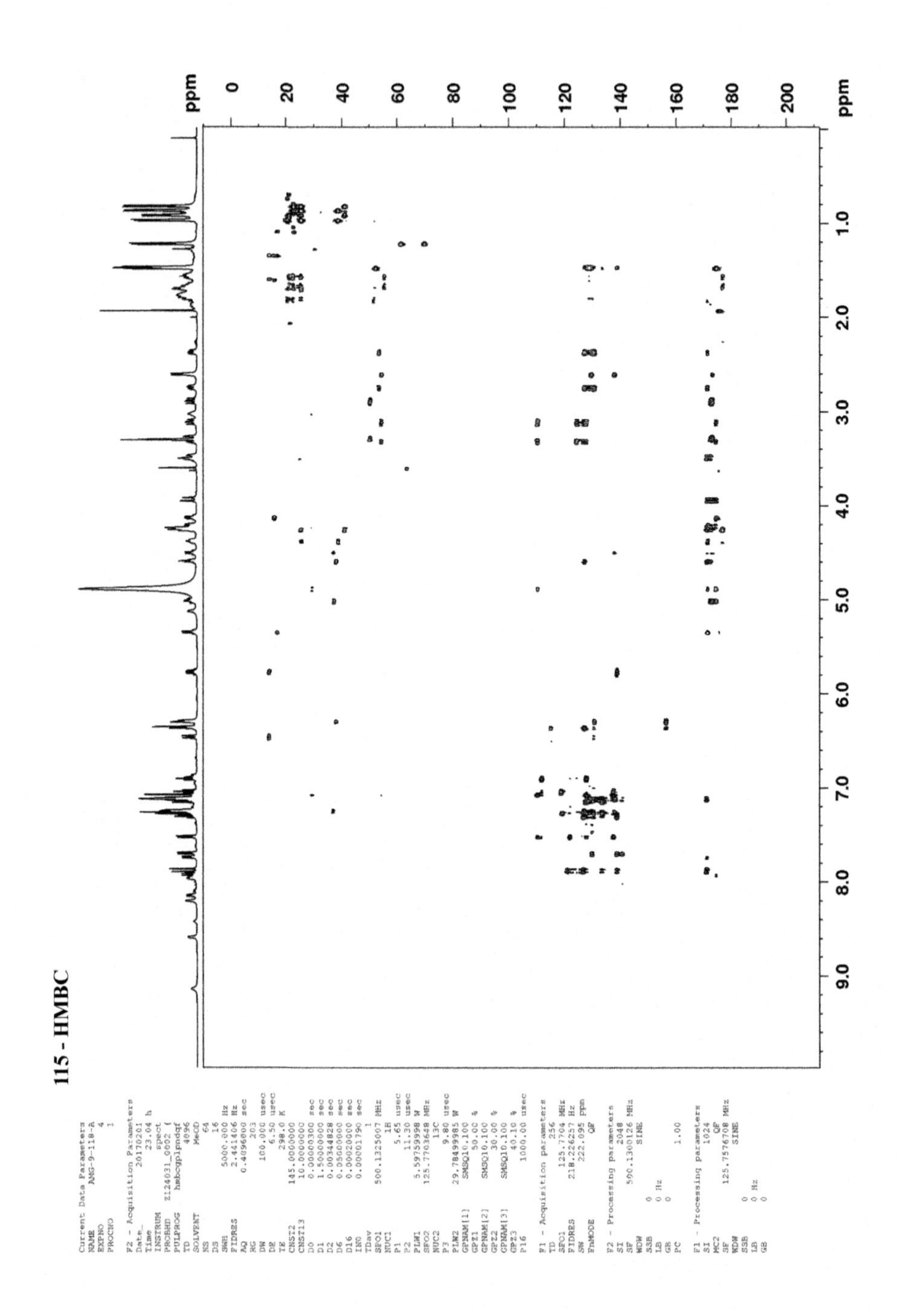
115 - HMBC

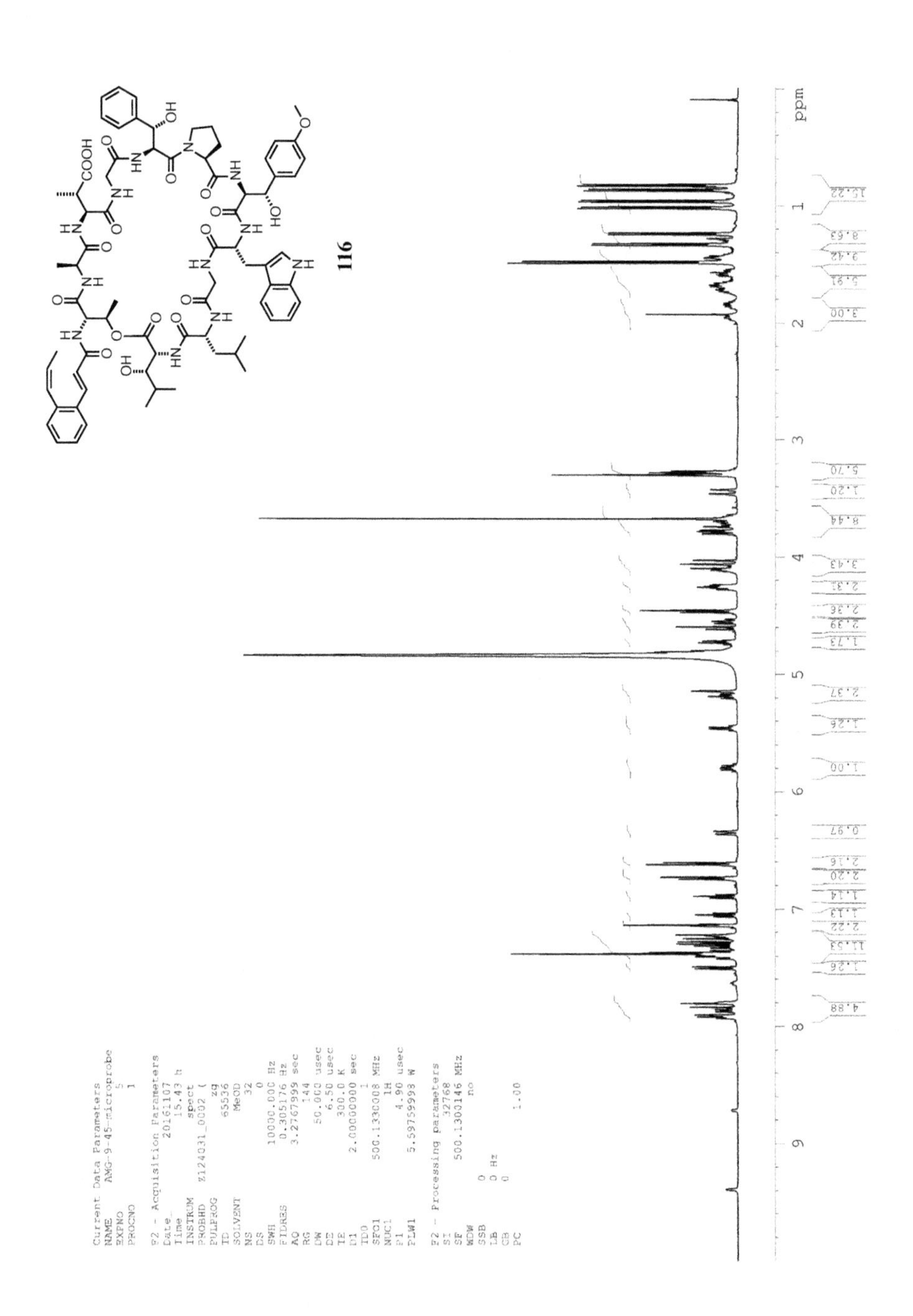

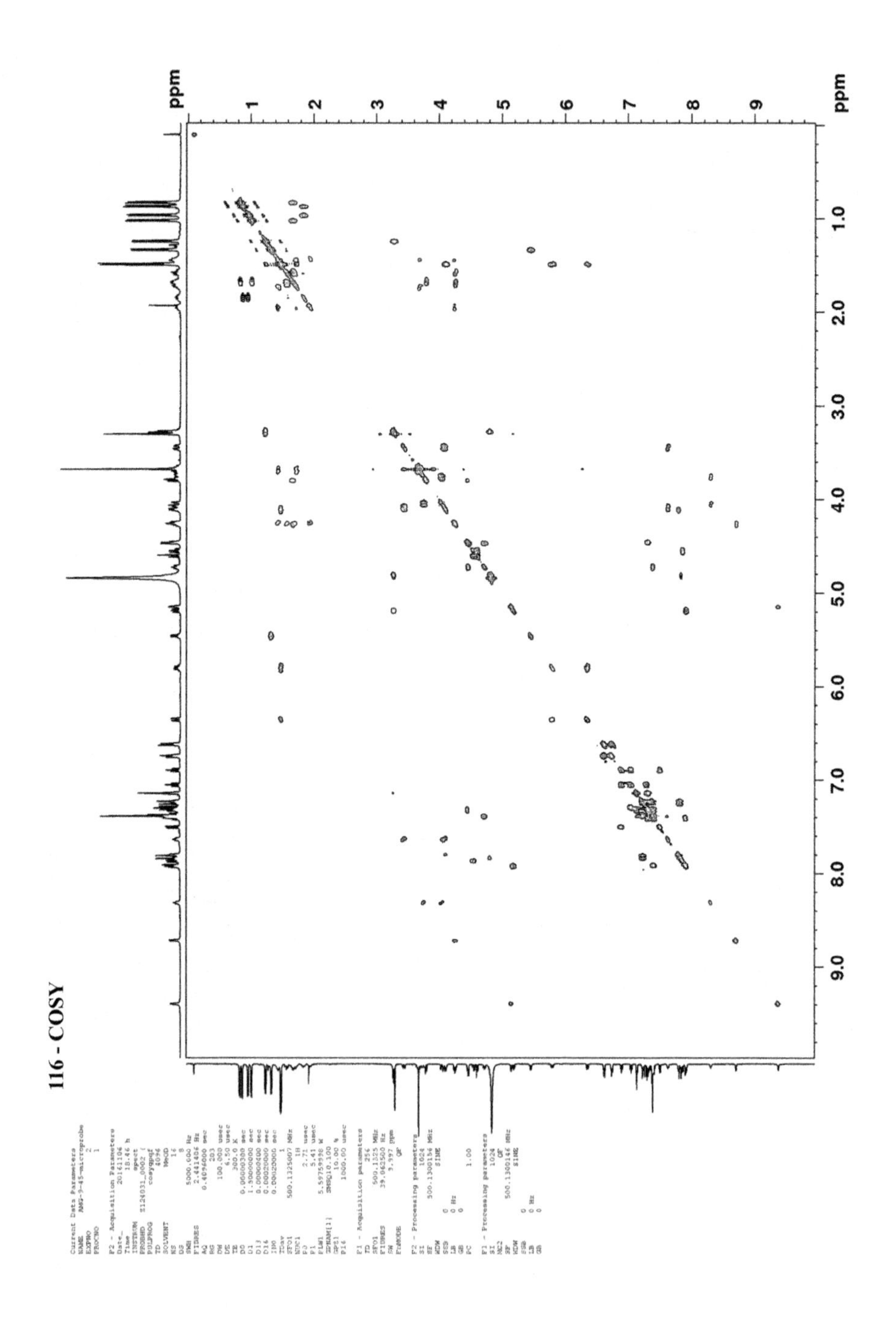
116 - COSY
ppm
1
2
3
4
5
6
7
8
9
ppm
1.0
2.0
3.0
4.0
5.0
6.0
7.0
8.0
9.0

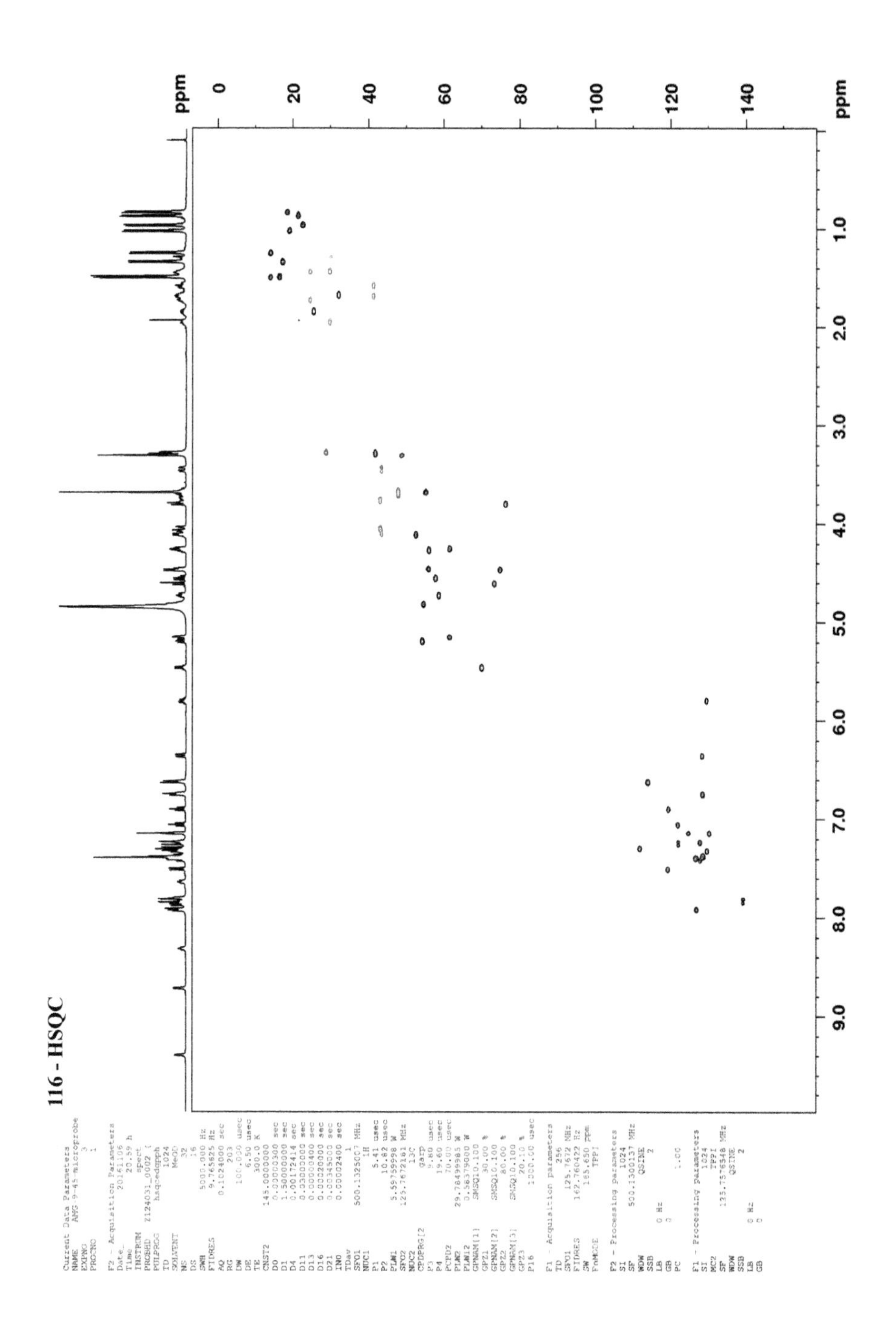
116 - HSQC
ppm
0
20
40
60
80
100
120
140
ppm
1.0
2.0
3.0
4.0
5.0
6.0
7.0
8.0
9.0

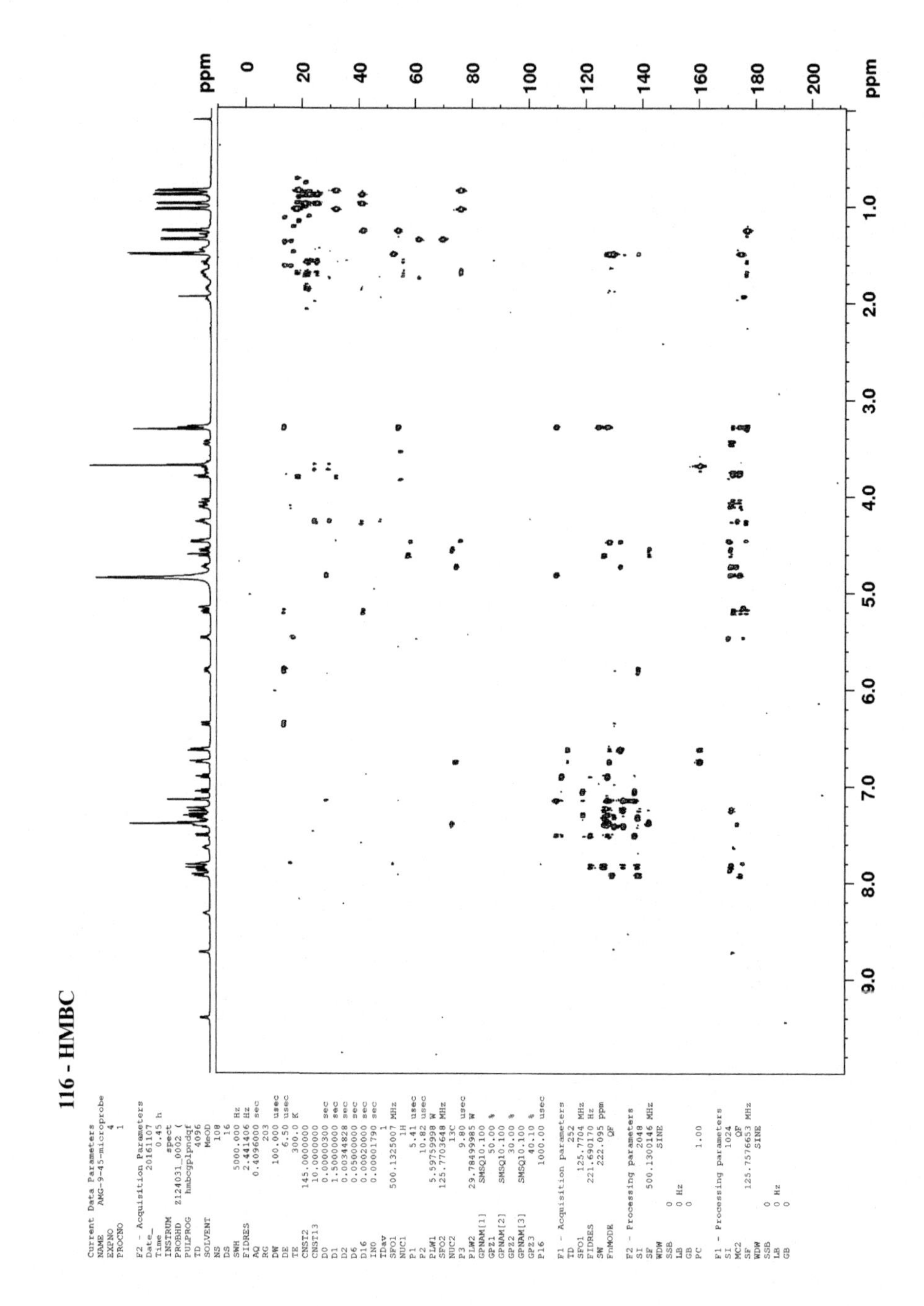
116 - HMBC
ppm
0
20
40
60
80
100
120
140
160
180
200
ppm
1.0
2.0
3.0
4.0
5.0
6.0
7.0
8.0
9.0

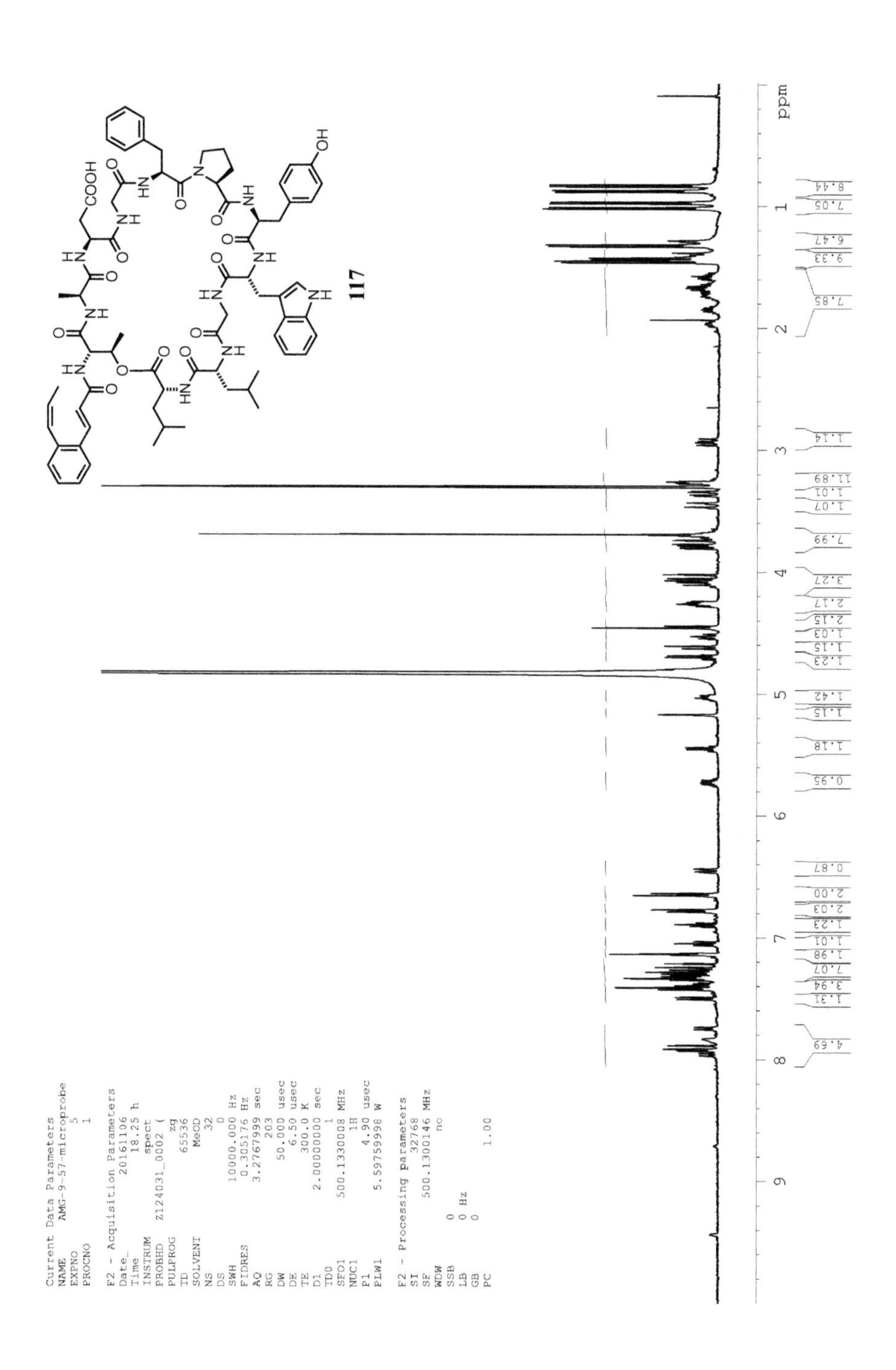
Current Data Parameters
NAME AMG-9-57-microprobe
EXPNO 5
PROCNO 1
F2 - Acquisition Parameters
Date_ 20161106
Time 18.25 h
INSTRUM spect
PROBHD Z124031_0002 (
PULPROG zg
TD 65536
SOLVENT MeOD
NS 32
DS 0
SWH 10000.000 Hz
FIDRES 0.305176 Hz
AQ 3.2767999 sec
RG 203
DW 50.000 usec
DE 6.50 usec
TE 300.0 K
D1 2.00000000 sec
TD0 1
SFO1 500.1330008 MHz
NUC1 1H
P1 4.90 usec
PLW1 5.59759998 W
F2 - Processing parameters
SI 32768
SF 500.1300146 MHz
WDW no
SSB 0
LB 0 Hz
GB 0
PC 1.00
COOH
OH
117
ppm
9
8
7
6
5
4
3
2
1

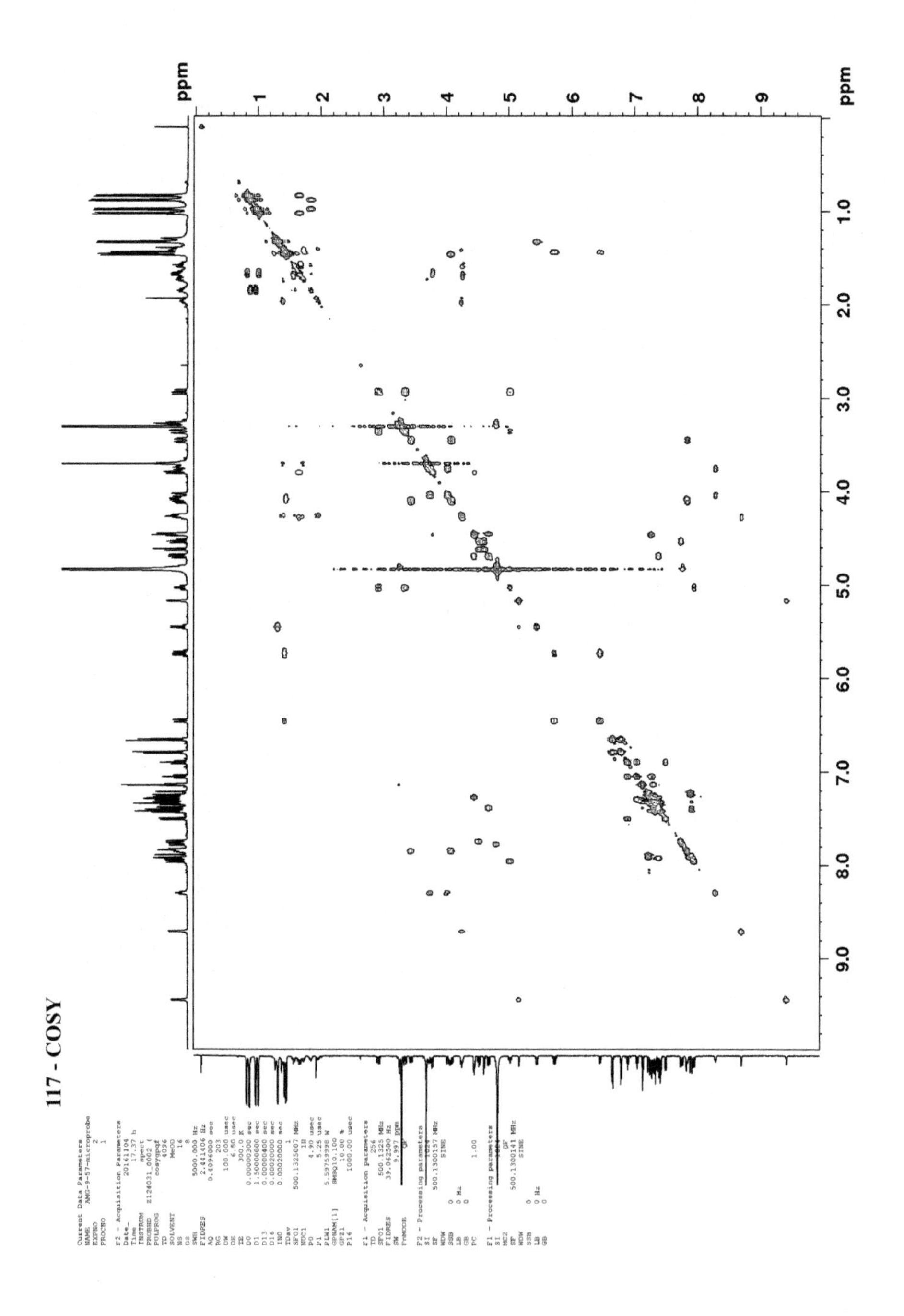
117 - COSY
ppm
ppm

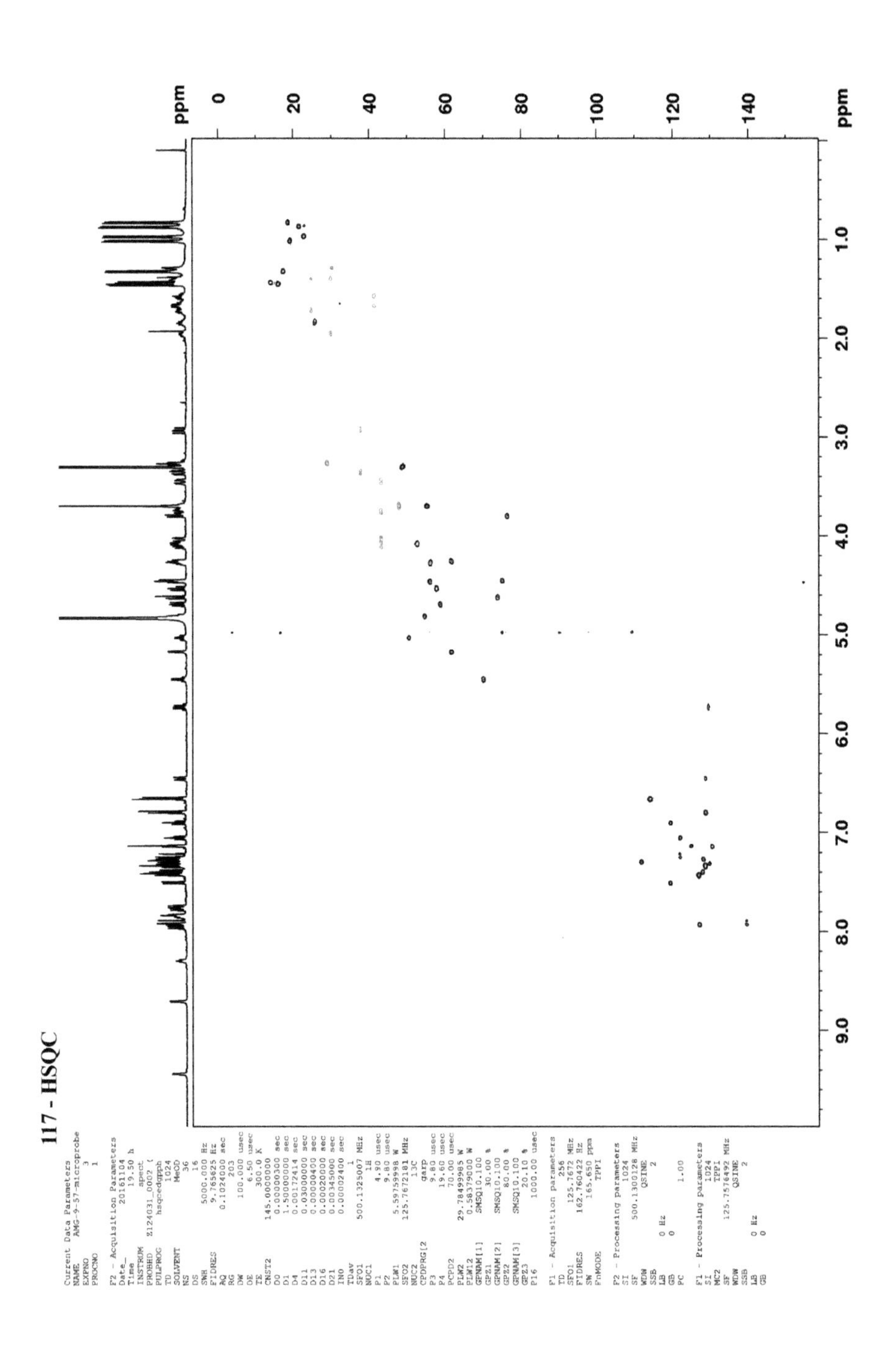
117 - HSQC
ppm
0
20
40
60
80
100
120
140
ppm
1.0
2.0
3.0
4.0
5.0
6.0
7.0
8.0
9.0

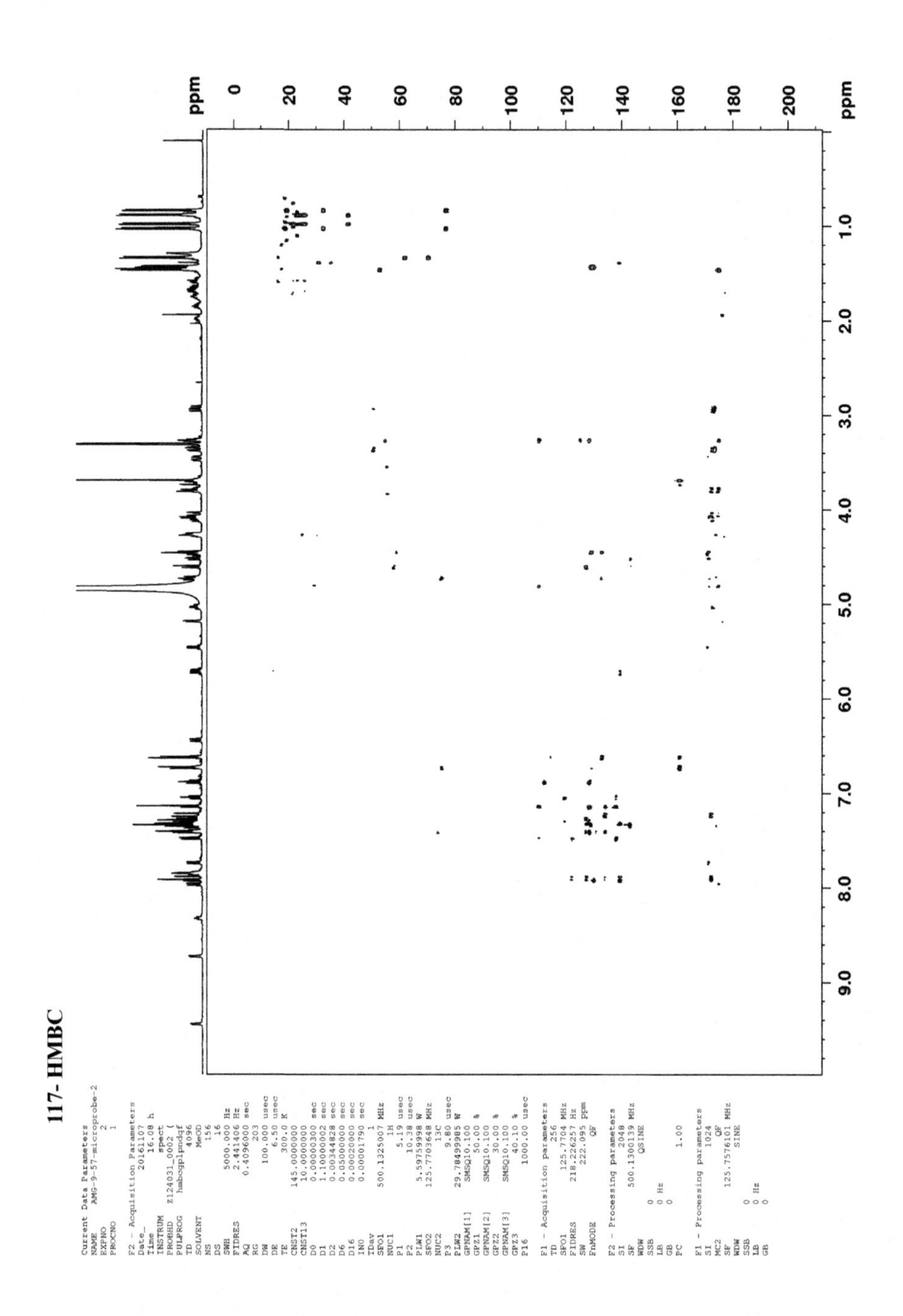
117- HMBC

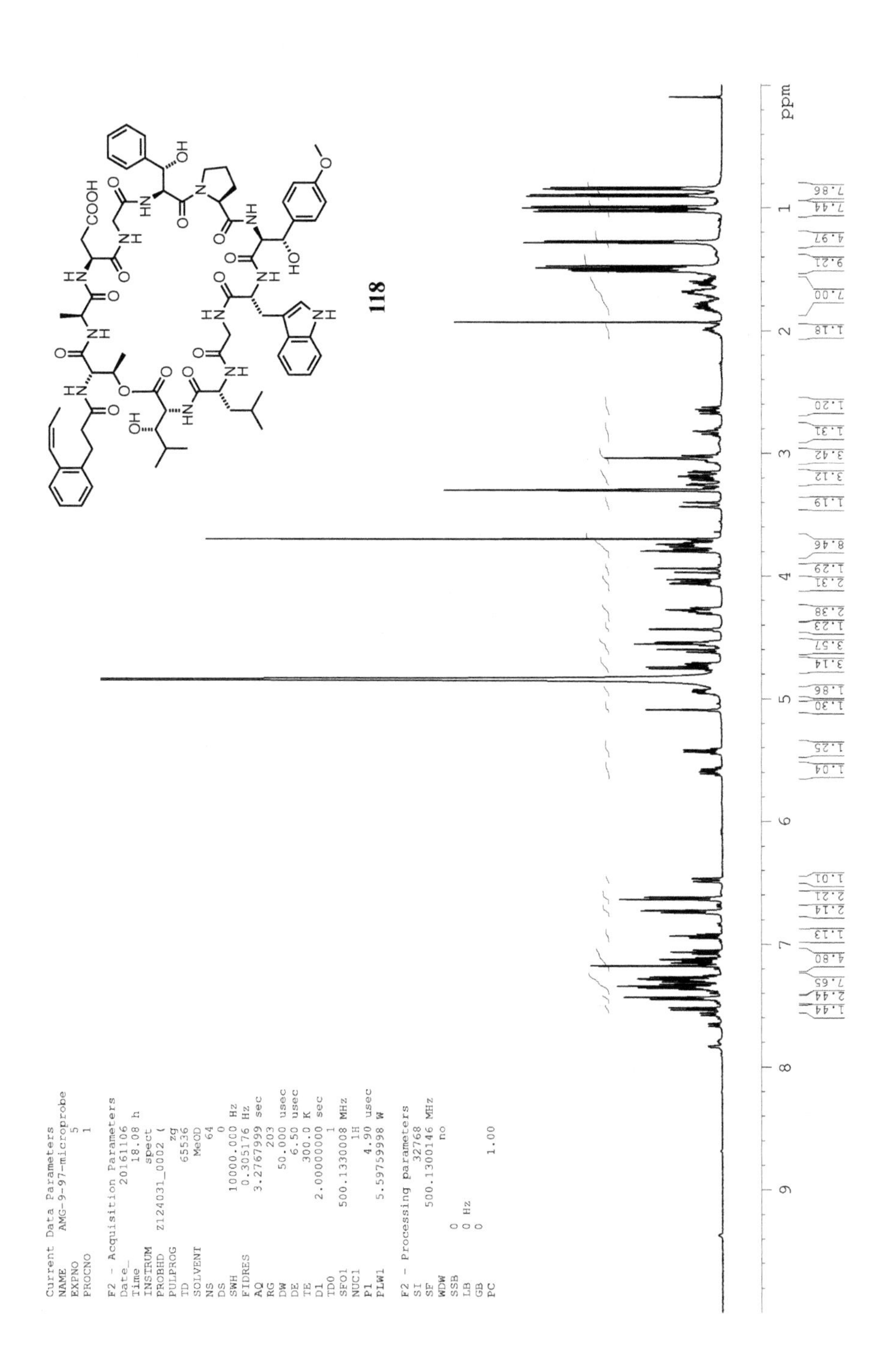
118
ppm
Current Data Parameters
NAME AMG-9-97-microprobe
EXPNO 5
PROCNO 1
F2 - Acquisition Parameters
Date_ 20161106
Time 18.08 h
INSTRUM spect
PROBHD Z124031_0002 (
PULPROG zg
TD 65536
SOLVENT MeOD
NS 64
DS 0
SWH 10000.000 Hz
FIDRES 0.305176 Hz
AQ 3.2767999 sec
RG 203
DW 50.000 usec
DE 6.50 usec
TE 300.0 K
D1 2.00000000 sec
TD0 1
SFO1 500.1330008 MHz
NUC1 1H
P1 4.90 usec
PLW1 5.59759998 W
F2 - Processing parameters
SI 32768
SF 500.1300146 MHz
WDW no
SSB 0
LB 0 Hz
GB 0
PC 1.00

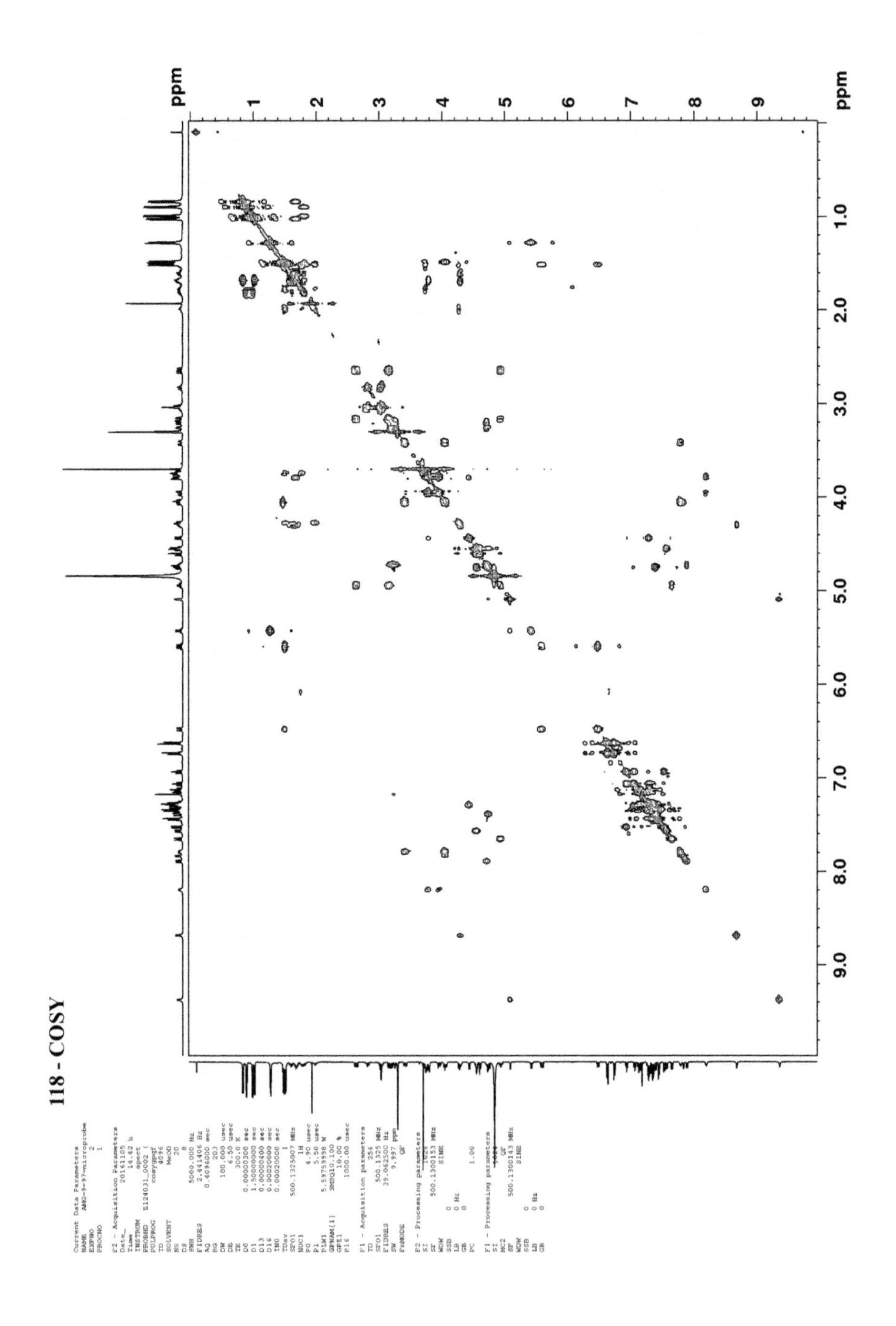
118 - COSY
ppm
ppm

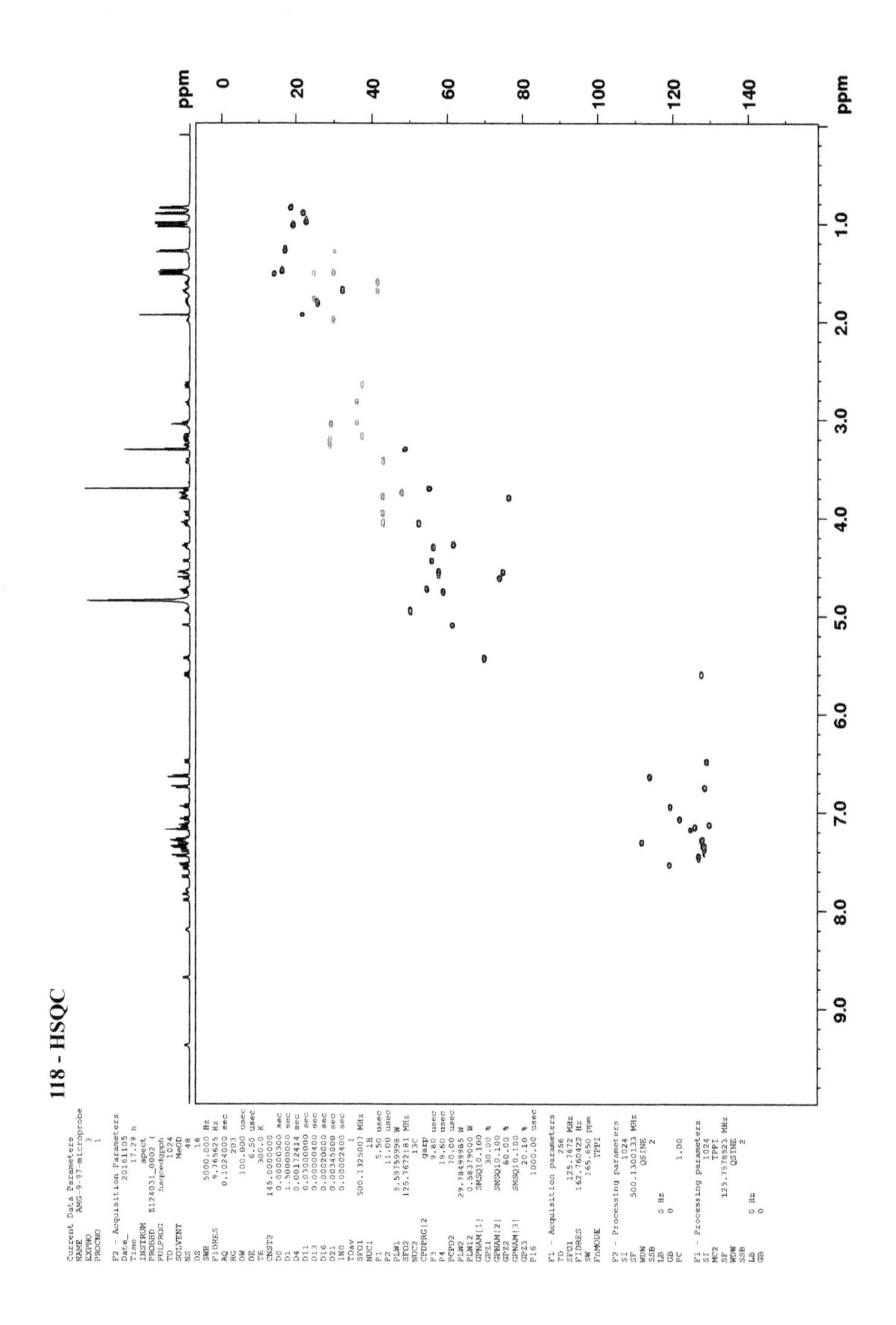
118 - HSQC
ppm
0
20
40
60
80
100
120
140
ppm
1.0
2.0
3.0
4.0
5.0
6.0
7.0
8.0
9.0

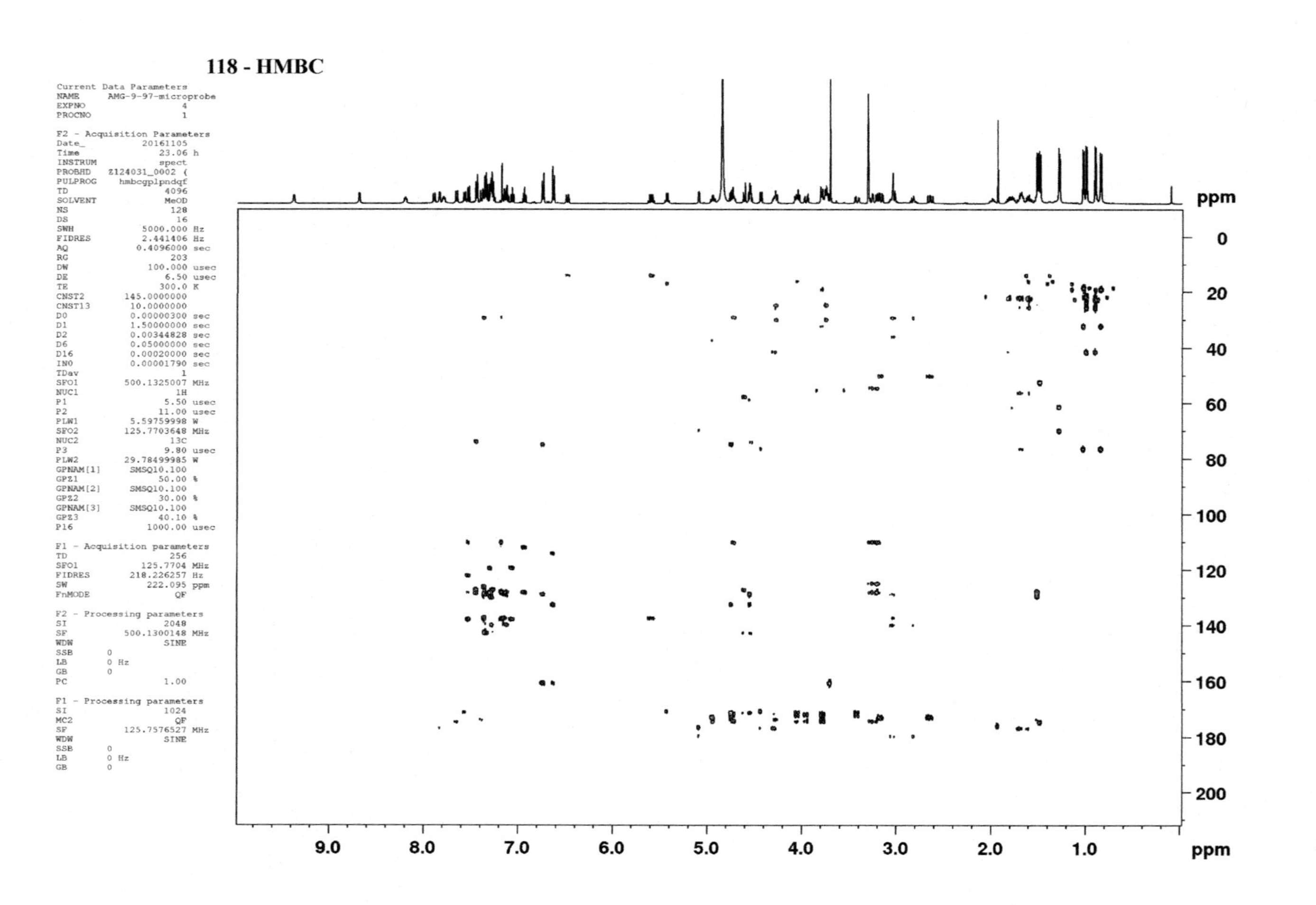
118 - HMBC
Current Data Parameters
NAME AMG-9-97-microprobe
EXPNO 4
PROCNO 1
F2 - Acquisition Parameters
Date_ 20161105
Time 23.06 h
INSTRUM spect
PROBHD Z124031_0002 (
PULPROG hmbcgplpndqf
TD 4096
SOLVENT MeOD
NS 128
DS 16
SWH 5000.000 Hz
FIDRES 2.441406 Hz
AQ 0.4096000 sec
RG 203
DW 100.000 usec
DE 6.50 usec
TE 300.0 K
CNST2 145.0000000
CNST13 10.0000000
D0 0.00000300 sec
D1 1.50000000 sec
D2 0.00344828 sec
D6 0.05000000 sec
D16 0.00020000 sec
IN0 0.00001790 sec
TDav 1
SFO1 500.1325007 MHz
NUC1 1H
P1 5.50 usec
P2 11.00 usec
PLW1 5.59759998 W
SFO2 125.7703648 MHz
NUC2 13C
P3 9.80 usec
PLW2 29.78499985 W
GPNAM[1] SMSQ10.100
GPZ1 50.00 %
GPNAM[2] SMSQ10.100
GPZ2 30.00 %
GPNAM[3] SMSQ10.100
GPZ3 40.10 %
P16 1000.00 usec
F1 - Acquisition parameters
TD 256
SFO1 125.7704 MHz
FIDRES 218.226257 Hz
SW 222.095 ppm
FnMODE QF
F2 - Processing parameters
SI 2048
SF 500.1300148 MHz
WDW SINE
SSB 0
LB 0 Hz
GB 0
PC 1.00
F1 - Processing parameters
SI 1024
MC2 QF
SF 125.7576527 MHz
WDW SINE
SSB 0
LB 0 Hz
GB 0
ppm
0
20
40
60
80
100
120
140
160
180
200
9.0
8.0
7.0
6.0
5.0
4.0
3.0
2.0
1.0
ppm

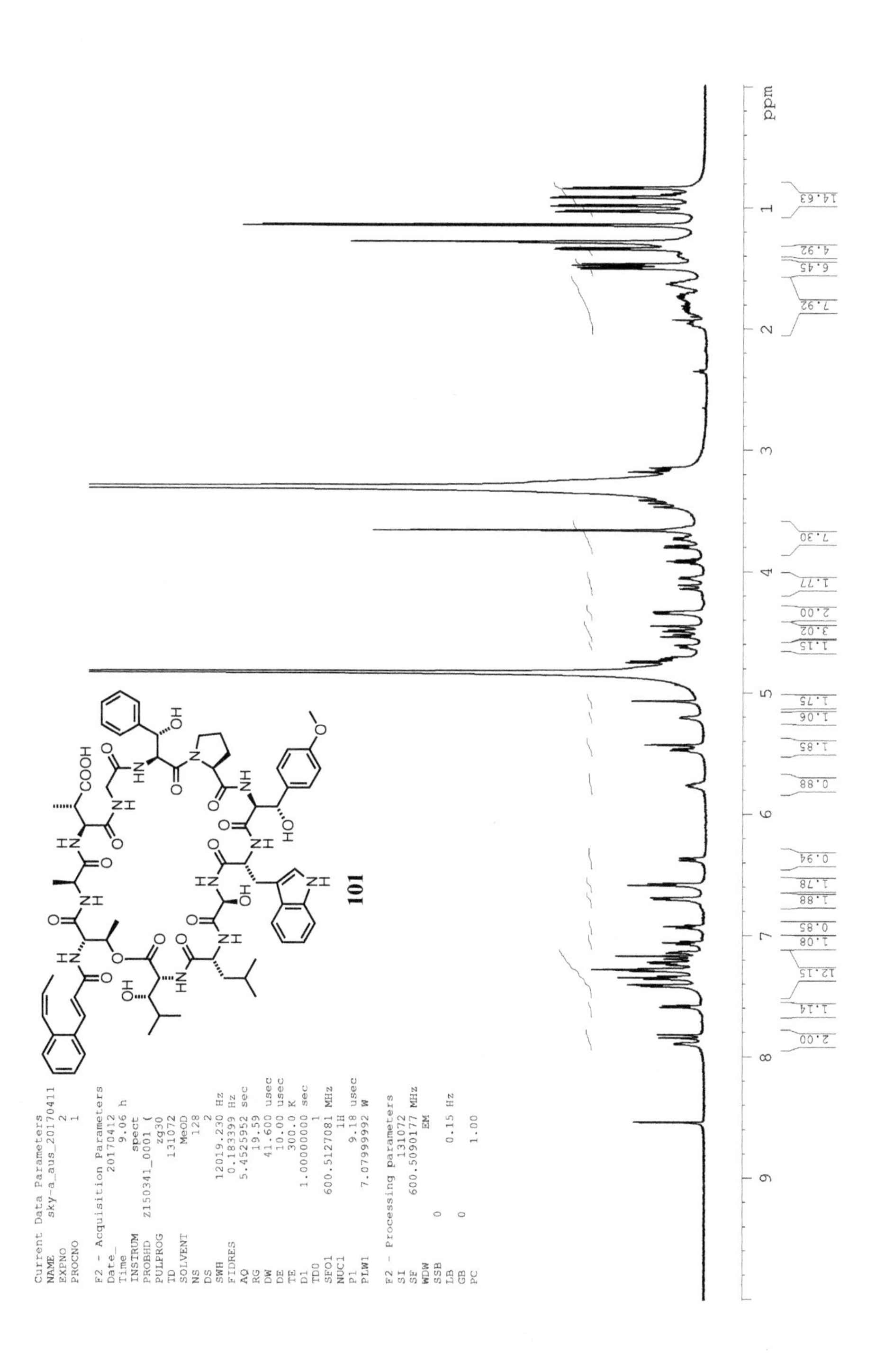
Current Data Parameters
NAME sky-a_aus_20170411
EXPNO 2
PROCNO 1
F2 - Acquisition Parameters
Date_ 20170412
Time 9.06 h
INSTRUM spect
PROBHD Z150341_0001 (
PULPROG zg30
TD 131072
SOLVENT MeOD
NS 128
DS 2
SWH 12019.230 Hz
FIDRES 0.183399 Hz
AQ 5.4525952 sec
RG 19.59
DW 41.600 usec
DE 10.00 usec
TE 300.0 K
D1 1.00000000 sec
TD0 1
SFO1 600.5127081 MHz
NUC1 1H
P1 9.18 usec
PLW1 7.07999992 W
F2 - Processing parameters
SI 131072
SF 600.5090177 MHz
WDW EM
SSB 0
LB 0.15 Hz
GB 0
PC 1.00
101
ppm

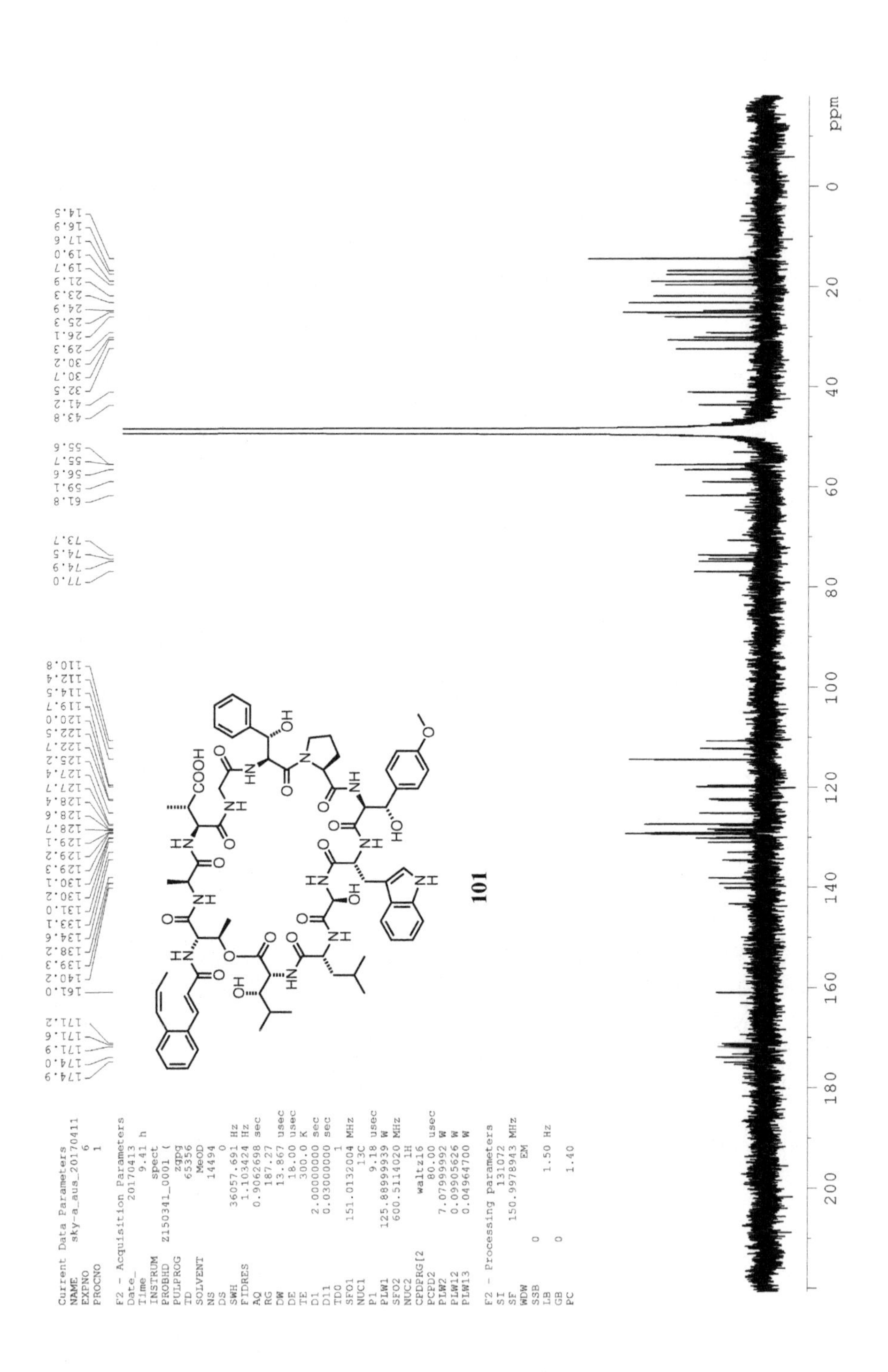
Current Data Parameters
NAME sky-a_aus_20170411
EXPNO 6
PROCNO 1
F2 - Acquisition Parameters
Date_ 20170413
Time 9.41 h
INSTRUM spect
PROBHD Z150341_0001 (
PULPROG zgpg
TD 65356
SOLVENT MeOD
NS 14494
DS 0
SWH 36057.691 Hz
FIDRES 1.103424 Hz
AQ 0.9062698 sec
RG 187.27
DW 13.867 usec
DE 18.00 usec
TE 300.0 K
D1 2.00000000 sec
D11 0.03000000 sec
TD0 1
SFO1 151.0132004 MHz
NUC1 13C
P1 9.18 usec
PLW1 125.88999939 W
SFO2 600.5114020 MHz
NUC2 1H
CPDPRG[2 waltz16
PCPD2 80.00 usec
PLW2 7.07999992 W
PLW12 0.09905626 W
PLW13 0.04964700 W
F2 - Processing parameters
SI 131072
SF 150.9978943 MHz
WDW EM
SSB 0
LB 1.50 Hz
GB 0
PC 1.40
174.9
174.0
171.9
171.6
171.2
161.0
140.2
139.3
138.2
134.6
133.1
131.0
130.2
130.1
129.3
129.2
129.1
128.7
128.6
128.4
127.7
127.4
125.2
122.7
122.5
120.0
119.7
114.5
112.4
110.8
77.0
74.9
74.5
73.7
61.8
59.1
56.6
55.7
55.6
43.8
41.2
32.5
30.7
30.2
29.3
26.1
25.3
24.9
23.3
21.9
19.7
19.0
17.6
16.9
14.5
COOH
OH
101
ppm
200
180
160
140
120
100
80
60
40
20
0

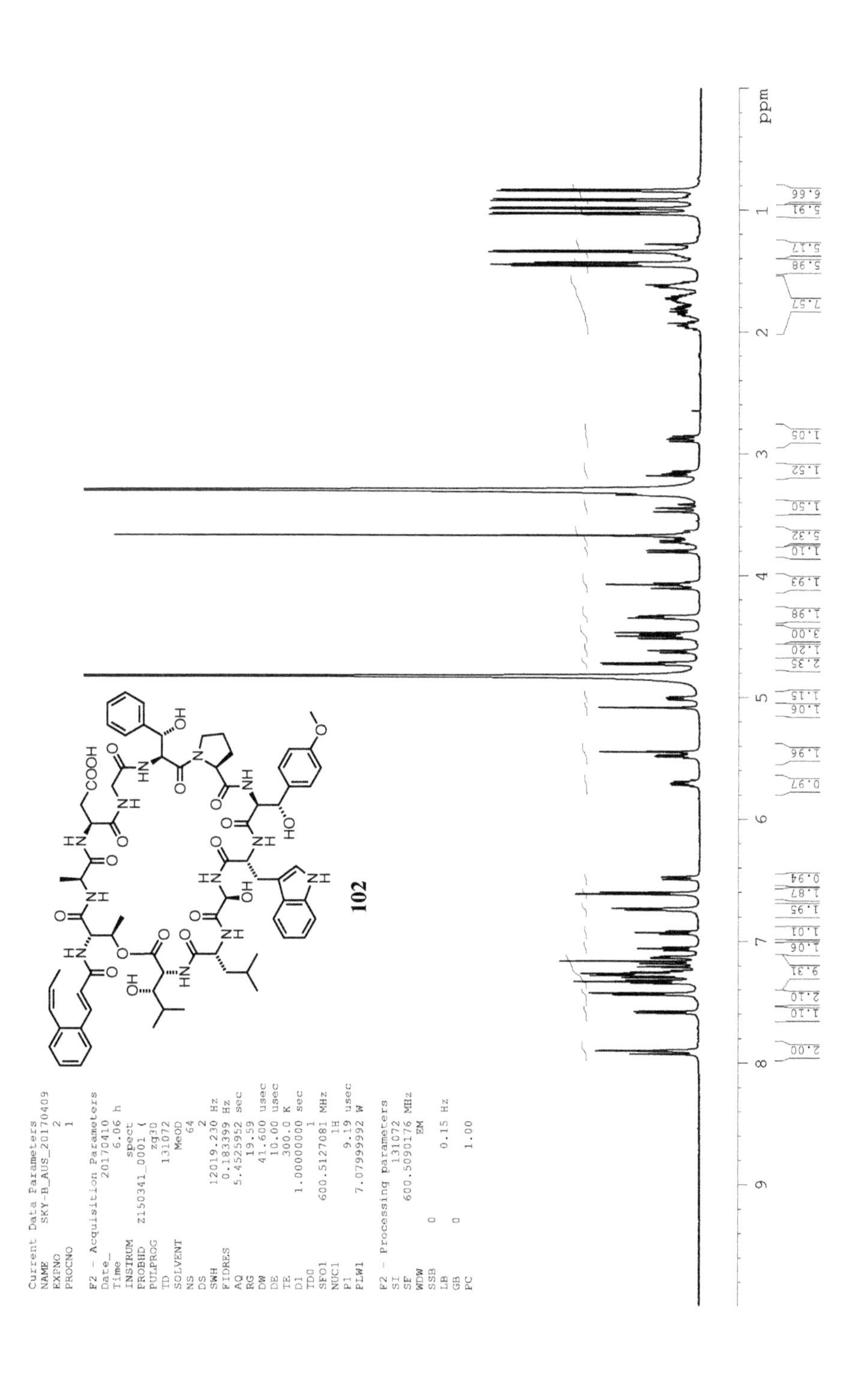
Current Data Parameters
NAME SKY-B_AUS_20170409
EXPNO 2
PROCNO 1
F2 - Acquisition Parameters
Date_ 20170410
Time 6.06 h
INSTRUM spect
PROBHD Z150341_0001 (
PULPROG zg30
TD 131072
SOLVENT MeOD
NS 64
DS 2
SWH 12019.230 Hz
FIDRES 0.183399 Hz
AQ 5.4525952 sec
RG 19.59
DW 41.600 usec
DE 10.00 usec
TE 300.0 K
D1 1.00000000 sec
TD0 1
SFO1 600.5127081 MHz
NUC1 1H
P1 9.19 usec
PLW1 7.07999992 W
F2 - Processing parameters
SI 131072
SF 600.5090176 MHz
WDW EM
SSB 0
LB 0.15 Hz
GB 0
PC 1.00
102
9
8
7
6
5
4
3
2
1
ppm
2.00
1.10
2.10
9.31
1.06
1.01
1.95
1.87
0.94
0.97
1.96
1.06
1.15
2.35
1.20
3.00
1.98
1.93
1.10
5.32
1.50
1.52
1.05
7.57
5.98
5.17
5.91
6.66

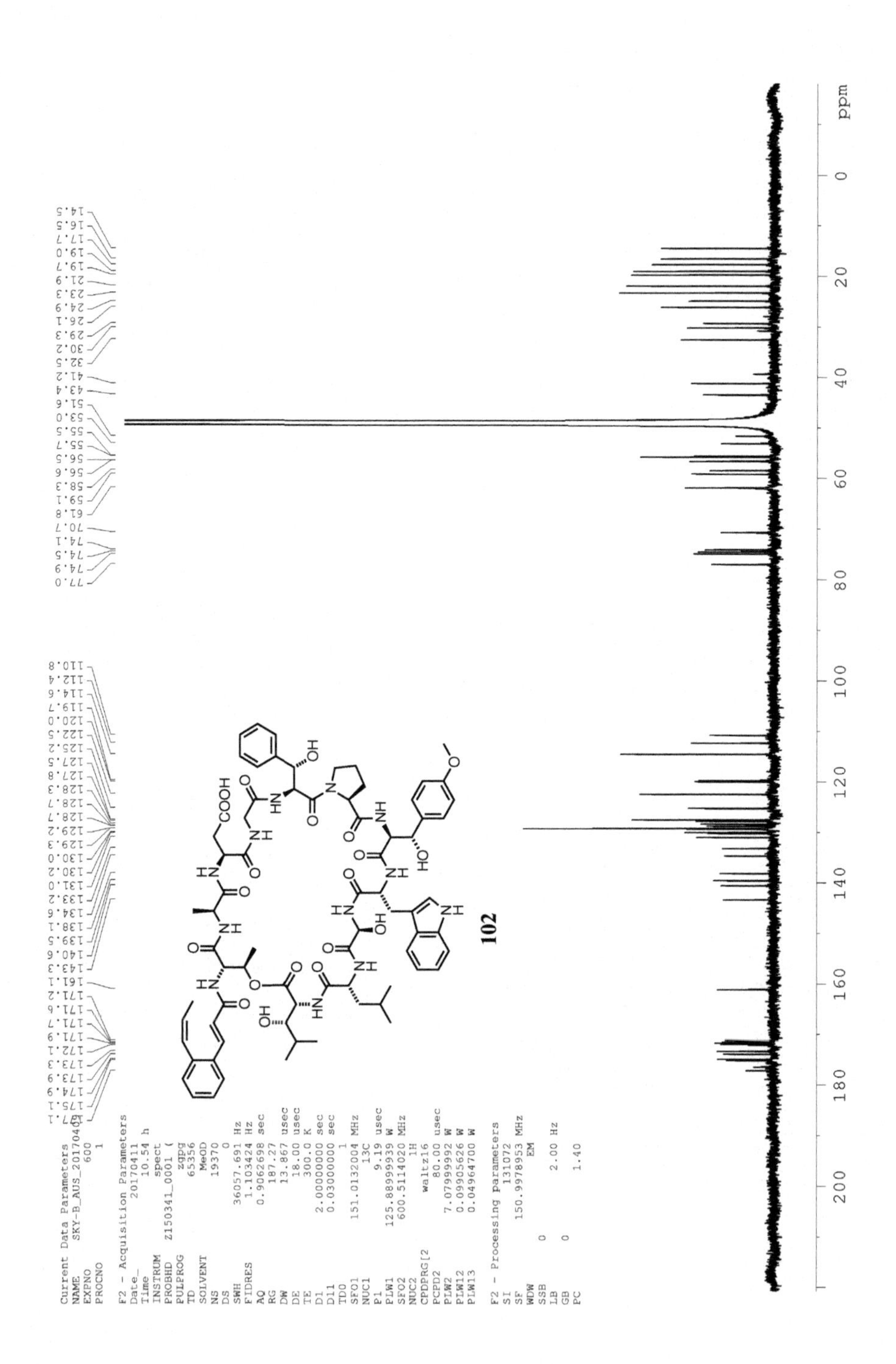
Current Data Parameters
NAME SKY-B_AUS_20170409
EXPNO 600
PROCNO 1
F2 - Acquisition Parameters
Date_ 20170411
Time 10.54 h
INSTRUM spect
PROBHD Z150341_0001 (
PULPROG zgpg
TD 65536
SOLVENT MeOD
NS 19370
DS 0
SWH 36057.691 Hz
FIDRES 1.103424 Hz
AQ 0.9062698 sec
RG 187.27
DW 13.867 usec
DE 18.00 usec
TE 300.0 K
D1 2.00000000 sec
D11 0.03000000 sec
TD0 1
SFO1 151.0132004 MHz
NUC1 13C
P1 9.19 usec
PLW1 125.88999939 W
SFO2 600.5114020 MHz
NUC2 1H
CPDPRG[2 waltz16
PCPD2 80.00 usec
PLW2 7.07999992 W
PLW12 0.09905626 W
PLW13 0.04964700 W
F2 - Processing parameters
SI 131072
SF 150.9978953 MHz
WDW EM
SSB 0
LB 2.00 Hz
GB 0
PC 1.40
102
ppm
200
180
160
140
120
100
80
60
40
20
0

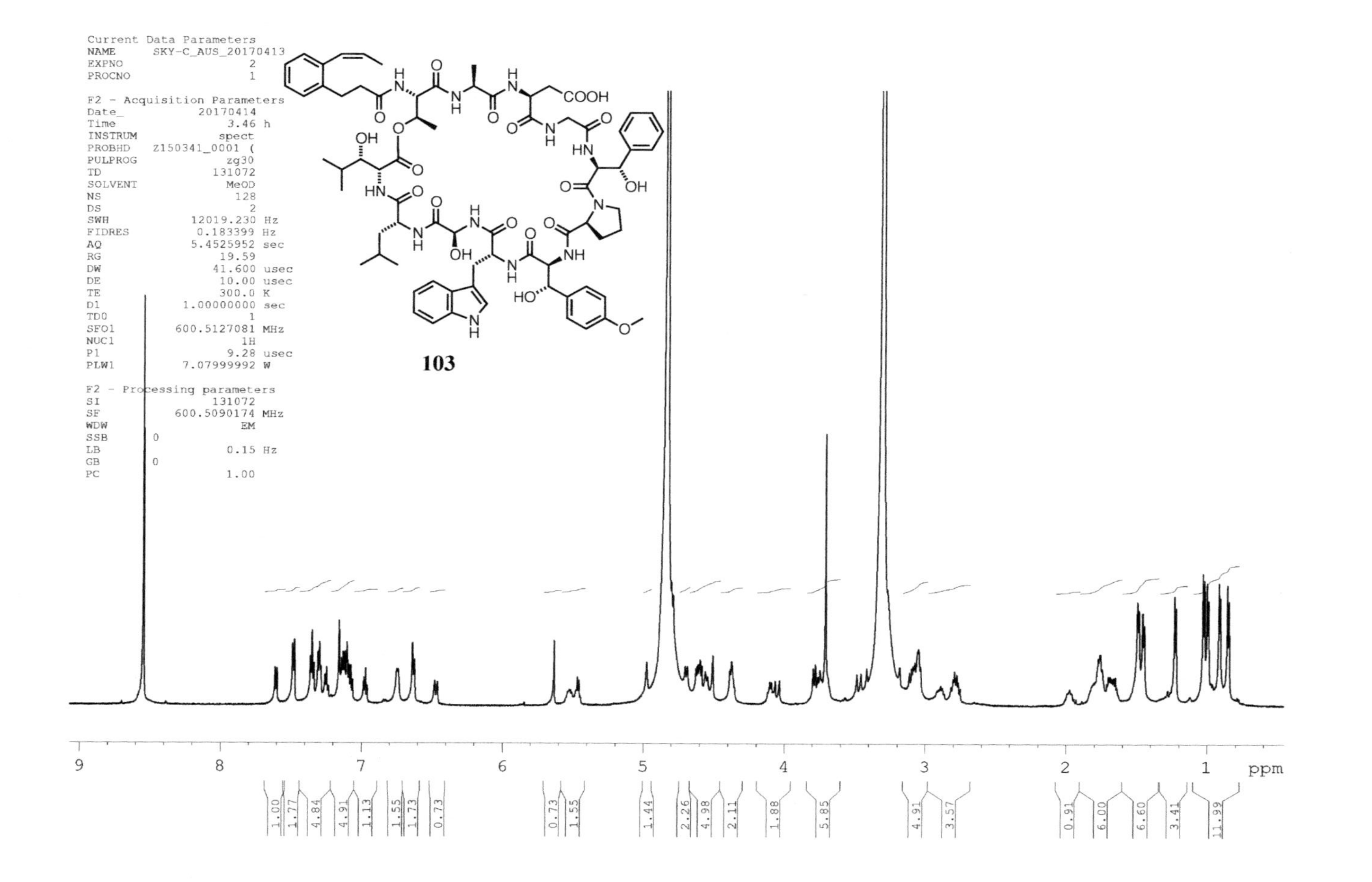
Current Data Parameters
NAME SKY-C_AUS_20170413
EXPNO 2
PROCNO 1
F2 - Acquisition Parameters
Date_ 20170414
Time 3.46 h
INSTRUM spect
PROBHD Z150341_0001 (
PULPROG zg30
TD 131072
SOLVENT MeOD
NS 128
DS 2
SWH 12019.230 Hz
FIDRES 0.183399 Hz
AQ 5.4525952 sec
RG 19.59
DW 41.600 usec
DE 10.00 usec
TE 300.0 K
D1 1.00000000 sec
TD0 1
SFO1 600.5127081 MHz
NUC1 1H
P1 9.28 usec
PLW1 7.07999992 W
F2 - Processing parameters
SI 131072
SF 600.5090174 MHz
WDW EM
SSB 0
LB 0.15 Hz
GB 0
PC 1.00
103
ppm

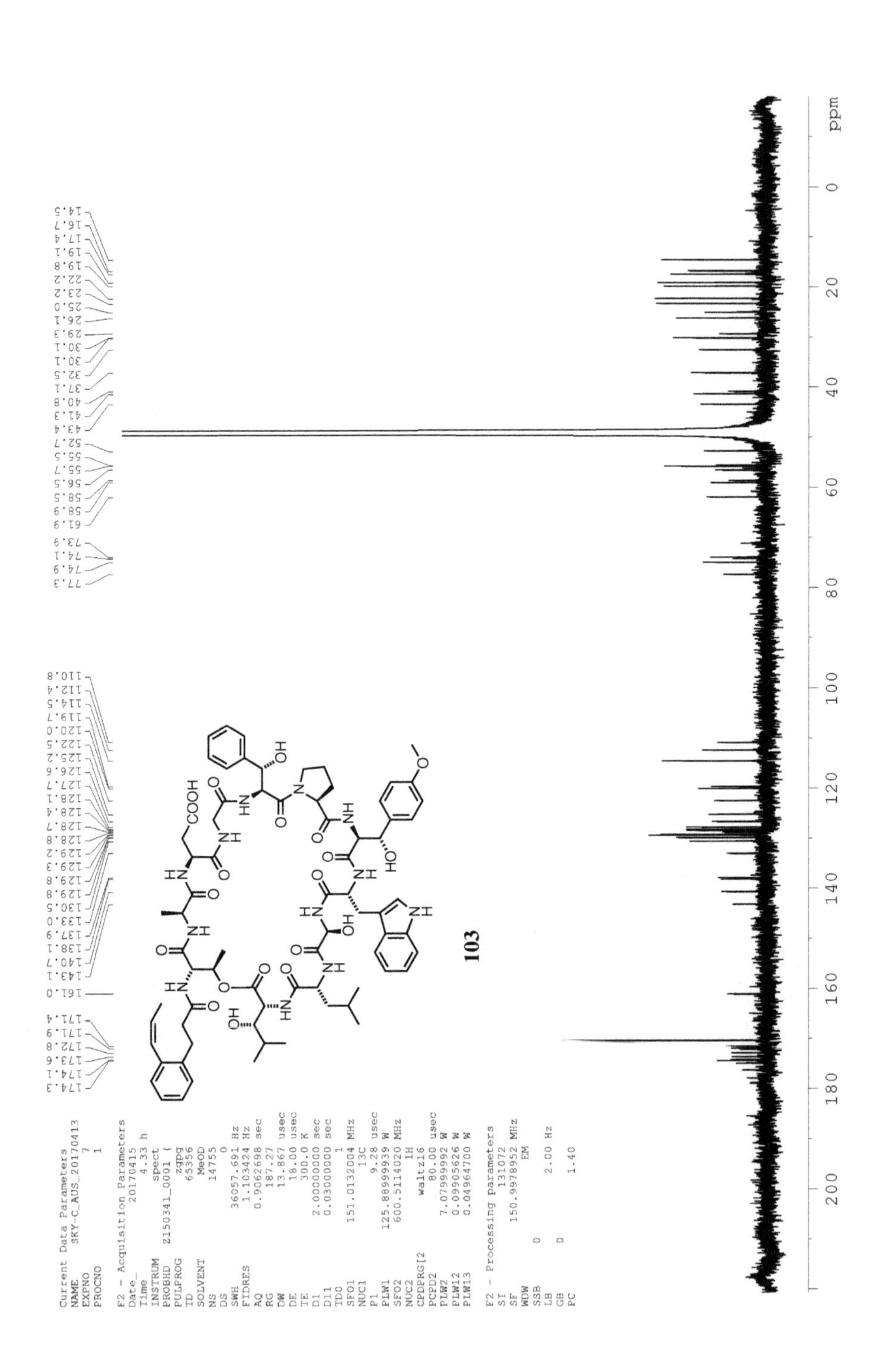